51.95

ATMOSPHERIC ECOLOGY

FOR DESIGNERS AND PLANNERS

WILLIAM P. LOWRY

Faculty of Landscape Architecture and Regional Planning,
University of Pennsylvania Faculty of Forest Science, Oregon State
University Faculties (Retired) of Ecology and Geography, University
of Illinois, Urbana

**with illustrations by
Samuel C. Lowry**

 VAN NOSTRAND REINHOLD
_____New York

WARNING & DISCLAIMER

This book is designed to provide designers and planners with information on climatic, bio-climatic, and meteorological aspects of the atmosphere that can be incorporated into their work. Every effort has been made to make this information as complete and as accurate as possible, but the author and publisher are not responsible for any errors or omissions, and no warranty or fitness is implied.

This book is sold with the understanding that the author and publisher are not engaged in rendering professional advice or services to the reader. Neither author not publisher shall bear liability or responsibility to any person or entity with respect to any loss or damage arising from information contained in this book.

Van Nostrand Reinhold
115 Fifth Avenue
New York, New York 10003

Chapman and Hall
2-6 Boundary Row
London SE1 8HN, England

Thomas Nelson Australia
102 Dodds Street
South Melbourne, Victoria 3205, Australia

Nelson Canada
1120 Birchmount Road
Scarborough, Ontario M1K 5G4, Canada

16   15   14   13   12   11   10   9   8   7   6   5   4   3   2   1

Library of Congress Cataloging-in-Publication Data

Lowry, William Prescott, 1927–
    Atmospheric Ecology for Designers and Planners
    Bibliography and Index
    1. Microclimatology 2. Ecology
    3. Environmental Planning 4. Architecture
    5. Environmental Physiology
    I. Title

Library of Congress Catalog Card Number 88-90598
ISBN 0-929002-07-5 (Paperback)
ISBN 0-442-00751-5 (Hardcover)

Errata

Page	Para beginning —	Line	Figure/Table	Change to (Old) **new** —
10	Between reservoir C—	1		Between **the** reservoir **to the right of** C —
17	How do these rules—	4		a depth of (0.1) **1.0** cm —
46	In most any—			Delete the paragraph — duplicated
47			Table 4-1	(See the little table in Fig. 4-**1**)
49			Table 4-2	(See the little table in Fig. 4-**1**)
—	Fig. 4-5 permits—	3		dots in fig. 4-**5**—
—			Table 4-2	cos Z = [(sin L)(sin **a**) + —
50		last		for (Feb 23/Oct **20**)
75	Footnote I	9		(spead) **spread**
76	You will notice—	6		from the **sand** into both
—	—	7		the **clay,** and from the loam into the **clay**
81	If we released—	5		look like Fig. 4-30**a**
—	—	6		look like Fig. 4-30**b**
89	Having just finished—	2		in the last sect**i**on
128	D_1 (migrating)—	1		and $\underline{\mathbf{D_2}}$ (static)
130	Fairbanks	last	Table 5-2	columns 3 and 4: (0.80) **1.57**; (22.8) **12.1**
146	as the LaPorte anomaly—	28		slightly **less, and then** more rainfall—
170			Fig. 6-11b	Arrowheads on "L" should be at heavy slant and vertical dashed lines
210	The technical consultant	2		game in it**s**elf
250	The value of the—	last		one part in **a** thousand
348			Fig. 10-41	"Flow aloft" spirals clockwise.
361	An environmental—	16		than its environment**.** Thus, this
425	Greenhouse effect			related to global climatic change **292**,383-385
426	Latent heat			67, 90

ATMOSPHERIC ECOLOGY FOR DESIGNERS AND PLANNERS
by
William P. Lowry

with illustrations by
Samuel C. Lowry

Introduction

Microclimate

Bioclimate

Atmospheric ecology and environmental design

Supplements: Mother Earth, Macroclimate, and Air Pollution

Conclusions

Preface

I want to tell you about my subject — the atmosphere. I want you to learn about it because it is interesting in its own right, and because you can make a lot of use, in the work you'll be doing, of the ideas you are going to read about.

In my experience*, designers and planners are not nearly as well equipped as they could be, and should be, to incorporate knowledge of the atmosphere into their work. Partly it's because they don't see how that knowledge fits into what they have to do, and partly it's because they don't know who to talk to about understanding the atmosphere. And partly, it's because they don't even seek out a specialist, thinking they'll never be able to understand the physics involved.

This book is intended to end that silence between designers and planners, on the one hand, and meteorologists, climatologists, and bioclimatologists on the other hand.

The book is laid out as follows. You start with a consideration of a few ideas about how designers handle scientific material, or could handle it (**Chapter 1**), and a few more ideas about physical systems in general (**Chapter 2**). A portion of Earth's atmosphere containing some living organisms and some man-made structures is the particular physical system considered in most of the book, but

---

*My experience has been mostly as a lecturer and consultant in the Department of Landscape Architecture and Regional Planning, Graduate School of Fine Arts, at the University of Pennsylvania. There I have worked with students and faculty from landscape architecture, architecture, urban, regional, and environmental planning. My Penn experience has been extended during several years working with landscape architects and planners at the University of Illinois in Urbana.

there are some useful ideas about general physical systems that you ought to be sure you know about.

Next, we go to a consideration of microclimates — the climates in small places, or very near the earth's surface, where people and most other organisms live. The plan here is to consider *first* the various patterns — time and space — that we observe in microclimates (**Chapter 3**). *Then* we consider the explanations for what we have observed (**Chapter 4**).

That sets a plan for the rest of the book: *"what" followed by "why."*

Next, we look at bioclimates — the energy connections between organisms and the atmosphere. *First* we follow the "what/why" pattern with human beings and their shelters (**Chapter 5**), and *then* we do the same with animals and plants (**Chapter 6**). Up to this point the effort is to deal with the immediate physical environments, and to place the living things within those physical environments, looking at how the organisms and the environments interact, and at the energetic similarities and differences among people, animals, and plants. These interactions *at the planning and design scale* are examined by means of the central concept of the energy balance.

It is likely that some instructors and students will believe the real meat of this book lies in its first half — through **Chapter 6** on animals and plants — dealing with process *at the planning and design scale.* That may be so for some, but for others, the sections on integrating the scientific consultant and his knowledge into the planning and design process (**Chapters 7** and **8**) and on atmospheric pollution and processes at the global and regional scale (**Chapters 9, 10,** and **11**) will be valuable extensions of the basic information in the first half of the book. In **Chapters 9, 10,** and **11** the "what/why" pattern is used again.

Finally, in **Chapter 12** we try to make clear, by reexamination of the major ideas of the middle portions of the book, the utility of the ideas and the information contained in the book.

Each chapter has three kinds of supplementary forms: (i) footnotes, (ii) notes, and (iii) peptalks. *Footnotes* are available on the same page where they are introduced in the text. The information they contain is briefly stated and of immediate utility in understanding the text. *Notes* at the end of each chapter contain information that would interrupt the flow of thoughts if they were to be included within the text. The information is useful, but it can "wait until later." *Peptalks* are also at the end of each chapter. They are comments meant to help readers who shy away from mathematical means for conveying information — "math anxiety" some educators call it.

Again, my personal interest here is seeing to it that *good scientific knowledge about the atmosphere* shows up in the built environment. My path to that objective, of course, is through the minds of those learning and practicing the planning and design professions — that's you. I hope the book will reach that objective. More than that, I hope our passage together along that path is a stimulating one.

Acknowledgments

I am pleased to acknowledge the assistance of the following friends and colleagues: Dean Burn, Brian Crissey, Aron Faegre, Lew Hopkins, Tom Kolba, Ian McHarg, John Reynolds, Gary Smith, Anne Spirn, Jim Thorne, Tom Worcester, and the Linfield College Departments of Physics and Computing Science.

W.P.L.
May 1988

1. What are WE doing here?
Why climatologists and designer/planners should meet

Have you ever had the thought, while walking at midday through the downtown streets of a large city in summer, "This is a sauna. I wonder why it's so hot and humid here?"

And have you wondered whether something might be done about it — some design and planning reponse to this effect? I imagine most designers and planners, at this point, would think of street trees as a means for cooling the urban canyon air evaporatively.

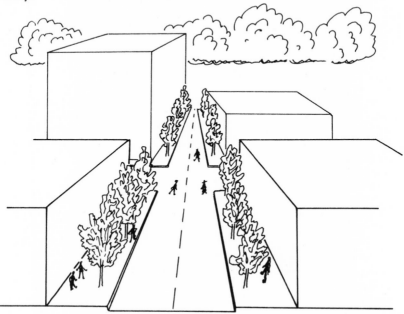

To my mind, exploration of this problem and suggested solution would make a perfect project for a design studio. I have no doubt that both the problem and the solution have appeared in studio presentations through the years. It seems straightforward enough: just plant trees and grass and try to make that little part of the city like the countryside. It's cooler in the country on a hot summer day, so try to make them alike. And the dozens of neat little urban parks, tucked away in Manhattan for example, are so cool and comfortable, there doesn't seem much left to study.

I wonder, though, whether these design analyses very often reflect the kind of analytical thinking you will find in this book. I doubt it, and I invite you to keep this problem in the back of your mind as you read. I present my physical analysis of it later.

For now, I invite you to keep in mind such ideas as the fact that trees can be introduced either in the original design or after development, in response to a problem.

But truly fundamental responses to the problem ought to be introduced at the beginning, since modification of major elements later is usually out of the question. Solutions making use of the shapes, materials, colors, and placement of elements involve both the architect and the landscape architect. Alas, this kind of inter-professional planning and design is too often omitted, and problems show up later. No doubt, that is why street trees, added later, so often come to mind to counteract thermal discomfort. The questions remain, however, do trees really solve the problem, and if so, why and how?

One of the reasons the doubts remain — another fact I invite you to keep in mind — is that one occasionally reads such reports as the article I saw recently on the science page of my regional paper. The article said that New York City loses about half the street trees planted there, because they are exposed to desert conditions in the streets — reflected light and extra heat. To me that sounds like a direct contradiction of the idea that street trees can cool their own immediate environments by their "breathing" — evaporative cooling.

I invite you to keep an open mind as you read this book, against the possibility that many other of your understandings may be shown to be doubtful, while many more will be reinforced with a new understanding of both "what" and "why." Among other things I hope you will recognize is that what "works" in one geographical region may be a disaster in another. And still another thing I hope you will recognize is that shapes and materials which one cannot even see from a certain place in an urban canyon —

for example at the roof level — nevertheless partly determine the microclimate at street level.

Finally, I want you to recognize that what you find out when you read — about things like trees to cool urban spaces by evaporation, about urban and rural temperature comparisons, and about the connections between temperature, humidity, and human comfort — depends very much on what you read and who you ask. These subjects may not be quite so simple as many seem to think.

And so we come to the purposes of this book.

Planners and designers have a *need to know* about the interactions of the atmosphere with the things they build in determining resulting microclimatic behavior and human comfort levels. At another scale, they have a *need to know* about how the parameters of microclimate and comfort change geographically — from one region or latitude to another. The best planners and designers will elect to satisfy that need rather than to ignore it and try to bury their ignorance, in a technological coffin, deep within their design. Instead, they will reach out to those with the specialized knowledge to satisfy their needs, and then manage, with the help of the specialists, to build the knowledge rationally and correctly into their design.

This book represents what I judge to be a catalog of basic information and concepts about the interactions of the atmosphere with the things you will build — that is, "the truth as I know it" about the parts of atmospheric ecology you need to know. The same basic information and concepts are to be found scattered here and there, but such a unified catalog as this does not exist elsewhere. What is more, I am certain there are many myths and half-truths scattered, along with correct information and concepts, in the writings of planners and designers who have tried — of necessity, because finding and talking with the right specialist isn't always easy — to be their own experts on microclimate and human comfort.

This book will enable you to *understand the interactive processes* involved in atmospheric ecology, and how they produce the various results of design alternatives. Thus, you will be able to *make rational comparisons* of those alternatives, and thereby to be able to choose the best one.

Once you have studied this book you will understand why the science article was correct about New York's street trees being in desert conditions, and why it is a myth that street trees will cool urban spaces by evaporation. And you will understand why you would be off the mark in thinking that temperatures at midday in summer are cooler in the countryside than in the city. It is another myth: you will understand why these temperatures are very nearly *the same*. And you will understand just how to relate the details of microclimate, produced by the design, to human comfort.

This book will enable you to *use the specialist-consultant intelligently* — how to approach him, query him, and listen to his feedback.

Finally, this book will enable you to *evaluate the work of others* in matters of atmospheric ecology. You will be able to tell if a consultant is "for real" and whether or not writers and other designers know what they are talking about in matters of atmospheric ecology.

The book is meant to take you by the hand through its catalog of information and concepts. The style is very informal, but don't let that blind you to the fact that you'll have to think hard about what you study. It's my intention that you will begin slowly, and then soon build up a momentum of comprehension because the topics build upon each other as you progress through the chapters.

I said the style is very informal, so before we embark, let me introduce myself. As an undergraduate I majored in chemistry, then engineering, and graduated in mathematics. I love to explain things, so at that time I intended to be a High School math teacher. Service in the Army had taken me to Alaska, where, though I didn't know the term, I met and loved "human bioclimatology." After the math degree I worked for the Corps of Engineers studying snow — more "human bioclimatology" — in the High Sierras of California, from which I went to Madison, Wisconsin, to study for an M.S. in meteorology.

After Wisconsin, I accepted a position with the Oregon Forestry Department, studying forest fire weather. That led me into ecology — forest ecology, in particular — from which I have happily never escaped. Since my doctorate at Oregon State University I have been engaged in teaching, as well as research and cooperative efforts of various kinds — consulting, graduate training programs, etc. — in atmospheric science, forestry, agriculture, air pollution control, and various design fields. My private pilot's license has let me see, from the air, many of the things I have been thinking, talking, and writing about. I hope my enthusiasm for my subject shows through.

Bon voyage.

P.S. I have told you about myself, and I would like to know about you, too. If you find this book to your liking, let me hear from you. Just a note will do, but I would very much enjoy knowing the ways the book serves you.

Wm. P. Lowry
Box 1264
McMinnville, Oregon 97128-1264

2. How does this thing work?
Systems and the energy balance

Most of the discussions in this book are about **physical systems**, and about the flows of **mass and energy** through those systems. *Mass* (or *matter*) is the collection of atoms and molecules that make up the "bulk" of a system, and *energy* (as the physicists define it) is that property of the mass that enables it to do work.

You can probably imagine the flows of energy — heat energy, for example — into and out of a brick, and not have any trouble thinking of that brick as a physical system. You might have a bit of trouble, however, thinking of a human being as a physical system. In the sense we are going to consider a human being, in particular the discussion of human comfort, the human being is a physical system with heat (energy) and water (mass) flowing into it and out of it. We will need to examine many kinds of systems in this book, so that the brick and the human body are just the two I've chosen to get us started.

To begin, think of a physical system in terms of the sketch in **Fig. 2-1**. The system itself has *boundaries* — it occupies a particular part of space. We can choose whatever boundaries we want when we discuss a physical system, depending on what processes we want to study.

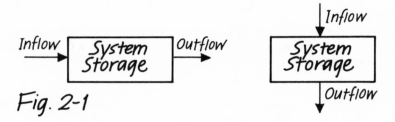

Fig. 2-1

The system has an inflow and an outflow. The particular '"orientation" — two are shown — doesn't really matter except as it might be chosen to represent the actual directions of flow in the real world. For example, the "left-to-right" flows might be through a wall, while the "top-to-bottom" flows might be through the horizontal surface of the earth. For the moment, it doesn't matter what is flowing in the arrows and through the system — mass or energy. In fact, as we will see in **Chapter 4**, there are several kinds of energy that might be flowing, and several kinds of mass. What matters right now is that the arrows represent **rates of flow** — how much *per unit of time*. Graphically, an arrow will be larger or smaller depending on whether its rate of flow is larger or smaller.

Fig. 2-2 suggests that a system may have several inflows and several outflows, each at a *different rate*. What we need to do now is relate the set of inflows and outflows to the SYSTEM STORAGE. We do that by using the **Law of Conservation** which says that neither mass nor energy can be created nor destroyed, so that *what flows in — in a certain period of time — must either flow out or result in a change in the storage in the same period of time.*

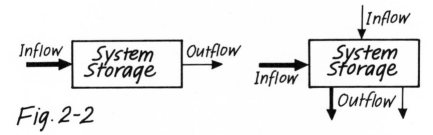

Fig. 2-2

We can represent that statement formally with an equation[P] like this:

$$\left[\begin{array}{c} Sum\ of\ all \\ Inflows \end{array}\right] = \left[\begin{array}{c} Sum\ of\ all \\ Outflows \end{array}\right] + \left[\begin{array}{c} Change\ in \\ Storage \end{array}\right]$$

In **Fig. 2-2**, if the total inflow exceeds the total outflow, there will be — in the time period being considered — a net increase in the system storage, of either energy or mass. We can suggest

---

(P) This "footnote" is marked **(P)** because it's a "Peptalk." Hereafter, when you see a **(P)**, refer to the end of the chapter for the peptalk.

graphically that flows are larger, for example, by arrows which are wider or longer, or both. But to show the actual balance, we must attach numbers, values, or *magnitudes* to the arrows. Before we do that, however, notice that we have said, in addition to the physical *boundaries* of the system (mentioned previously), we must also specify the *time period* we are talking about in order to make a study of the **energy or mass balance** of the system.

In **FIG. 2-3** we have attached numbers to the flow arrows, so that the balance, using the equation above, is: (10 + 4) = (7 + 1) + (Change in Storage). The change in storage must be (+6) to make the equation balance (14 on each side of the equal sign), so that *at the end of this time period* the system storage will have been increased by (6) and be equal to (36) as the next time period begins.

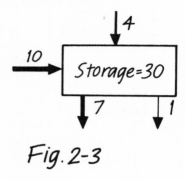

Fig. 2-3

Seen in this way, it is clear that *the storage in a system **now** is the net result of all the flows that have taken place in all the preceding time periods. In any time period during which the total outflow exceeds the total inflow, the change in storage will be negative*, and the system storage will be *decreased* during the time period.

In more complex analyses, the whole system may be shown as having more than one box (also called a "reservoir"), the several boxes being connected by a set of arrows, as in **Fig. 2-4**. If the system is in **equilibrium**, the flow rates will be such that **no changes in storage** will take place: *(Inflow)* = *(Outflow) for each reservoir*. The system in **Fig. 2-4** is in equilibrium.

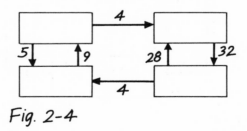

Fig. 2-4

Before we leave these systems, notice that *there need be no correlation between the magnitudes of the flows and the magnitudes of either the system storage or the rate of change in that storage*. As shown in **Fig. 2-5**, either may be large or small, independent of the others, as long as the equation (representing the Law of Conservation) balances. Can you spot the fourth case that has been omitted in **Fig. 2-5**?

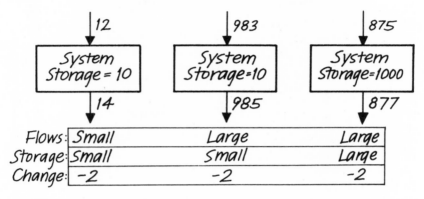

Flows: Small	Large	Large
Storage: Small	Small	Large
Change: −2	−2	−2

Fig. 2-5

Finally, we need to examine the concept of **feedback,** which is the term used to say that something within the system functions as a *regulator,* so that what is going on at one point in the system determines — at least in part — what goes on at another point in the system. This connection between two points in the system is called a **feedback loop,** and the loop may connect any two points. **Fig. 2-6** shows two examples, indicated by dashed lines.

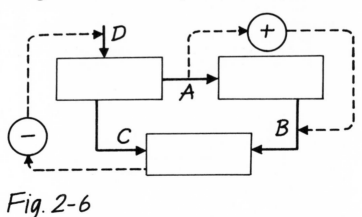

Fig. 2-6

Between *flow* (**A**) and *flow* (**B**) there is **positive feedback**. This means that any increase in the flow rate in (**A**) will result in an increase in (**B**), while a decrease in (**A**) will result in an decrease in

(B). Flow **(A)** regulates flow **(B)**. *Positive feedback means an enhancement* of whatever is occurring, either increase or decrease.

Between *reservoir* **(C)** and *flow* **(D)** there is **negative feedback**. This means that any increase in the storage in **(C)** will result in a decrease of flow in **(D)**, while a decrease in **(C)** will result in an increase in **(D)**. Storage in **(C)** regulates flow **(D)**. *Negative feedback means a suppression* of whatever is occurring, either increase or decrease.

Not shown in **Fig. 2-6** are the **feedback rules**, which are the quantitative relationships governing the form of the regulation taking place. As very simple examples, the response in **(B)** might always be twice as large as the change in **(A)**; and the response in **(D)** might not begin until the storage in **(C)** decreased to a certain prescribed amount.

Systems analysis is a large and complex subject, but these few ideas will serve as a basis for study of many climatological and biological systems in later chapters. We can leave the subject at this point by remembering that what flows in the system may be either mass or energy, and that the Law of Conservation is to be obeyed — the *balance* is to be maintained — as one of the operating rules of the systems we will study.

Peptalks

(P₁) In peptalks, I mean to try to take care of any problems you might have with numbers or ideas in the various forms of math and equations. Maybe after a while you will see that "math and equations" are just symbols, and that when we use them, they are just a shorthand, used to get on with our thinking about some complex ideas. If you don't have trouble with "math and equations", so much the better.

3. What is it like HERE?
Observed microscale patterns

In this chapter we will examine the time and space patterns of several weather variables as they are observed on the local, microclimatic scale. The variables are *insolation* (solar energy), *temperature* (and heat), *moisture* (and humidity), *wind*, and *clouds*. Finally, we will look within plant canopies, at how to express the three-dimensional structure of the plant parts there, and very briefly at how the time and space patterns of the same weather variables (except clouds) change within the canopies.

Local patterns of insolation

Insolation is the rate of arrival of solar energy on a unit area of the earth's surface. In the definition we will use, the surface *need not be horizontal*, though in some things you will read "horizontal" is part of the definition. In the terms of physical units, that is (Energy divided by Area divided by Time), or (energy/area-time)$^{(P_2)}$. There are several parameters of insolation that need to be specified in a discussion: (a) *latitude*, (b) *date*, (c) *hour*, and (d) the *orientation of the surface* on which the insolation falls. Sometimes the discussion involves *integrated values* of insolation: values of the instantaneous insolation at one place, on one surface, that are then "summed" according to the needs of the discussion. For instance, we may talk about the instantaneous insolation (a) at 40° North latitude, (b) on April 12, (c) at 3 p.m., (d) on a northeast-facing slope, inclined at 30°. One might then integrate for all the daylight hours of April 12 to get a "daily sum", or integrate over all the daylight hours of a year to get the "annual sum." In the extreme case, one could integrate for both time and space and get the annual sum for the whole earth. You get the idea.

While insolation pertains to the amount of solar energy arriving at the surface, much (but not all) of the variation in insolation is due to the changes in **solar geometry**, which is concerned with the position of the sun in the sky. Most of the rest of the variation in insolation is due to changes in dirtiness and cloudiness of the air. **Fig. 3-1** shows the commonest form of presentation about solar geometry. The **sun path diagram**[1] is for one latitude: 40° North. As will be explained in greater detail in the next chapter, reading the sun path diagram will give you information on the sun's location in the sky, but not its location with respect to any except a horizontal surface.

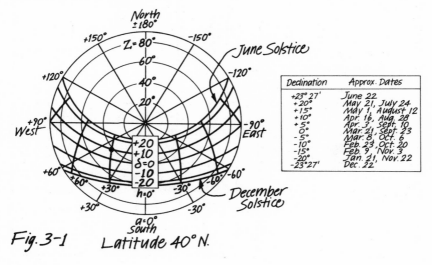

Declination	Approx. Dates
+23° 27'	June 22
+20°	May 21, July 24
+15°	May 1, August 12
+10°	Apr. 16, Aug. 28
+5°	Apr. 3, Sept. 10
0°	Mar. 21, Sept. 23
-5°	Mar. 8, Oct. 6
-10°	Feb. 23, Oct. 20
-15°	Feb. 9, Nov. 3
-20°	Jan. 21, Nov. 22
-23°27'	Dec. 22

Fig. 3-1   Latitude 40° N.

Fig. 3-2 shows two kinds of presentation for insolation: (a) daily sums on a horizontal surface, for each day of the year, for each of several latitudes; and (b) instantaneous values on each of five slopes, at one latitude on one date. The two presentations, obviously, are useful in two different kinds of discussion. Both are useful for "instructional purposes", but in practice their uses are different. One is to compare climates at different latitudes, while the other is to compare different slopes at the same place on one day.

The number of combinations of the four parameters and their integrated forms is practically endless. One could present a huge variety of diagrams about insolation.[P_3]

About all we can do in this Chapter is talk about the things all the presentations have in common, and I can mention a few places

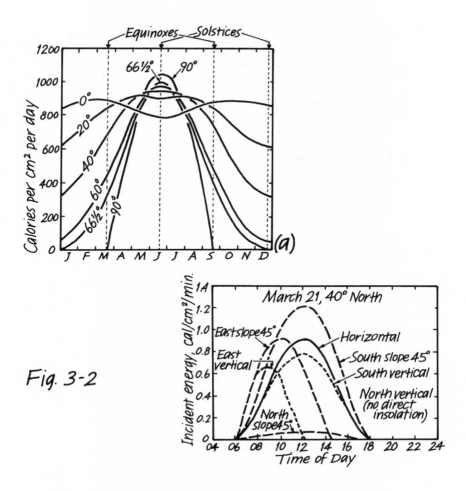

Fig. 3-2

to look (in the library) where people have already answered some questions about insolation[2].

On the subject of insolation, we ought to note that presentations may be based either on *theory* (discussed in **Chapters 4 and 9**) or on *observations*. In the case of the latter, for example, one may obtain, directly and with confidence but at some expense, the effects of dirty air or of clouds on insolation. In the case of theory, on the other hand, one may **estimate** these effects, with less confidence but also with less expense. This *trade-off between theory and observation* is another point that will keep showing up in this book.

Fig. 3-3 shows two other kinds of presentation about insolation. **Fig. 3-3a** is based on measured values, on clear days at Tucson, Arizona (Sellers, 1965, page 36). Each panel contains all combinations of hour and slope steepness (called the "inclination" of the slope) for a single date and a single slope direction (called the "aspect" of the slope). This kind of presentation is clearly meant for comparisons of insolation on different parts of the local terrain, but we are stuck with the dates and directions selected by someone else.

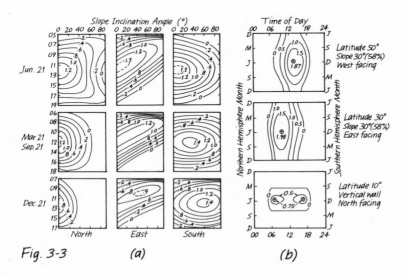

Fig. 3-3   (a)   (b)

Fig. 3-3b is based on theoretical calculations. Each panel contains all combinations of date and hour for a single combination of latitude and slope. Among other things it shows clearly the times of *sunrise and sunset* on the slopes presented (where the lines have the value of zero), but we are stuck with the limited number of combinations chosen by someone else. Obviously, this presentation is only good for "instructional purposes."

Local patterns of temperature

As with insolation, a good way to start this section is with a few definitions. First, it's good to note that *there is a difference between heat and temperature*. The difference will be discussed more completely in the next chapter, but you should know now that, while temperature differences determine which way heat will flow, two objects with the same temperature can contain quite different amounts of heat.

The next thing to do is to take note of the different **temperature scales** used by ecologists. Often, on public signs, both
Fahrenheit and **Celsius** (formerly "Centigrade") temperatures are
displayed. Both are found in technical ecological literature as well.
The following two tables will give you enough basic information
about these two scales, but when you need to convert from one to
the other, it will require a little bit of algebra[P.].

Table 3-1

Celsius	0	10	20	30	40
Human comfort	Cold	Cool	Comfort	Warm	Hot
Fahrenheit	32	50	68	86	104

Table 3-2

Scale	Freezing	Boiling	Number of degrees (Boiling - freezing)
Celsius	0	100	100
Fahrenheit	32	212	180

Consideration of microenvironmental patterns of temperature
involves three parameters: (a) *time*, (b) *temperature*, and (c) *distance from the interface* between the air above and the soil or
water beneath. As with graphical presentations of insolation, we
have a choice of combinations of these parameters in our presentations of temperature patterns. In the discussion here we are concerned only with the microclimatic temperatures of air and soil (or
water), leaving discussion of the temperatures of living things
within the microenvironment until **Chapters 5** and **6**, and of the
temperatures on the larger regional and global scales until **Chapter
10**.

Incidentally, in case you still aren't quick at relating distances in
meters and centimeters to feet and inches, here's a little scale to
help you get used to the scientist's use of the **metric scale** of
lengths.

Fig. 3-4 presents probably the most common
combination of parameters,
in which the two coor-
dinates are time (in this
case the hours of a day)
and temperature (degrees
C), and each line repre-
sents one distance from
the interface. The comparisons here are between (a)
air and (b) *soil* during one
24-hour (**diurnal**) cycle.

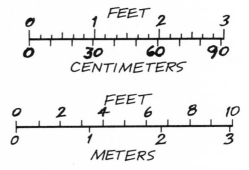

The case of *water* below the interface will be considered presently.

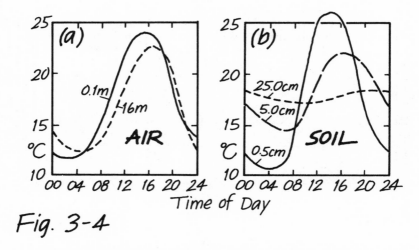

Fig. 3-4

In **Fig. 3-5** the presentation is of the same kind, except here the comparisons are between (a) the *diurnal* and (b) the *annual* cycles of soil. The numbers — the values of the parameters — are different, but the basic patterns are very nearly the same.

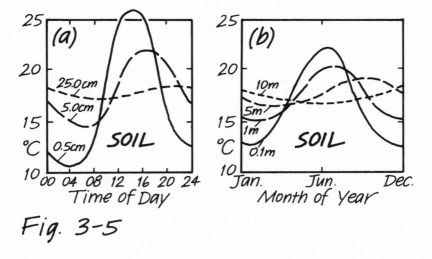

Fig. 3-5

In **Fig. 3-6** the presentations are of the other two combinations of parameters, with both air and soil shown together. In **Fig. 3-6a** the coordinates are temperature and distance, with each line (called a **temperature profile**) representing one hour. In **Fig. 3-6b** the coordinates are time and distance, with each line (called an **isotherm**) representing one temperature.

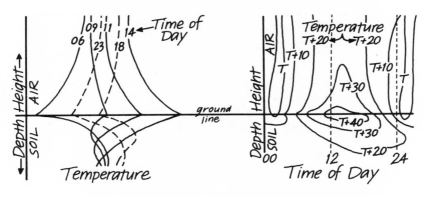

Fig. 3-6

Careful study will show you that the information, though presented in three different ways, is qualitatively the same in **Figs. 3-4** and **3-6**. Further, careful study of these figures will show you that we can deduce several *"rules of behavior"* of these microenvironmental temperature patterns, and that both the diurnal and the annual patterns in soil follow the same rules. Here are the rules, *which apply in both air and soil.*

1) The greatest variability of temperature, through a day or a year, is found at the surface;

2) the **delay** in time of occurrence of any temperature event (such as the maximum or the minimum value) increases with increasing distance from the surface;

3) the **decay** (reduction) in amplitude of a temperature "wave" — the difference between the maximum and the minimum — increases with increasing distance from the surface; and

4) the distance from the surface at which one finds *a given amount of delay or decay* is much greater in air than in soil, because air is a light fluid and soil is a dense solid.

As we shall see later, the values of various physical characteristics of the soil determine the particular temperature patterns in the soil, and one can change — manage — these patterns by judicious changes in the values of these physical characteristics.

How do these rules of behavior apply to water? **Fig. 3-7** suggests the answer. In **Fig. 3-7a** the diurnal temperature behavior at the surface and the bottom of a shallow 30cm-deep pool (from Geiger, 1965, p.190) is compared with that at a depth of 0.1 cm in soil on a clear, sunny day (the same as the soil temperatures in **Fig. 3-4b**). Because of (a) evaporative cooling at the pool's surface, (b) the ability of the water to mix and distribute its heat downward, and (c) the fact that the sun can shine through the water to the bottom, the amplitude of a temperature wave at the surface in the pool is much less than it is at the surface in the soil. The original records for the pool show that the wind was blowing during the night and morning, so that the water was well mixed, and the surface and the bottom were at nearly the same temperature. When the wind was calm, on the other hand, the water near the surface could warm above the bottom temperature. Summarizing, because of the water's transparency, its mixing ability, and the variability of the wind even on a clear day, the rules of behavior do not apply as well as they do in soil.

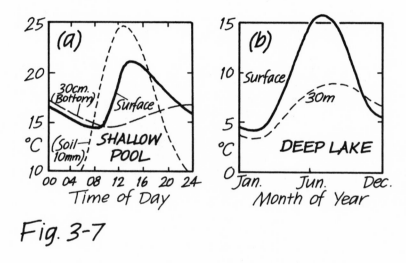

Fig. 3-7

In **Fig. 3-7b** the *annual* cycle of temperature at the surface of a deep lake (from Geiger, 1965, p.196) is similar to the *diurnal* cycle in the shallow pool. At a depth of 30 meters — 100 times the depth of the pool — the amplitude for a whole year is similar to that found at the pool's bottom during a single day. If we examined a typical *diurnal* cycle of temperature at the surface of a deep lake, we could scarcely detect any variation at all — the amplitude would be something like 1°C, so it's not even worth graphing the diurnal cycle to go with the annual cycle in **Fig. 3-7b**.

Fig. 3-8 shows dramatically the results of water's ability to be mixed by the wind, a primary difference from soil. The figure describes the annual cycle of events in a northern lake[A]. Before following the events clockwise around the cycle, you should know that water has the interesting property of being most dense — heaviest — when it is at a termperature of $+4\,°C$, above freezing. That is, when it is *either warmer or colder* than $+4\,°C$, it will float above any water at $+4\,°C$.

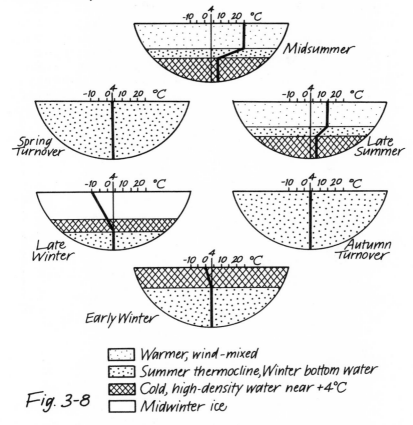

		Warmer, wind-mixed
		Summer thermocline, Winter bottom water
		Cold, high-density water near +4°C
		Midwinter ice

Fig. 3-8

Beginning at the top in midsummer, notice that the surface water has been warmed by the sun, and then mixed by the wind through a deep layer that has reached a temperature of $+25\,°C$. (In each sketch there is a temperature *profile*, drawn according to the temperature scale on which $+4\,°C$ is marked prominently). The deep water is near the temperature of greatest density, $+4\,°C$, and between these two layers is a transition layer, called the **thermo-**

(A) I realize I'm breaking my own rule about "*What* in this chapter and *Why* in **Chapter 4**", but this is the only place in the book where we'll discuss microclimates of water.

cline, in which the temperature becomes very rapidly colder as you pass downward through it.

As you go clockwise around the cycle, notice that the surface water, in contact with sun and wind, follows the seasonal cycle of air temperatures — from +25 to -10 degrees — and that the bottom waters are seldom far from +4 °C, the temperature at which all other water floats above. As autumn comes, there is a time when the wind-mixed surface waters reach +4 °C, and then, for just a short while, *the whole lake has the same temperature*. Since then all the water has the same density, it does not stratify any longer. In fact, surface and bottom waters, together with all their oxygen and nutrients, all mix together in the **autumn turnover**. Into winter, the surface waters grow colder and less dense, and stay on the surface. If the winter is cold enough, an ice layer grows downward from the surface, leaving a layer of unfrozen bottom water, part of which is between freezing and +4 °C. Unless the winter is too cold and long, this provides an unfrozen, but oxygen-poor, winter refuge for fish. With spring, there comes another *turnover*, and the mixing of oxygen and nutrients is repeated as the annual cycle is complete.

Fig. 3-9 introduces the **temperature envelope**, which is the plotting of the *maximum and minimum values of temperature* versus distance. There are two plots (a maximum and a minimum) for each "substrate" — kind of material or medium — and the two look as if they are temperature profiles. A moment's thought will tell you, however, that these plots are not true profiles, because the observations plotted do not occur at *one time*, as they do in a profile. **Fig. 3-9a** shows diurnal envelopes for air and a typical soil

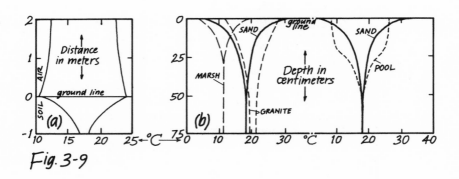

Fig. 3-9

substrate. They are equivalent to the "outer limits" of all the profiles in the generalized diagram of **Fig. 3-6a**.

Fig. 3-9b shows sub-surface, diurnal envelopes for three kinds
of soil and the water in the shallow pool. The envelope for sand is
repeated in both parts of the figure for comparison. Notice that
the height (distance) scales are not the same for the two parts of
the figure. In **Fig. 3-9b** the distances are exaggerated.

So what's special about an envelope? What good is it? Of course,
it shows the amplitude at each distance within a particular substrate, thereby showing, among other things, how far the diurnal
temperature variation at the surface penetrates into the soil or
water. More than that, it shows how much heat flowed into and
out of the soil or water during the course of a 24-hour period.

For example, in the right hand part of **Fig. 3-9b**, the envelopes
for sand and water show that the wind has mixed the heat farther
down into the pool than the heat could penetrate into the soil. But
did more heat enter and leave the water than the soil? In which
substrate was the exchange between the surface and the depths
greater during the course of the whole day? The total heat
exchange is proportional to the *area within the envelope* —
between the two curves and below the surface. Judging by the two
envelopes in the right side of **Fig. 3-9b**, the total exchange was
slightly greater in the water than in the soil, and it was distributed
quite differently.

Local patterns of humidity and soil moisture

When you hear "humidity" you probably think about the term
relative humidity (RH), because that is the measure of
atmospheric moisture most often encountered in ordinary conver-
sation. Unfortunately, that term has limited utility when it comes
to ecological concepts. True, you can, by experience, predict you'll
feel one way when you go outside on a sunny summer day knowing the Weather Man has said "the relative humidity is 68%", and
another way if he had told you "the relative humidity is 35%."
You'd know the former would leave you feeling "sticky" and the
latter perfectly comfortable. *Knowing the approximate temperature*, you can predict differences in your *comfort* on the basis of
statements about RH. But, as you'll see, that's not good enough for
a reliable scientific statement about *how much water vapor
(moisture) is in the air.*

Fig. 3-10 shows a typical pair of diurnal patterns — one for air temperature and one for RH —during a summer day in the central United States[B]. It is almost certain that on that day the amount of moisture in the air was fairly constant, but the RH swings widely from night to day. In fact, as one goes up the other goes down — they are *negatively correlated*. We will examine this relationship more fully in the next chapter.

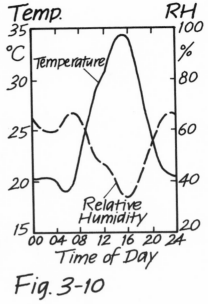

Fig. 3-10

Rather than RH to express the amount of moisture in the air, it is much more appropriate and useful to use the **dewpoint temperature**, which will be discussed in the next chapter[P]. **Fig. 3-11** shows air temperature, dewpoint, and RH, all plotted side by side. **Fig. 3-11a** is for a cloudy winter day at a city in the western U.S.[B], while **Fig. 3-11b** is for a clear summer day at the same city.

About the only things similar between the two days is that the *dewpoints are nearly the same on both days, and both vary only a little bit from hour to hour* no matter what the temperature and the RH are doing. Since the dewpoint is about 10 °C at all times in the figure, we know that the amount of moisture — water vapor mass — was about the same. Clearly, using the dewpoint is very reliable for describing atmospheric moisture. As we will see in **Chapter 5**, however, combinations of temperature and RH are customarily used to specify measures of *human comfort*.

Fig. 3-12 shows temperatures and dewpoints (but not RH) at the same city and through the same year[B] as in **Fig. 3-11**. The values plotted are for midday on the 1st and 15th of each month. Again, the dewpoint varies much less than the temperature. That by itself isn't so important as the fact that the record of the dewpoint discloses that, through the year, there are about three "kinds of day" that show up — three levels of atmospheric moisture: with dewpoints near 0 °C, 5 °C, and 10 °C. These are related to the concept of **airmass**, as will be discussed in **Chapter 10**.

(B) Fig. 3-10 is from central Nebraska. Figs. 3-11 and 3-12 are from Salem, Oregon, in 1972.

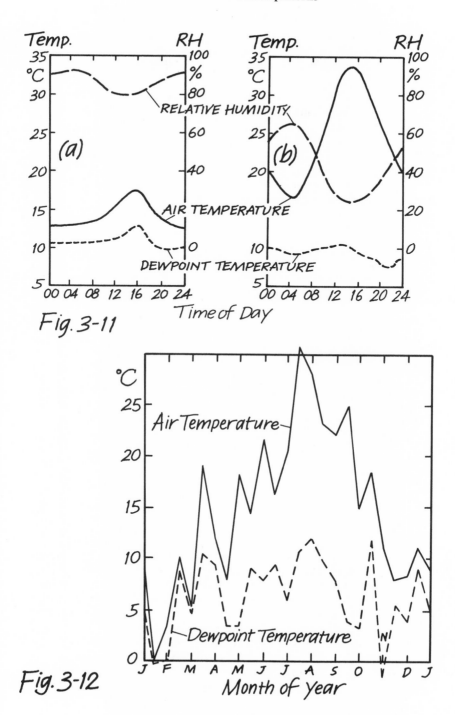

Fig. 3-11

Time of Day

Fig. 3-12

Month of year

In making a presentation, at the planning stage, about the climate of a locality, it is preferable to use the dewpoint temperature, rather than the RH, to represent atmospheric moisture in the presentation.

What can we say about local patterns of *soil moisture?* There are no simple, recognizable *diurnal* and *annual cycles* of soil moisture. The space and time patterns are much more variable and less predictable, and of a greater variety, than those for temperature and humidity, since they depend not on the daily and seasonal passage of the warming sun through the sky, but on the irregular coming of rainfall associated with storms. And if the **precipitation** comes as snow rather than as rain, the effect on soil moisture is delayed until melting occurs.

Typically, *we* experience soil moisture in terms of the wetness of the soil surface, while *plants* experience it mainly in terms of the moisture content in their **root zone**. The two need not tell the same story at all.

On the surface, we can trace a cycle starting with sloppy, puddly mud right after a downpour, followed by drainage and the disappearance of the puddles. The physical characteristics of the soil largely determine how much puddling and how fast the drainage. Then, as the sun shines and the wind blows, the surface dries out some more — no longer simply by drainage, but by evaporation. If the dry period continues, the surface begins to form a "crust" — how thick and how crusty also depends on the physical characteristics of the soil.

A shovel may disclose plenty of soil moisture beneath a dry crust, but then again the soil may be bone dry. It is difficult to tell the status of the soil moisture just by looking at the surface. On the other hand, the plants can tell us, by curling or drooping leaves, if the soil is dry beneath the surface.

The point of this word picture is simply to say that *local patterns of soil moisture cannot easily be described in terms of regular, general cycles and trends* as temperature and humidity usually can. In **Chapter 4** we will discuss the details of the relationships between the physical characteristics of soil and the soil's wetness, as experienced by us and by plants.

Basic characteristics of windflow

Up to this point we have described local patterns of three environmental variables: sun, temperature, and moisture. The last of the "Big Four" is **wind**. *As a general rule, unless you include all four in any description you give for a microclimate, your description is incomplete, usually in a major way.*

To begin consideration of windflow, here are scales to compare the various *speed* systems you will encounter. This is a good place to note that people who study wind make the distinction between **speed** and **velocity** this way: velocity has two components — speed and **direction**, and a wind is named for the direction *from which it comes*.

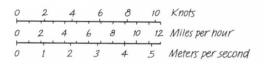

A British admiral in the days of "tall ships" devised a scale of wind speed in terms of the things it does — what you can see and feel. The **Beaufort Scale** bears his name. Here is the Admiral's scale translated into "landlubber" terms. It can often be helpful when you are making field observations, or when you need to translate, quickly and in your mind, the relationship between wind speed and its effect on objects.

Beaufort Number	Wind Speed (mph)	Description
0	Under 1	Calm; smoke rises vertically.
1	1-3	Smoke drift indicates direction.
2	4-7	Leaves rustle; wind vanes move.
3	8-12	Leaves and small twigs in constant motion.
4	13-18	Small branches move; leaves, dust, paper lifted from ground.
5	19-24	Small trees in leaf sway.
6	25-31	Larger tree branches move; whistling heard in wires.
7	32-38	Whole trees in motion; difficulty in walking upright.
8	39-46	Twigs and small branches broken from trees; walking impeded.
9	47-54	Slight damage to structures, roof tiling blown off.
10	55-63	Trees broken or uprooted; damage to some structures.
11	64-72	Widespread damage begins.
12	Over 72	Widespread damage.

Experience tells us that wind is sometimes "smooth" and sometimes "gusty." You can feel the gustiness and see it in the erratic flight of leaves and paper scraps. To the component of wind that makes the gustiness, we give the name **turbulence**. Smooth-flowing wind has only a little bit of it, and gusty wind has a lot of it. In both cases there is motion, but one has more of the turbulent component added in than the other does. In the form of an equation the idea looks like this:

(**Instantaneous** velocity) = (**Average** velocity) + (**Turbulent** velocity)
(Actual flow) = (Steady flow) + (Turbulent flow)

Graphically, we can think of this turbulent flow as shown in **Fig. 3-13**. The average velocity has a constant speed and direction, as in **Fig. 3-13a**. It is the time-averaged component that would be registered by any instrument that didn't respond to turbulence. Then there is the turbulent component, which we further break down into three components: *downwind, crosswind, and vertical.* If we were to observe the *actual flow* a large number of times, and then mathematically separate out from each observation the turbulent component[P6], the large number of turbulent components in our sample would plot like the familiar *bell-shaped curve* in **Fig. 3-13b**. This means these turbulent components are random in nature. Note that some kinds of turbulence have larger components than others (see footnote (C) later).

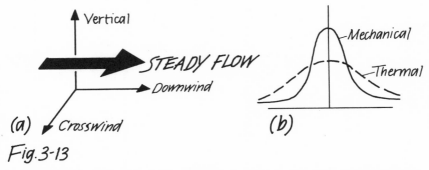

Fig. 3-13

Fig. 3-14 shows what happens when we plot the *average speed* at various heights against *height.* The small figure standing in this **wind speed profile** — he's about 2 meters tall — has his feet in relatively calm air and his head in air that's moving, on the average, at about 9 meters per second (mps). That's a big change of speed within a small (vertical) distance. Above his head, the speed increases only to a little bit more than 12 mps all the way up to 8 times his height — 16 meters. These height differences — difference in speed divided by difference in height — are called **vertical gradients** of speed. In fact:

(Gradient of anything) = (Difference of that thing) ÷ (Height difference).

Here are the two gradients just mentioned:

below the man's head:

(9 - 0) ÷ (2 - 0) = 4.5 mps per meter

above the man's head:

(12.2 - 9) ÷ (16 - 2) = 0.23 mps per meter

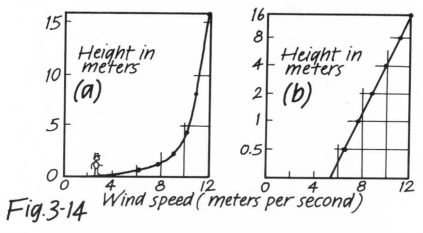

Fig. 3-14 Wind speed (meters per second)

But back to the profile in **Fig. 3-14a**. It is very curved, because the frictional drag of the surface on the passing wind stream slows it down right next to the surface — around the man's legs and feet. It is a lot less curved above the man's head *where the gradient is small* and the frictional drag is less. The layer of air in which the wind's behavior is affected by the presence of the surface — where the surface drags on the wind — is called the **boundary layer** of the atmosphere.

The wind profile in the boundary layer is **logarithmic**, as we can see in **Fig. 3-14b** where replotting the speeds in a certain way makes the curved profile into a straight line. The new plotting has the heights on a logarithmic scale where each *doubling of height* is the same distance on the scale[P.].

Fig. 3-15 connects the ideas of *turbulence* and *wind speed profile*. In the upper part of the figure, we see the **trajectories** of air parcels — like tiny balloons — as they move along in the turbulent windstream. All the parcels left the same place — let's say different heights on the same pole — at the same time, so they have all travelled the same length of time. Notice that the ones farther from the surface have travelled farther — their *average speed* is greater — than those closer to the ground. In fact, if we connect the end points of these trajectories, they form a *logarithmic wind speed profile*. The figure is like a *time exposure* of the parcels.

Notice also that the parcels have not followed straight line paths. They have wandered up and down (and also back and forth, if we can imagine the dimension into and out of the page) as they moved on the average, steady flow. The departures from a straight line are due to the turbulence. There is more of it farther from the surface and less of it near the ground.

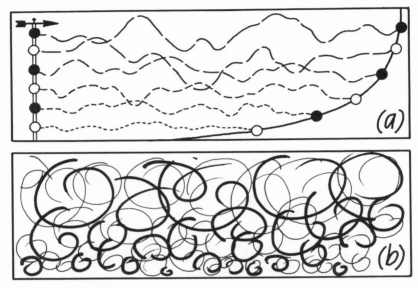

Fig. 3-15

In **Fig. 3-15b** is another view of the turbulent boundary layer. Here we see **streamlines** of the parcels in motion at one instant — more like a *snapshot* of the motion. The quasi-circular elements of flow are called **turbulent eddies**, and their diameters are larger above the surface than near the ground. So, to summarize one idea about windflow, a *trajectory* tells you where the parcel has been, while a *streamline* tells you what it is doing right now. And all the turbulence and variability in the boundary layer is due to the presence of the surface interacting with the passing airstream[C].

(C) The discussion might leave you with the impression that all the turbulence is due to the frictional drag of the surface. If the wind speed is moderate or more, it is in fact this **mechanical turbulence** due to friction that accounts for most of the turbulence. However, if the surface is warmer than the airstream, the heating of the air by the surface produces buoyant forces which appear as an additional component called **thermal turbulence**. If the wind is light, most of the turbulence is thermal; if the wind is stronger, most is mechanical, though there is almost always a mixture of the two. Refer to the two curves in **Fig. 3-13**.

Before we go on to the subject of local differences in wind behavior, we need to note the fact that, in addition to the kind of boundary layer we have examined just above the earth's surface, *any object protruding into the windstream will have its own boundary layer.* For example, any person, bush, branch, twig, or leaf will have its own **microscale boundary layer** in which the presence of the object affects the motion of the air. **Fig. 3-16** suggests the forms of these microscale boundary layers, in terms of the two most-studied shapes: *flat plates* (simulating leaves) and *cylinders* (simulating things like a man standing). Without going any more into the details of this complex subject, it is enough to recognize that the *boundary layer is thicker* if the wind is lighter and the object is larger.

WIND

Fig. 3-16

Boundary Layer

Boundary Layer

Responses of wind to locally varying surface properties

In the previous section we considered some of the basic characteristics of wind and of the boundary layer associated with a simple, smooth surface. In **Chapter 4** we will examine the basic *cause* of wind. Let's look now at how wind flows over and around different kinds of surfaces and different objects on the surface, and how the boundary layer responds as the airstream moves across the boundary from one kind of surface to the next one downstream.

Boundary layers come in many sizes — around an individual leaf and between Earth and space. Even at one size, there may be several boundary layers, caused by the atmosphere flowing first over one kind of surface and then over another kind. The basic feature of the transition from one to the next is the **leading edge** of a newly encountered surface type — at the boundary between two "patches" on the earth's surface. **Fig. 3-17** shows this leading edge and its relationship to the transition.

Downstream from the leading edge, the accomodation made by the airstream is the formation of an internal boundary layer as the primary layer thickens upward. If, again downstream, the air meets another leading edge — another "patch" — the internal boundary layer appears as a **plume** reaching up into the primary boundary

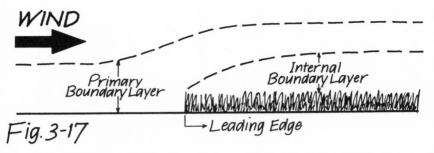

Fig. 3-17

WIND

Primary
Boundary Layer

Internal
Boundary Layer

Leading Edge

layer. The air within a plume, of course, takes on the characteristics (heat, moisture, turbulence, dust, etc.) of its underlying patch. The downwind length of the patch is called its **fetch**, and the greater the fetch, the larger and deeper its plume. **Fig. 3-18** shows how these plumes appear in cross section.

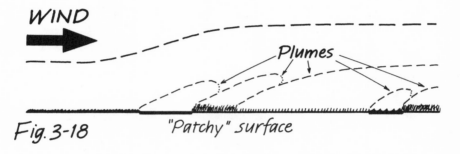

Fig. 3-18

WIND

Plumes

"Patchy" surface

In addition to the airstream's accomodation to underlying patchiness, it must also respond to underlying *obstacles:* individual features such as isolated hills and buildings. That response depends on several things about the obstacle: (1) its height-to-length ratio (h/d), (2) the "sharpness" of its edges — whether or not it is streamlined — and (3) whether or not it is at the same temperature as the air — whether buoyancy is present. **Fig. 3-19** shows typical flow lines around an obstacle that has a small ratio (h/d), is streamlined, and is at or below air temperature. The flow is "like a rock in a stream" — relatively unperturbed.

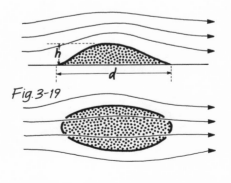

Fig. 3-19

In **Fig. 3-20**, on the other hand, the building has a large ratio (h/d), sharp edges, and is heated above air temperature. Here the flow is more complex, and breaks down into several generalized flow zones, or sub-volumes, all of which are readily recognizable in actual cases, such as those shown in **Fig. 3-21** based on well known photographs. Though it is not strictly true, it helps understanding to think of the boundaries (surfaces) between the sub-volumes as being **surfaces of separation**, across which there is no flow — compartments within the modified boundary layer.

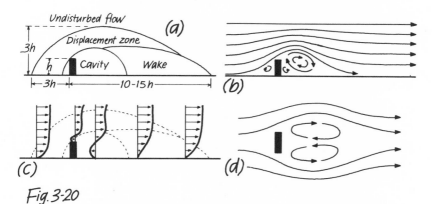

Fig. 3-20

There is a lot of information in the four parts of **Fig. 3-20**. The first panel shows the various sub-volumes and their names, together with the size dimensions expressed in multiples of the building height, (h). The second panel, **Fig. 3-20b**, shows the flow lines associated with the sub-volumes, while **Fig. 3-20c** shows the wind speed profiles in different parts of the boundary layer. Notice the *reverse flows* within the cavity. Finally, the last panel shows a plan view of **Fig. 3-20b**.

Fig. 3-21

Fig. 3-22 suggests some of the various flow patterns associated with *groups of buildings*. The number of variations of building heights, lengths, and spacings is so large, only a few examples are possible here[3].

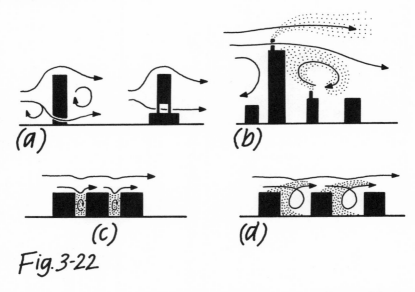

Fig. 3-22

Fig. 3-23 shows the basic information about the effects of *hedgerows and shelterbelts*, both natural and cultivated, on windflow. In general, the principles which emerge from such examples are that (1) a planting of low to intermediate density produces an effect farthest downstream, and (2) as in the case of the building wake in Fig. 3-20, the downstream distance is in the range of 10-15h.

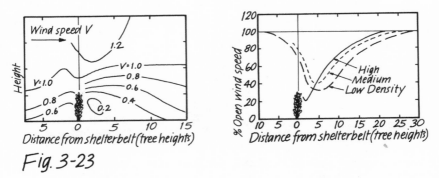

Fig. 3-23

Effects of local terrain features on airflow

In a different category from obstacles are *terrain features* that affect local windflow. As with building clusters, the possible combinations of terrain features — shorelines, hills, mountains, valleys, and the regional wind directions relative to those features — is so large as to make it necessary here to illustrate basic concepts with only a few simplified examples.

Because of temperature differences, as will be explained in **Chapter 4**, *shorelines* produce important local windflow patterns. **Fig. 3-24** illustrates these patterns, which consist of "onshore flow" by day and "offshore flow" at night when the skies are clear and regional winds relatively calm. When the regional weather is stormy, these shoreline effects do not become well developed. Following the rule that winds are named for the place from which they come, these day-night, shoreline patterns are among the **local wind systems: land-sea breezes and lake-shore breezes**. Chicago is "The Windy City" because of the lake breeze from Lake Michigan, so you know these flows can be vigorous when conditions are right. It is the sea breeze that brings the fog into San Francisco Bay.

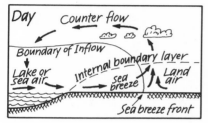 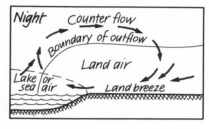

Fig. 3-24

Also because of temperature differences, *hillsides and mountainsides* produce important local windflow patterns. **Fig. 3-25** illus-

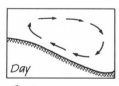 

Fig. 3-25

trates these patterns, which consist of "upslope flow" by day and "downslope flow" at night when the skies are clear and regional winds relatively calm. The comment above about storminess suppressing these local winds applies here. These day-night, slope patterns are also among the **local wind systems: mountain-valley winds**.

When the slopes are associated with isolated *hills and basins*, the flow patterns appear as closed, self-contained systems, as shown in **Fig. 3-26**. In three dimensions, they are shaped like giant doughnuts. Flows in the basin, when the regional wind is calm, are not the simple reversals of flow that they are for the hill. In particular, on calm nights cold, dense air settles in the basin bottom[(D)] to form a **frost pocket**. This air gets colder and deeper the longer is the night, as more cold air flows in and the air already collected there loses radiant heat to the clear night sky. Unlike the lake water in **Fig. 3-8**, air is *not* most dense above freezing — it keeps getting more dense as it cools below freezing; hence, on long clear nights the basin earns the name of "frost pocket." It is important to note that the *most efficient cold air drainage takes place on gentle slopes. A steep slope promotes turbulent mixing with warmer air, as the flow tries to accelerate, and reduces the efficiency of the downslope sliding of cold air.*

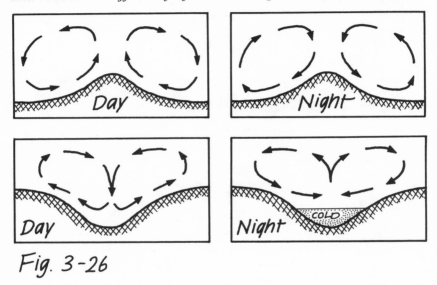

Fig. 3-26

Fig. 3-27 describes basic patterns of flow in a set of terrain features one step more complex than the isolated hill and basin. There are five daytime cases out of the many possible from the different combinations of basic factors: (a) day or night, (b) ridgelines E-W or N-S, and (c) regional wind perpendicular or parallel to the ridgelines. The sketches assume we are in the *Northern Hemisphere* where the midday sun is from the south, so that north-facing slopes are not warmed and so do not experience upslope flows. In **Figs. 3-27c and 3-27d**, the dashed lines represent the same kinds of *surfaces of separation* found in **Fig. 3-20**. The

---

(D) Frost can form by this kind of flow in basins both large and small — for example, from the Central Valley of California down to a minor depression in a hilly region.

two sketches of a basin in **Fig. 3-26** can be added as nighttime examples if the basin is considered as a valley instead.

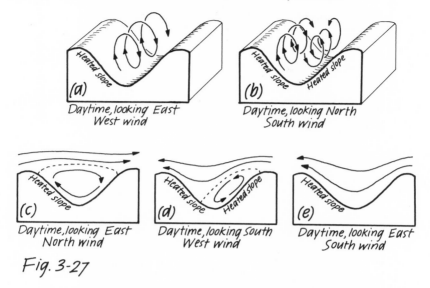

Fig. 3-27

In a more particular case involving the same ideas, if the valley *faces toward the midday sun and closes at the upper end* — as in a transverse valley of an E-W mountain range — a daily cycle of winds occurs, as suggested in **Fig. 3-28**, taken from a classic study[4] in the south-facing Inn Valley of Austria.

Again, these sketches concerning terrain features are *only simplified versions of what actually happens,* and the terrain they picture is also simplified. The purpose, of course, is just to try to illustrate the ways in which combining basic principles yields insights into wind flows in increasingly complex terrain.

Clouds and terrain

Technically, clouds are *colloidal suspensions of water droplets in air.* That means that the droplets are so small they fall only very slowly, with the result that they are effectively suspended in the air — millions of them floating near each other but only rarely colliding. There is a lot more to the story than that, of course, as we'll see in the next chapter. For now it is worth noting that clouds respond to terrain in ways that *make visible* some of the flow patterns we have just been considering.

As you've probably already seen for yourself, moist air rising vertically will often produce a cloud or a cloud cluster at some

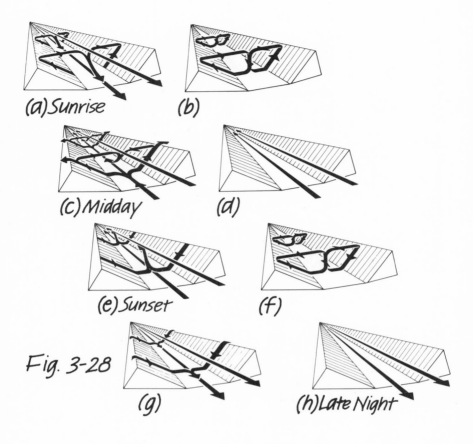

(a) Sunrise   (b)

(c) Midday   (d)

(e) Sunset   (f)

Fig. 3-28   (g)   (h) Late Night

level above the base of the rising air column. The more moist is the air, the lower the cloud base, and "the drier the higher."

Fig. 3-29 shows clouds in relation to some of the flow patterns just discussed. It is only rarely that smaller, local clouds such as these will produce precipitation, but, being tied to stationary terrain features, they are reliable producers of *localized sun shading* under the right conditions of wind direction and humidity.

Local cloud groups such as these are of the **cumulus** variety of cloud. When a broad airstream is forced up the upwind side of an equally broad range of mountains or hills, the clouds that form are of the **stratus** variety, as suggested in **Fig. 3-30**. If clouds form in the cool air trapped at the bottom of a basin, they are also **stratus** clouds. "Cumulus" simply means the vertical dimension is greater than the horizontal, while "stratus" means the opposite.

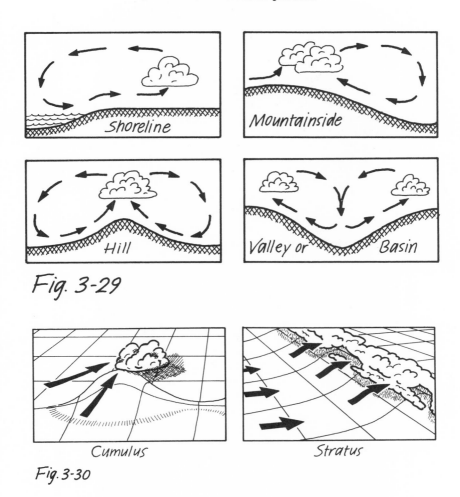

Fig. 3-29

Fig. 3-30

Most precipitation (rain, snow, and the various forms of each) is associated with *larger scale storm systems* rather than with the local clouds described here. A great deal more will be said in **Chapter 4** about cloud types and the processes of precipitation. Larger scale storm systems are discussed in **Chapter 10**.

Before setting aside the subject of local clouds, we ought to note that clouds form locally where large amounts of water are being injected steadily into the atmosphere, such as at factory chimneys and power plants, industrial cooling ponds, and (in cold, moist air) along major highways. The point of that remark is simply that the *injection of moisture* and *vertical motion* are separate reasons for local cloud formation.

Plant canopies: their structure and atmospheric processes within them

A logarithmic wind speed profile over grass, like the one in **Fig. 3-14**, is repeated in **Fig. 3-31a**. The individual clumps of turf sticking up from the surface and dragging on the windstream are called **roughness elements** and in grass they are considerably shorter than those in a major plant canopy such as that shown in **Fig. 3-31b**[E], where the elements have a height (h). Depending on such things as the spacing between elements and their flexibility in bending with the wind, a canopy will modify the wind speed profile in a manner suggested in **Fig. 3-31b**.

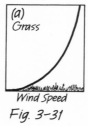
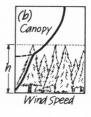

Fig. 3-31

In general, the modification will be as if the wind were responding to the presence of a second surface located somewhere near the canopy top — dragging along a surface above the true ground surface. Again in general, the profile will be logarithmic above that surface, which tends to be located near a height of about (0.8h). The surface is *lower or higher* as the spacing between roughness elements is *greater or smaller*. Below the surface, there is little change of speed — a very small gradient, except immediately next to the ground — in what are relatively low winds speeds.

Mention of spacing and flexibility brings up the subject of *canopy structure* and how to express it quantitatively. Based on what has just been said, one measure of the structure is the height of the "second surface" above the ground. But that's pretty crude as a measure of canopy structure. A far better measure is the **Leaf Area Index (LAI)**.

In calculating the LAI, researchers sample *each layer of the canopy above a unit of ground area.* Within each layer the *amount of leaf area* is measured to get the **Leaf Area Density,** given the mathematical symbol **a(z)** where 'a' means "area density", and '(z)' means "at a height of (z) above the ground."[Ps] Then the values of a(z) are plotted against the heights of the layers which they represent, as in **Fig. 3-32a**. Thus, the value of a(z) is

---

(E) The qualitative remarks made here about "canopies" will pretty much hold true whether the plants are low-growing field crops or giants of the rain forests or the redwoods. A canopy can be thought of as a collection of vertical elements which are too close together to act like individual obstacles to the wind, but that form a pretty much homogeneous spatial network — part of a "patch" of landscape.

the average leaf area within the layer, and above a unit of ground area.

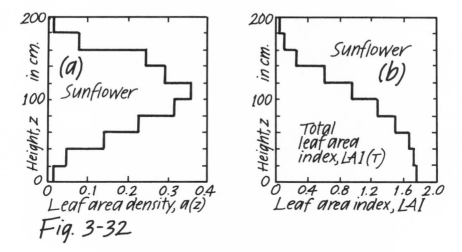

Fig. 3-32

To get the value of the LAI from those of a(z), simply *add up the a(z)* starting at the top of the canopy — height (h) — and going to the ground line, z = 0. **Fig. 3-32b** shows what the result is like, using the values of a(z) from **Fig. 3-32a**. This is a very useful quantitative description of the canopy structure. The value of the LAI added from top to bottom is called LAI(T) — the *Total LAI*[F]. For examples, **Fig. 3-33** shows LAI(T) through one growing season for an annual plant (sunflower), and through a lifetime for a perennial plant (oak).

Fig. 3-34 shows the graphs of LAI(h-z) for three types of canopy often encountered. Clearly, grasses are not as tall as trees, but in these sketches the heights are **normalized**, which means all heights are expressed *as multiples of (h)*. From these sketches you can deduce the rule that *the LAI increases most rapidly downward — the profile is most nearly horizontal — where the densest foliage occurs.*

(F) Since the distance travelled from the canopy top — in adding up the LAI — is (h-z) by the time you get down to height (z), the result is designated as LAI(h-z), so that *at the ground line where z=0, LAI(h-z) = LAI(T)*. The value of LAI(h-z) at each level can be thought of as the number of times a rod would have encountered a leaf if the rod were lowered vertically to that level from the top of an *average* location within that canopy. If you had, for some reason, wanted to do your adding upward from the ground line, you would have a result called LAI(z), but we don't need that one here. The remark is simply to suggest the flexibility of the LAI concept in representing canopy structure.

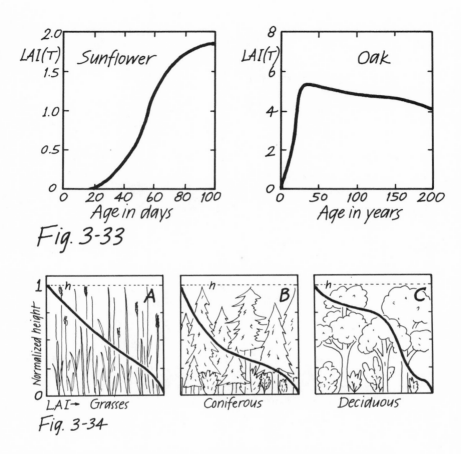

Fig. 3-33

Fig. 3-34

In **Fig. 3-35a**, the *normalized wind speeds* from the three canopy types are plotted against *normalized height*. That means, for example, that at the top of all of the canopies the wind speed is shown as being the same even though the speeds would be different in "absolute numbers" — in the real world. Wind speeds don't have to be normalized, but it is a "trick" often used to help bring out the effect of the *canopy structure* more clearly.[G]

(G) For the mathematically inclined reader, I am able to point out a good example of how the LAI is so useful in representing the canopy structure. If instead of just the height (z) to represent the vertical dimension in a profile, you used instead the value of LAI(h-z), you would then get a *linear graph* of (logarithm of wind speed) *versus* LAI(h-z). Thus wind speed is related mathematically to the foliage density that slows it down — a result exactly in agreement with intuition. This linear graph will often result for a plot of an atmospheric element, such as solar radiation (see **Fig. 3-36**) that decreases more and more — never increases — downward into the canopy. In this respect *air temperature* (see **Fig. 3-35b**) does not qualify.

Midday air temperature profiles in the three canopy types are plotted in **Fig. 3-35b**. Unlike the wind profiles, temperature is not normalized, though height is. The maximum temperatures occur within the canopies *at a level where light penetration (energy added) is large enough while ventilation (energy removed) is small enough* for that to occur. In grasses, that happens much nearer the ground than in a forest, where light penetration is not so great and ventilation is greater nearer the ground.

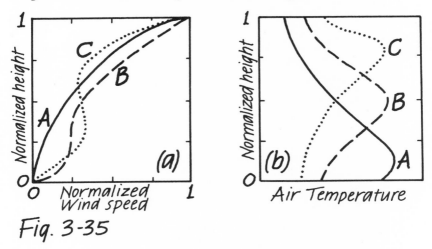

Fig. 3-35

Fig. 3-36 shows how profiles of solar radiation in a grassy canopy (Type A in **Fig. 3-34**) typically change through the hours of the day. In order to show the effect of the hour (solar position) on the profile, the values of solar radiation are *not normalized*.

Just a word or two more about canopy profiles. Profiles of moisture require use of the variable called *vapor pressure* (see **Chapter 4**), and they show only very small variations from the top to the bottom — hardly worth sketching here. In general, *normalized* wind profiles look about the same at night as they do in **Fig. 3-35a**, but temperature profiles are reversed — maxima become minima, and *vice versa*. Solar radiation, of

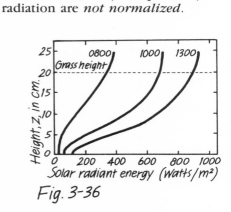

Fig. 3-36

course, is absent at night. A great deal more about what the various canopy profiles can tell us is explained in **Chapter 4**.

As a reminder about the way this book is organized, I have tried in this chapter to present *what* we observe about microclimates, and in **Chapter 4** I will try to explain more of *why* things are that way.

Notes

(1)   Sun path diagrams are presented in many places, among them: Architectural Graphic Standards, Smithsonian Meteorological Tables (Table 170), and D.M. Gates (1980, Appendix 4).

(2)   A few good, readily available sources I know about for graphs on insolation are: T.R. Oke (1978, Chapter 5), D.M. Gates (1965, Chapter 1), R. Geiger (1965, Chapter VII), and W.D. Sellers (1965, Chapter 3). Any good library, of course, will have others.

(3)   For a series of excellent illustrations of flow patterns around building clusters, see the work of T.R. Oke, in particular the figures and references in Chapter 8 of his book "Boundary Layer Climates", (1978, 1987b). In addition, the literature on air pollution control is a good source of studies such as those shown in Fig. 3-22.

(4)   This figure is often encountered in American textbooks. It first appeared in our technical literature in a chapter by Defant (1951).

Peptalks

(P₂)   Often, as in this case, the meaning of the definition of a physical variable becomes more exact when you examine the group of "physical units" that make up the variable: (energy/area-time). The four most basic physical units, of which all others are combinations, are (i) mass, (ii) length, (iii) time, and (iv) temperature. The major use of these units, however, is for *quantifying* variables. That is, specifying the physical units is like specifying in which monetary system — Dollars, Dinars, Francs, or Yen — the transaction will be carried out. For example, area can be in square inches, square meters, or hectares. It usually doesn't matter much — except for pure convenience — as long as everyone knows which.

(P₃)   The thing to realize — and I will keep bringing this point up throughout the book — is that *knowing the theory behind a concept permits you to make the presentation you need for your project, rather than to have to depend on presentations someone else has prepared, that might not be exactly what you need.* Said another way, *the library is full of*

answers to questions they hoped someone would ask. If your presentation requires an answer to some other question, don't settle for the "next best thing" in the form of a pre-packaged answer — learn how to ask the right questions and prepare your own answers. **Chapters 4, 5, 6, 9, 10,** and **11** all present theory of the kind you will need. If you don't understand the theory, or want to check your understanding with an expert, Chapters 7 and 8 discuss the use of a consultant.

(P₄)   You need the following simple formulas to convert:

Formula   **Example**

Celsius = (5/9)x(Fahrenheit - 32)   Convert 90 °F to °C:
                                                          (5/9)x(90 - 32) = 32.2 °C.

Fahrenheit = (9/5)x(Celsius) + 32   Convert 10 °C to °F:
                                                          (9/5)x(10) + 32 = 50 °F.

Kelvin (also known as "Absolute") and *Rankine* temperatures are also used in ecological writings, but you can always look these up in a physics book.

(P₅)   Most published weather records for major cities will have values for air temperature, dewpoint, and RH; so they are all readily available to you. Even so, in **Chapter 4** you will learn how to convert graphically from one to the other.

(P₆)   No, you don't need to do that, except in your mind. It's just that to do it actually one needs to use mathematics.

(P₇)   What is important here is NOT the details of how you plot something logarithmically. What is important is the fact that, *in a boundary layer profiles of most atmospheric properties are logarithmic*, because all these different properties are affected by the frictional drag that produces the turbulence in the windstream. We hadn't talked about it earlier, but the temperature profiles (in the air) in **Fig. 3-6a** are also logarithmic.

(P₈)   I'm not going to expect you to follow any complicated math. This is just an example in which using mathematical symbols is a good "shorthand" for making the discussion clearer.

4. Why is it like that here?
Explaining what we observe

In this chapter, following the descriptions of microclimates just completed, we need to explain *why* the solar radiation, temperature, moisture, wind, and clouds behave *locally* as they do. In the course of doing that we will need to use several ideas from physics, so this chapter will be more in the nature of a physics lecture than were **Chapters 2** and **3**$^{(P_y)}$.

Solar radiation: geometry

In **Chapter 3** we had a look at the sun path diagram as the basis for solar geometry, and noted that each diagram is for one latitude. Most published diagrams are for the *Northern Hemisphere only*. **Fig. 4-1** shows sun path diagrams for 40 °N and 40 °S latitudes. We're about to explain the anatomy of the diagram, but you can see from a comparison that *changing a published diagram to the opposite hemisphere* involves two steps: (a) rotating it around the horizontal axis, and then (b) reversing (top to bottom) the numbers attached to the arcs.

Fig. 4-2 provides you with the "anatomy" of the sun path diagram, and **Table 4-1** defines all the angles referred to. In the figure, there are two sets of coordinates *superimposed* — one for the *input* and one for the *output* of information. Here's how it works. There are three kinds of input data, and one of them — the latitude — is already specified once you've chosen your diagram. The other two are the *date* and the *hour*. **Fig. 4-2a** shows you the "web" of lines used here: the horizontal set is for the date, and the "vertical" set is for the hour.

In most any sun path diagram you find, the key dates (also see **Chapter 10**) are the **equinoxes** and the **solstices**. The equinoxes

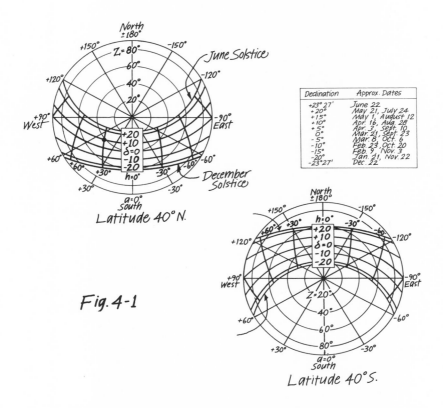

Declination	Approx. Dates
+23°27'	June 22
+20°	May 21, July 24
+15°	May 1, August 12
+10°	Apr. 16, Aug. 28
+5°	Apr. 3, Sept. 10
0°	Mar. 21, Sept. 23
-5°	Mar. 8, Oct. 6
-10°	Feb. 23, Oct. 20
-15°	Feb. 9, Nov. 3
-20°	Jan. 21, Nov. 22
-23°27'	Dec. 22

Fig. 4-1

are 21 of March and September, when the days and nights are *equal* — 12 hours. The solstices are 21 of June and December — the longest and shortest days of the year.

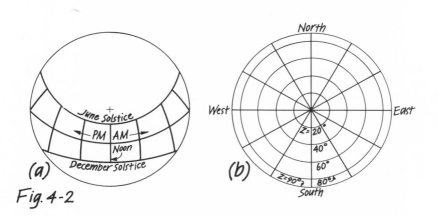

Fig. 4-2

In most any sun path diagram you find, the key dates (also see **Chapter 10**) are the **equinoxes** and the **solstices**. The equinoxes are 21 of March and September, when the days and nights are *equal* — 12 hours. The solstices are 21 of June and December — the longest and shortest days of the year.

There are two kinds of output data: the *solar zenith angle*[(A)] and the *solar azimuth angle*, shown on the polar coordinates of **Fig. 4-2b**. To check your understanding of the diagram, the *dot* in **Fig. 4-1** for 40°N is at 2 p.m. on April 16 (+10° on the declination scale). It tells us that the sun is *40° down from directly overhead, and about 50° west of south* in the sky.

What is the sunpath diagram, exactly? **Fig. 4-3** shows the answer. It is the *horizontal mapping, or projection, of a set of sun path arcs* for a particular latitude. If you need to, you can construct your own set of three arcs for a latitude of your choice. Here's how. (a) The central arc for the equinoxes is tilted *from the*

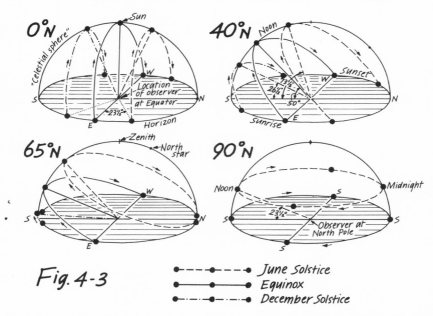

Fig. 4-3

●- - -●- - -● June Solstice
●——————●——————● Equinox
●—·—·—●—·—·—● December Solstice

(A) In some published diagrams, the scales for the two output variables may be different from the ones used here. The solar zenith angle (Z) may be the **solar elevation angle** (90° - Z) instead. Not to worry — the difference is one you can handle easily. The solar azimuth angle may be expressed with some other scale; for example, the scale may go from 0° at North through 180° at South to 360° at North again. You can handle it if you understand what is being represented. I've chosen the scales I use because they work better with the math formulas coming up shortly.

TABLE 4-1   Angles used in calculations of Solar Geometry

NAME	SYMBOL	DESCRIPTION
Latitude	L	+90° at North Pole; 0° Equator; -90° South Pole.
Solar declination	∂	One value for each date: the latitude at which today's noon sun is directly overhead. +23.5° June 21; 0° equinoxes; -23.5° Dec 21. (See the little table in **Fig. 4-4**)
Hour	h	(15° x number of hours from Noon): -30° at 10 am; +75° at 5 pm; etc.
Solar zenith	Z	0° when sun is directly overhead; +90° when it is on the horizon.
Solar azimuth	a	angle between South and horizon position directly below the sun: -90° at East; 0° at South; +45° at SW; etc.
Slope inclination	i	The slope is inclined this many degrees from horizontal: 0° to +90°.
Slope azimuth	a'	The direction toward which the slope points (its *aspect*); measured the same way as the Solar azimuth, **a**.
Solar incidence	B	The angle between the solar beam and a perpendicular to the local slope.
Sunrise/Sunset	H	The hour angle at the times of sunrise and sunset; measured the same way as **h**.

vertical by an angle equal to the latitude. (b) That means the line from the central position on the horizontal plane to the noon position of the sun at equinox forms a zenith angle (Z_0) equal to the latitude. (c) The lines from the central position on the horizontal plane to the noon positions of the sun on the solstices form zenith angles of 23.5° greater and less than (Z_0). (d) The planes of the three arcs are always parallel to each other, and the same distance apart. (e) The set of three arcs for the *opposite hemisphere* is obtained by grabbing the set you have drawn and rotating it — as is, date labels and all — around the line connecting East and West on the horizontal plane.

From the sets of arcs in **Fig. 4-3** you can deduce some useful bits of information. (1) At the equator (0°) the sun is always in the northern half of the sky for one half of the year, and in the southern half the rest of the year. (2) At either pole, the sun is below the horizon for 6 months each year. (3) The closer you get to a pole — the *higher the latitude* — (a) the longer is the longest day and the shorter is the shortest day, and (b) the greater is the difference between east and the position of sunrise (and between west and the position of sunset) on the solstices.

Fig. 4-4 gives you a set of sun path diagrams for four latitudes of the *Northern Hemisphere*. The diagram for 40°N is repeated from **Fig. 4-1**. For more detail or for other latitudes, either construct your own or refer to one of the publications in **Note 1 of Chapter 3**.

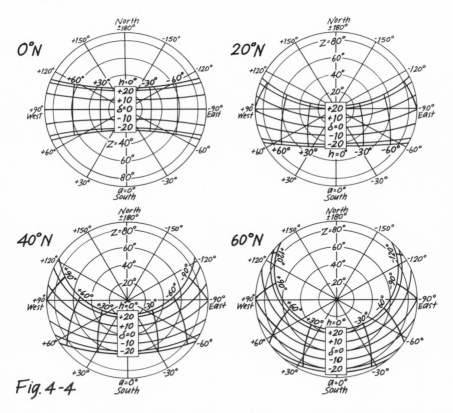

Fig. 4-4

Table 4-2 gives the mathematical essentials for making calculations about solar geometry. This information is the basis for the diagrams in **Figs. 4-1** through **4-4**, all of which are found in a variety of readily available books. To try to emphasize the point, again, that the formulas enable one to construct diagrams otherwise unavailable, I include **Figs.** 4-5 **and** 4-6, neither of which is readily available elsewhere.

Fig. 4-5 permits you to find the times of sunrise and sunset — and thus the **daylength** — for any combination of latitude and date. The *pair of dots* in **Fig.** 4-4 are for 40° (N or S) on April 16 (+10° on the declination scale) the same combination as for the dot in **Fig.** 4-1. The dots tell you that *in the Northern Hemisphere* (left scale) the sunrise is a little after 5 a.m. and the sunset is a little before 7 p.m. However, *in the Southern Hemisphere* (right scale) the sunrise is a little after 7 a.m. and the sunset is a little before 5 p.m.

In **Fig.** 4-6, the sketch shows that the information concerns the directions *shadows* are cast at different combinations of latitude,

---

Table 4-2   Formulas used in the calculations of Solar Geometry

---

Calculating the sun's position in the sky:

Input variables		Formula
Latitude	L	
Date/Solar declination	∂	(see the little table in **Fig.** 4-4)
Hour	h	

Output variables

Solar zenith angle	Z	$\cos Z = [(\sin L)(\sin d)] + [(\cos L)(\cos \partial)(\cos h)]$
Solar azimuth angle	a	$\cos a = [(\sin L)(\cos Z) - (\sin \partial)] \div [(\cos L)(\sin Z)]$
	or	$\sin a = (\cos \partial)(\sin h) \div (\sin Z)$

Calculating the times of sunrise and sunset:
$$\cos H = -(\tan L)(\tan \partial)$$

Calculating the angle of incidence of the solar beam on a slope:

Input variables

Solar zenith angle	Z
Solar azimuth angle	a
Slope inclination angle	i
Slope azimuth angle	a'

Output variables

Solar incidence angle	B	$\cos B = [(\cos Z)(\cos i)] + [(\sin Z)(\sin i)\{\cos(a - a')\}]$

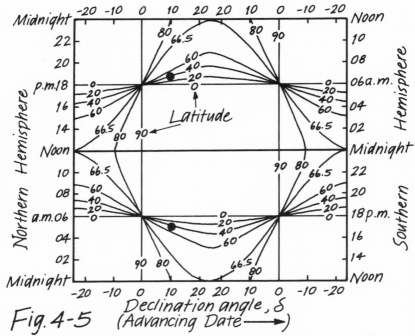

Fig. 4-5 (Advancing Date ⟶)

date, and hour (the three input variables for calculation of the sun's position). In the sketch, the *direction and relative length* (relative to the height of the wall) *of the shadow edge* is indicated by the line A-B. That idea is transferred to the diagram for latitude 60°N, where A-B runs from the "intersection of the cross hairs" to the 2 p.m. (1400) spot on the curved line for (Feb 23/Oct 10) —

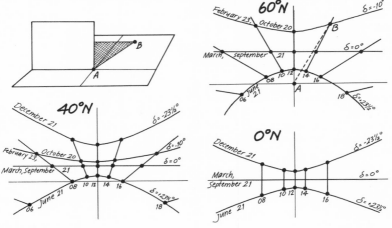

Fig. 4-6

∂ = -10°. If you have trouble picturing what the diagram represents, think of a pole casting a shadow from A to B rather than a wall casting a shadow edge. The information is the same. The length scales in the diagrams are *relative to the wall (or pole) height*.

Solar radiation: insolation

Solar geometry describes the sun's position in the sky and in relation to the surface on which it is shining. **Insolation** connects the geometry with the *rate of energy arrival* on the surface. As we saw in the last chapter, the number of combinations of solar position and surface orientation is so large we can present here only a sampling of the combinations. But there is more to insolation than just the geometry. The condition of the atmosphere through which the sun shines — its dirtiness and its cloudiness — is nearly as important in determining the rate of energy arrival.

Radiant energy — solar radiation — leaves the sun, travels outward through space in all directions, and a very small fraction of it is intercepted by Earth. The rate of arrival of the solar energy *on a perpendicular surface at the outer limits of Earth's atmosphere* is called the **solar constant**. It isn't really constant (it varys a few percent through many years) and we can't know exactly its value (precise measurements aren't that easy), but the value usually used is about 2 calories per square centimeter per minute (cal/cm²min) = 1.4 kilowatts per square meter (kW/m²). It is the value against which most other energy transfer rates are compared in various meteorological and climatological studies.

Fig. 4-7 suggests, in very simplified form, the various pathways the sunshine takes on its way to a particular surface on Earth. The arrival rate at (**W**) is the *solar constant*, while the rate at (**Y**) is the *insolation*. There are three broad factors operating here:

(a) energy is removed from the direct solar beam (D) by the processes of **absorption** (A) and **scattering** (S) — both by atmospheric molecules and particles — *en route* to the surface,

(b) *part* of the energy removed by scattering (S) arrives indirectly at the surface from the bright part of the sky surrounding the sun, and

(c) the longer the path through the atmospere (the lower the sun is in the sky) the more chance there is for removal from the direct beam, so the insolation is less. The shortest path through the atmosphere is (**XY**) if the sun is directly overhead, and otherwise the **solar path length** is equal to (**WY**) = (**XY**)÷(coz **Z**), where **Z** is the solar zenith angle obtained in the last section.

The scattered energy that does not reach Earth returns to space. It is most of the light we see coming from Earth in an Earth-photo taken from space by an astronaut. What we see on Earth is (on a clear day) mostly in the direct beam (D) and partly in the indirect (S) light from around the sun. *The direct light decreases and the indirect light increases as the sky becomes dirtier or more cloudy.* In round numbers, these effects are as follows (as multiples of the direct beam in clean air):

	Clean air	Dirty air	Overcast
Direct sunlight (D)	1.00	0.5	0.02
Indirect sunlight (S)	0.06	0.3	0.5
Ratio (S/D)	0.06	0.6	25.0

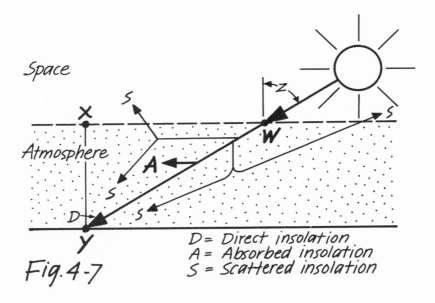

Fig. 4-7

D = Direct insolation
A = Absorbed insolation
S = Scattered insolation

We pretty much have to be satisfied with "round numbers" here, because the mathematical theory is complex and the measurements are so various[1]. Just because insolation arrives at the earth's surface doesn't mean, of course, it will be absorbed there. If it strikes water, a little bit will penetrate and be absorbed below the surface. Mostly, however, the insolation is either *absorbed* or *reflected*, on both water and land. The *fraction of the arriving sunlight that is reflected* is called the **albedo**. Shiny, reflective surfaces have large albedos. Here are some sample values.

Table 4-3   Typical values of surface albedo

Surface	Albedo	Surface	Albedo
Thin clouds from above	0.40	Thick clouds from above	0.70
Fresh snow	0.80	Old snow	0.60
Dry sand	0.50	Wet sand	0.25
Open water (high sun)	0.05	Open water (low sun)	0.25
Plowed fields	0.20	Forests	0.10

Mathematically, if the albedo is (**a**), then the fraction of the insolation that is absorbed is (1 - **a**). So, if the insolation (energy) is **S**, then the absorbed energy is (1 - **a**) **S**.

Now let me tell you why the sky is blue and sunsets and sunrises are red. **Fig. 4-8** suggests the reasons. The sunlight traveling through space (coming from the left) is made up of radiation in all wavelengths of the visible spectrum — only red (longer wavelengths) and blue (shorter wavelengths) are shown. Passing through the atmosphere scatters the blue light but the red scarcely at all[B]. Scattering occurs in all directions, but only those toward the eye are shown.

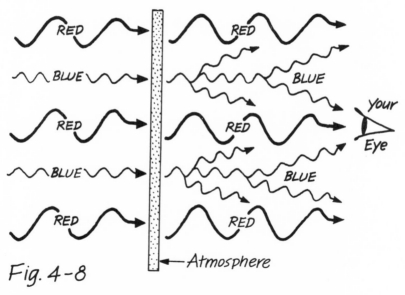

Fig. 4-8

(B) Physically, this differential scattering between the wavelengths near the blue end of the color spectrum — violet, indigo, and blue — and the red end — yellow, orange, and red — is because the particles and molecules doing the scattering are very small, so that their diameters are more nearly equal to the blue wavelengths than to the red wavelengths.

The net result is that the red reaching your eye comes directly from the sun, while the blue comes from elsewhere in the sky. The longer the *solar path length* (see **Fig. 4-7**) the more blue is removed and the redder the sun seems to be; the path is longest near sunrise and sunset. The farther away from the sun you look, the more red is removed and the bluer the sky seems to be.

Radiation physics

The ideas in this section will be a bit harder for most readers to follow than the ideas just considered, because they are less directly and obviously related to common experience. On the other hand, understanding them will give you greater confidence when subjects such as *radiation inversions* and *leaf temperatures* come up.

First, we need to have a couple of definitions.

Radiation is the word used for both the *process* of energy transfer and the *energy* that is transferred. **Incident radiation** is that arriving at a material.

Absorptivity is the *fraction of incident radiation absorbed* by a material.
Reflectivity is the *fraction of incident radiation reflected* by a material.
Transmissivity is the *fraction of incident radiation passing on through* a material. "Transmits" means "allows to pass through."

A black body is a material with the greatest absorbing and emitting ability physically possible. It absorbs all radiation, of all wavelengths, that falls upon it. Its *absorptivity* is equal to 1. It reflects and transmits *no incident energy*. Most materials are not black bodies — their efficiencies are a little or a lot less than 1 — and true black bodies are specially constructed in physics laboratories.

Emissivity is the efficiency with which a material emits — sends out — radiation *relative to* the efficiency of a black body at the same temperature. Since the black body does it best, it is natural to compare all other efficiencies to that one.

Electromagnetic spectrum — or just "spectrum" — the full range of wavelengths in which radiation occurs as it is emitted and transmitted in physical systems. In **Fig. 4-8** red and blue are shown with different wavelengths. As we are about to see, there is a much wider range than the range of wavelengths in light — the "visible spectrum."

Fig. 4-9 shows the spectrum, along with some of the terms used to describe parts of it. In ecology we are concerned mostly with the parts called *ultraviolet (UV), visible (light), shortwave, near infrared (NIR), far infrared (FIR), and longwave.* As you can see in the figure, "shortwave" and "longwave" each include several of the other kinds within them. Radiation that does *biological damage* is at the very shortwave end, near "Gamma rays", but is not part of our discussion here.

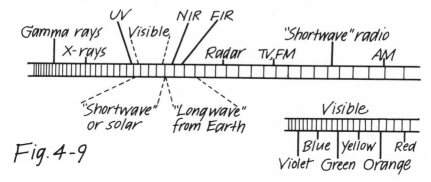

Fig. 4-9

After the definitions, we need to know about four **Laws of Radiation Physics**[C]. The first three laws describe the behavior of a *black body*, with which all other materials are compared. The last law describes any material.

1) Any *black body* not at absolute zero (degrees Kelvin) emits radiation at all wavelengths, but with most of it near a single maximum wavelength somewhere in the middle of the spectrum.

2) The higher a *black body's* temperature, the more radiation it emits.

3) The higher a *black body's* temperature, the shorter is the wavelength at which it emits most of its energy.

4) A material that emits efficiently also absorbs efficiently, and in the same wavelengths for both processes.

Fig. 4-10 connects the first three laws. The *shapes* of the black body curves are described by Law #1. The fact that the *area under a curve* increases as the temperature rises agrees with Law #2. The fact that the "*peak*" of a curve shifts to shorter wavelengths as the temperature rises agrees with Law #3.

Here's what the curve in Law #1 means physically. If you point a radiation detector at a black body, and the detector has *a filter*

(C) The laws are named after their discoverers: (1) Planck, (2) Stefan and Boltzmann shared the discovery, (3) Wien, and (4) Kirchhof. Mathematically, the Stefan-Boltzmann law says the rate of radiation emitted is proportional to the *fourth power of the Kelvin temperature:* T^4, which means "T multiplied by itself 4 times." Thus, a 1% increase in Kelvin temperature means a little more than a 4% increase in radiating power. Water freezes at Kelvin 273 and boils at Kelvin 373.

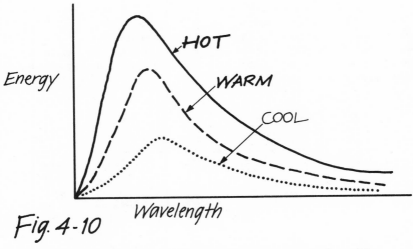

Fig. 4-10   Wavelength

that allows only certain wavelengths to pass through and be detected, then the meter on the detector will register a certain amount of energy. In **Fig. 4-11** that energy is shown by the small rectangle located at a particular place on the wavelength scale — the spectrum — where the filter let the radiation pass through.

If you keep changing filters, the amount of energy detected will keep changing, and you will plot the curve of Law #1. Notice there is a wavelength at which the maximum amount is detected, referred to in Law #3. The curve is called the **black body emission spectrum** for that temperature.

We know the temperature of Earth is, averaged over the whole surface, about 288 °K, about 15 °C. That is a whole lot cooler than

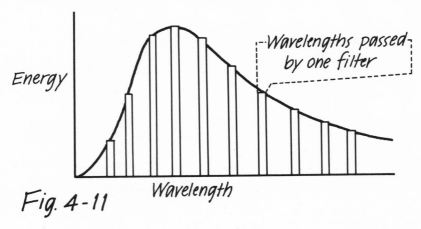

Fig. 4-11   Wavelength

the surface temperature of our sun, which we estimate to be about 6000 °K. According to Laws #1 and #3, the sun emits a great deal more radiant energy than does Earth, *and at much shorter wavelengths*. Although the sun emits a lot more energy than Earth, by the time it travels through space and spreads out, its intensity is much less when it reaches Earth.

Fig. 4-12 shows the spectra of radiation arriving at Earth from the sun — we call it *"shortwave"* or *"solar"* radiation — and of radiation departing from Earth in all directions toward space — we call it *"longwave"* or *"terrestrial"* radiation.

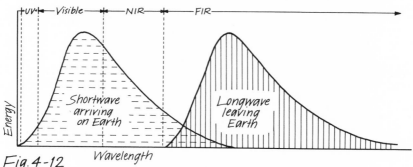

Fig. 4-12

Note several things in **Fig. 4-12**:

a) the travel through space has reduced the total amount of the sun's energy arriving — the area under its curve — to very nearly the same amount as is leaving Earth headed toward space;

b) the sun and the earth act very much like black bodies, *but of very different temperatures*[D], and so their curves take on the characteristic shape of the black body emission spectrum;

c) sunlight is mostly visible, but also contains UV, near infrared (NIR), and a little bit of FIR (see **Fig. 4-9**);

d) longwave radiation is nearly all far infrared (FIR); and

e) despite a little bit of overlap in the curves near where NIR and FIR meet, they are essentially separate, and are usually discussed ecologically as if they were two different kinds of energy.

(D) In fact, with an error of only a couple of percent, we can treat them as if they are black bodies. For the reader interested in numerical data, the energy leaving the sun is a large multiple of the energy leaving Earth, *per unit area per unit of time (see the footnote (C) above):*

$$\text{(Energy leaving sun/Energy leaving Earth)} =$$
$$(T_s/T_e)^4 = (6000/288)^4 = 188,380 \text{ approximately.}$$

Clearly, when we note that the sun's surface area is so much larger than that of Earth, the travel through space has made a very large reduction.

All this about black bodies is fine, but didn't I say most materials are *not* black bodies? Yes, I did, so it's about time to discuss the behavior of some non-black bodies in our environment. **Fig. 4-13** is closely related to **Fig. 4-12**, as you can see. Both contain the shortwave and the longwave emissions of the sun (as it arrives at Earth) and Earth. In **Fig. 4-13**, however, we also have information on the radiative behavior of typical *living leaf materials* and the *gases of Earth's atmosphere.*

Here's how you read these curves in **Fig. 4-13**. To the left you can see that when sunlight strikes a living leaf, about half the energy is reflected or transmitted, and only half is actually absorbed by the leaf. Clearly, the *absorptivity* is different from one wavelength to another, so the transmitted and reflected energy consists mainly of green light (G) — that's why leaves look green from above and below — and of *NIR (F)* — that's the part of the spectrum that "infrared *film*" responds to, so living leaves look white in IR photoprints.

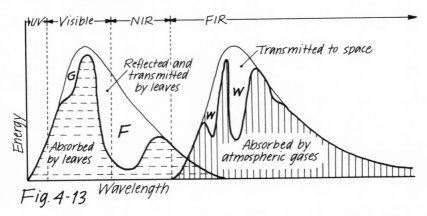

Fig. 4-13

To the right you can see that, though invisible longwave radiation leaves the surface of the earth in all wavelengths, it is only selectively absorbed by the gases of the atmosphere[E]. A small fraction of it passes on out, directly to space in what is called the **atmosphere's radiation window** — (W) in the figure. *According to Law #4*, the area marked "absorbed by atmospheric gases" can also be marked *"emitted by atmospheric gases — toward space and toward Earth."* If you were an astronaut, pointing an infrared detector back at *a cloudless part of Earth*, you would know that whatever you were detecting in the wavelengths of the window

---

(E) The absorption-emission spectrum for atmospheric gases refers to clear skies — no clouds. When the water vapor condenses to form clouds, the cloudy part of the air acts pretty much like a black body.

(W) had come from the surface rather than from the atmosphere. In the other wavelengths, the radiation would have come to you from the gases of the atmosphere. That is so because — according to Law #4 — those gases, having absorbed in the other wavelengths, also emit in those same wavelengths[2].

It may be nice to know why leaves are green, you say, but what has all this stuff about astronauts and detectors got to do with planning and design? Ok, here's what it has to do with it. The business about astronauts and detectors was only to let you check your understanding of these complex ideas of radiation physics before we talk about *temperature inversions*. And temperature inversions are very important factors in microclimate, not to mention in problems of air pollution control (see **Chapter 11**).

The soil and water of the earth's surface act very much like a black body, while the gases of the cloudless part of the atmosphere let some of the surface energy escape directly to space without being absorbed by, and heating, the air. Because of this, **on a cloudless night** *the surface loses radiant energy at a faster rate than the air near it sends radiant energy back to the surface.* This being so, and recalling the discussions of changes in stored energy in **Chapter 2**, you can see that the surface will cool off, on a cloudless night, faster than the air just above it. That condition — cold beneath warm — is called a **temperature inversion**. In particular, this kind is called a **radiation inversion** because of the major process forming it.

Energy, heat, and temperature

In **Chapters 2 and 3** we noted several things about energy, heat, and temperature. To discuss them, first think about a group of molecules all spinning, vibrating, and moving about. These molecules are not only individually *in motion*, in the three ways just mentioned, but are also bound together and held in particular *positions* by chemical bonds. That's at the atomic and molecular size scale, and the energy is **internal** to the molecules.

Now think about a much larger group of molecules — a visible sample of material — like a rock or a volume of air, in which the energy is **external** to the molecules. That group also has *motion* and *position*. The motion, or lack of it, can be detected by our eyes. The energy of position can be represented by the energy it takes to lift a rock and place it on a ledge. It has more energy on the ledge than it had before being lifted.

The following table summarizes these four combinations and the names for the kinds of energy associated with them. There are some kinds of energy, such as electrical, not represented in the table, but it contains all the ones we need to know about ecologically.

Size scale	Energy of Motion	Energy of Position
Molecular: internal	Heat energy	Chemical energy
Visible: external	Kinetic energy	Potential energy

Energy can be changed from one kind to another, and from one place to another, but the total amount of it cannot change as long as we are still talking about the same group of molecules — the same *physical system*. For example, when a rock collides with a wall, the kinetic energy it had appears as heat, with a small, temporary warming of both rock and wall; but the total amount of energy in the rock and the wall is the same. For another example, if the rock falls off the ledge, its potential energy is first converted to kinetic energy, and then to heat energy when it hits the ground; but the total amount of energy in the rock and the ground is the same.

In **Chapter 3**, we noted that there is a difference between heat and temperature, and that heat flows, not necessarily from the place where there is more of it, but from the place where the temperature is higher, regardless of heat content. For a group of molecules, *heat measures the total internal energy of motion* (of the three kinds mentioned above), while *temperature measures the average internal energy of motion per molecule*. So that a mass may have a large or a small heat content, and, quite separately, a high or a low temperature.

An air sample may have a high temperature but a low heat content, because it contains (as gases usually do) very few molecules. Nevertheless, if the air comes in contact with a rock of lower temperature but a much larger number of molecules (as solids usually have), heat will flow from the warmer air, which has less stored heat, to the cooler rock, which has more stored heat.

To sum up, knowing whether a part of the microenvironment is hot or cold need not tell you very much about the amount if heat stored in it. But you can be sure that, regardless of the locations of stored heat, given a chance it will always flow from higher to lower temperatures.

Heat flow and storage in soil, air, and water

In **Chapter 3**, after examining several sets of data on temperature behavior in soil, air, and water, we came up with four rules of behavior found in all the examples. The rules were about daily and seasonal variability being greatest at the surface, about *delay* and *decay* of amplitudes with distance from the surface, and about differences among air, soil, and water. Now we need to explain those rules as a basis for really understanding the behavior.

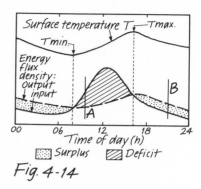

Fig. 4-14 connects the time variations of temperature with the flow-storage ideas of **Chapter 2**, and with the small system — probably best seen as a shallow block of soil located at the surface — in **Fig. 4-15**. **Fig. 4-14** repeats for us that temperatures rise and fall in response to the differences between inflow and outflow that produce *changes in storage*. For example, the temperature *maxima* and *minima* — when the direction of temperature change reverses — occur when the inflow and the outflow are equal, usually *not* at the times of the maxima and minima of *inflow*, as intuition might suggest.

Fig. 4-15 adds the idea that *warming and cooling* — changes in storage — *are separate from the directions of heat flow.* **Fig. 4-15a** is occurring at *Time A* in **Fig. 4-14**, while **Fig. 4-15b** is occurring at *Time B* in **Fig. 4-14**.

Fig. 4-15 also introduces the idea — the concept of **conduction** — that *heat flows from higher to lower temperature, and at a rate proportional to the difference in temperature along the path.* From this we can infer that the temperature on the upper

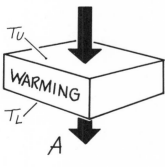
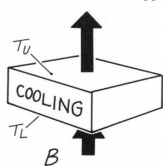

surface (T_U) of the layer in **Fig. 4-14a** is higher than the temperature on the lower surface of the layer (T_L). In **Fig. 4-14b** (T_L) is higher than (T_U). If the temperature difference is doubled, for example, the rate of flow — the amount of heat energy flowing in a unit of time — will also double. The flow is proportional to the difference. No big surprise. Less obviously, *if the temperature difference remains the same, but the distance between the upper and lower surfaces — the length of the path — is only half as great*, the rate of flow will also double.

Clearly, both the length of the path and the temperature difference are involved in determining the rate of energy flow along that part of the path. Both of these variables are combined in the **temperature gradient**, which is

(Temperature difference) ÷ (Path length).

The idea that *changes in storage* and *direction of heat flow* are separate is continued in **Fig. 4-16**, where the flow arrows in the temperature profiles are always from higher to lower temperature. Also, *the arrows are longer when the temperature profiles are least vertical* — that is, when the temperature gradients are greater.

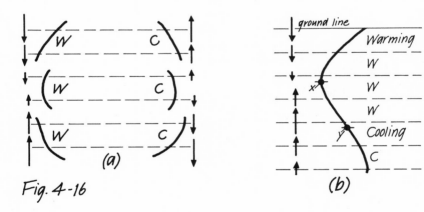

Fig. 4-16

On the left are six combinations of difference and direction — three each for warming and cooling. As it turns out, we can also associate warming and cooling with **curvature of the temperature profile**, and not have to imagine little flow arrows at all. The *positive (+) curvature associated with warming* is called "concave to the right" while The *negative (-) curvature associated with cooling* is called "concave to the left." It is also true, but we need not go into it any farther now, that *the greater the curvature — the less straight the profile — the faster the warming or cooling is taking place*.

In **Fig. 4-16b** we apply these rules of curvature and differences between flow arrows to a typical midday soil temperature profile (of about a foot in depth). Compare it with **Fig. 3-6.** The new idea here is that the *zones of warming and cooling* and the *zones of upward and downward flowing heat* are also separate. In particular, the division between warming and cooling occurs at (**Y**) — the level of the **point of inflection** where the curvature changes from (+) to (-) On the other hand, the division between upward and downward flowing heat occurs at (**X**) — the level of temperature maximum (or minumum).

In **Fig. 4-17** the soil temperature profiles of **Fig. 3-6** have all been laid out separately; that is, though they occur in the correct sequence, you must imagine that each has its own temperature scale. The dynamics of soil heat flow and storage is such that, when we apply to **Fig. 4-17** the rules about curvature, and the differences between warming-cooling and upward-downward flow, we see several interesting outcomes. They are evident in the lines slanting downward from the upper left to the lower right. The lines connect — from one profile to its neighbor — (a) temperature maxima and temperature minima, in the case of the division between directions, and (b) points of inflection, in the case of the division between warming and cooling. Here are the outcomes.

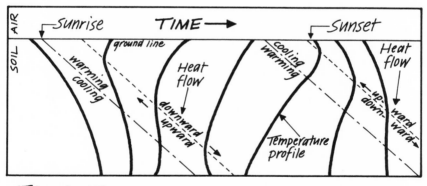

Fig. 4-17

1) The rising and setting of the sun initiate *waves — layers — of warming and cooling* that move from the surface downward into the soil. Not much of a surprise.

2) However, *layers of upward-flowing and downward-flowing heat* also move from the surface *downward* into the soil. The idea of a downward-moving layer of upward-flowing heat is a little bit strange — counter-intuitive, as we say — but it's what happens.

On a longer time scale, these same two outcomes occur as the surface changes temperature due to changes in season. Also, the same kinds of waves move downward into water and upward into the air, though at very different speeds than those for soil.

Up to this point we haven't said anything quantitative about "*how much* heat is moving and being stored", or about "*how fast* the heat and the layers are moving." To do that we need to define and discuss several **thermal properties** of the different media — soil, air, and water. First, a few more definitions.

Thermal conductivity — or just **conductivity** — tells us *how much heat flows for a given temperature gradient*. Clearly, since the rate of flow depends on something else besides the gradient, that something else is a *physical property of the medium* through which the flow is taking place. You can think of the conductivity — call it (**k**) — as measuring the *efficiency of conduction* of the medium. It answers the question "How much heat flows?"

Heat capacity — call it (**C**) — measures *the temperature change produced in a volume by the addition or subtraction of one unit of heat energy*. Seen another way, it also measures the amount of additional energy required to produce a given amount of temperature change. You can think of the heat capacity as measuring the *efficiency of heat storage* of the medium. It answers the question "How much heat is stored?"

I know those definitions are a lot of information coming at you in a short time. Let's hasten your comprehension of them by using them at once in a description of the interactions of flow and storage in a small volume, such as that in **Fig. 4-15**. Recall the statement from **Chapter 2** that

(Inflow) = (Outflow) ± (Change in Storage), which is also

(Inflow) - (Outflow) = (Change in Storage).

One can look at the process in either of two ways. (1) The difference (Inflow) - (Outflow), but not the magnitudes of the flows, depends in part on the *heat capacity* of the material — on how much heat is "put in storage" along the path between the inflow and the outflow. (2) The amount of heat flowing through the system — the magnitudes of the flows, but not their difference — depends in part on the *conductivity* of the material in the flow path. Clearly, *both (k) and (C) are involved in partitioning the inflow between storage and outflow* in the little system of **Fig. 4-15**.

We can now see why the amplitudes of temperature become less and less (decay) farther from the surface. It is simply because *the*

*heat flowing into one layer is the heat flowing out of the layer
next closer to the surface*, so that *the heat flowing into one layer
is the heat not stored in any of the layers closer to the surface*. It's
like a series of hanging waterfalls: the amount that falls into a
basin depends on how empty all the basins are above it, and upon
how much enters the top basin.

Furthermore, we can now see the reason for differences in temperature behavior between soil and air. Because *soil has a large
heat capacity*, not much heat flows into lower layers, so the amplitudes are reduced rapidly with increasing depth. The opposite is
true for air, which has a *small heat capacity*.

Here are two more definitions of thermal properties.

The **thermal diffusivity** — (k/C) — measures the *time it takes*
for the layers of warming and cooling to pass down into the soil
(and up into the air).

The **thermal admittance** — Square root of (kC) — measures
the amount of heat involved in both flow and storage — the
amount that goes into the soil and back out again during the
course of a day or a year. Much more can be said about the
dynamics of heat flow and storage in soil, air, and water[3]. **Table
4-4** summarizes, in numbers *relative to the smallest* in each category, the various thermal properties and their differences among
various materials found in a microenvironment. From the table we
can deduce many useful results, among them:

Table 4-4   Relative magnitudes of the thermal properties of materials

Material	Conductivity (k)	Heat Capacity (C)	Diffusivity (k/C)	Thermal Admittance $\sqrt{(kC)}$
Still air	1.	1.	139.0	1.
Gently moving air	1000.	1.	139,000.0	32.
Still water	30.	4165.	1.0	353.
Gently moving water	300.	4165.	10.0	1118.
New snow	4.	200.	2.8	28.
Old snow	20.	1000.	2.8	141.
Ice	110.	1960.	7.8	464.
Dry soil	15.	1200.	1.7	134.
Wet soil	90.	3000.	4.2	520.
Granite	220.	2165.	14.2	690.

a) the best *practical* material in which to store heat efficiently is water (**C** = 4165), with granite (**C** = 2165) being second;

b) as seen in **Fig. 3-6b**, there is only a negligible delay in heat movement through air (**k/C** = 139,000) as compared with the delay in soil (**k/C** = about 3); and

c) still air (**k** = 1) is the best natural insulator, though dry mud blocks (**k** = 15) and snow blocks (**k** = 20) — as used by "primitive" builders — are also good natural insulators.

Water and humidity

On this planet Earth, water exists in all three of its *states*: solid (ice), liquid (water), and gas (water vapor). As we will see in **Chapter 9**, this fact, that we take so for granted, is rare and the basis for life as we know it. For a group of water molecules to pass from one state to another — a **phase change** —requires that a certain amount of energy change form, between heat energy and chemical energy. For example, as shown in **Fig. 4-18**, it takes 80 calories for each gram of ice (80 cal/gm) to melt to water. *Heat enters the process.* Freezing the same gram of liquid again releases 80 calories. *Heat leaves the process.* About six times as much heat for each gram is involved in the phase changes of evaporation and condensation. These amounts of heat are called the **latent heats** of water. Any substance that changes phase, or state, has a set of latent heats for its changes. We are concerned only with those of water.

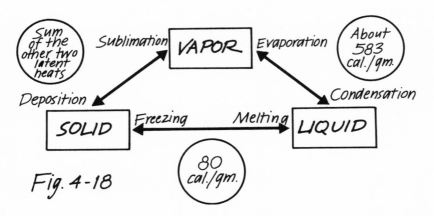

Fig. 4-18

Fig. 4-19 places the latent heats of water in a different perspective, related to the temperature of the water. Starting at the left with ice below freezing (0 °C), increases in temperature are associated with increases in the *internal energy* of the water, except that *during phase changes the changes of energy are not accompanied by changes in temperature.* That's why they are called "latent" heats: they don't "show up" when all you watch is temperature. Going from left to right shows heat is added — some of it latent — as

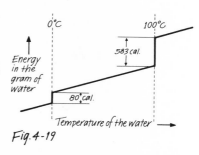

Fig. 4-19

temperature rises[F]. You can just as well go from right to left and see the energy released (or removed) as temperature falls.

The little experiment in **Fig. 4-20a** will help us define the **vapor pressure** of water. Under the air-tight bell jar are a pan of water and a pressure gauge (called a manometer).

The temperature of all the apparatus is kept constant. At first the water is covered so no evaporation can take place. In **Fig. 4-20b** the pressure on the gauge(P_{lo}) with the water covered is shown as a horizontal trace. When the pan is uncovered, the pressure rises, at first rapidly, then slowly, until it levels off at a higher value. *Call the pressure increase (∂p).* Then:

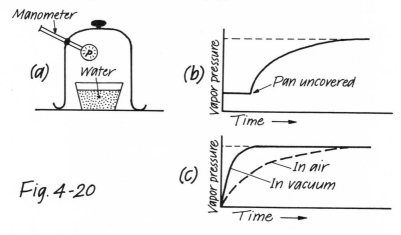

Fig. 4-20

a) if there were *a vacuum* under the bell jar at the beginning, and the cover remained off until the pressure leveled off, then (∂p) *and the final pressure* would both be the **saturation vapor pressure (s.v.p. = e_s) of water** at the temperature of the experiment.

b) if there were *perfectly dry air* under the bell jar at the beginning, and the cover remained off until the pressure leveled off, then (∂p) would be the **(s.v.p. = e_s)** at the temperature of the experiment.

c) *whatever the conditions* under the bell jar at the beginning, if the cover remained off until the pressure leveled off, then the final pressure would be the sum of the pressure due to the dry air *plus* the **(s.v.p. = e_s)** at the temperature of the experiment.

d) if there were *perfectly dry air* under the bell jar at the beginning, and the cover were replaced before the pressure leveled off, then (∂p) would be the **ambient vapor pressure (v.p. = e) of water** at the resulting conditions of temperature and humidity.

e) if we made pressure measurements during conditions (a) and (b) — vacuum and dry air — the results would look like **Fig. 4-20c**. The final pressure would be the s.v.p., but it would be reached more slowly when the vapor molecules had to "fight their way among the air molecules" that were already there. Thus, *the value of the s.v.p. does NOT depend upon what other gases are present.* That is **Dalton's Law of Partial Pressures**. It's part of the reason that hot food cools more rapidly (by evaporation) at higher altitudes — lower pressures. The other part of the reason is that, usually, the air is cooler at higher altitudes.

So we see that vapor pressure is a way to express the amount of water vapor in the air sample (under the bell jar). The more water vapor molecules, the higher the vapor pressure. Finally, we have to note that *if the whole experiment had been carried out at a higher (lower) temperature, the saturation vapor pressure would have been higher (lower).* That relationship between temperature and the **(s.v.p. = e_s)** is shown in **Fig. 4-21a**.

In **Fig. 4-21** we begin discussion of the ways to convert from one humidity variable to another, by means of a chart called the **psychrometric diagram**. It involves mainly the three variables temperature (T), relative humidity (R), and vapor pressure (e), so we'll just call it the **TRe diagram**. In the first panel, **(s.v.p. = e_s)** is given as a function of air temperature. As a *rule of thumb* for this curvilinear relationship, no matter what pressure units are used, *e_s doubles for each increase of $11°C$ ($20°F$).*

If we draw a vertical line at each of several temperatures, as in **Fig. 4-21b**, then the lengths of the lines will each be (e_s) for that temperature. If we divide each vertical line in half, and connect those points, *we will have drawn a line whose label is (50% relative humidity: R = 0.5)*. By this rule, then: a) the curve of temperature and the **(s.v.p. = e_s)** is labelled (100% RH); and b) **relative humidity** (R) is the *fraction of moisture (water vapor) in the air compared with the amount that **could be** in the air at saturation, at the present temperature of the air.* Said another way, R = (ambient v.p.) ÷ (s.v.p.) = (e / e_s).

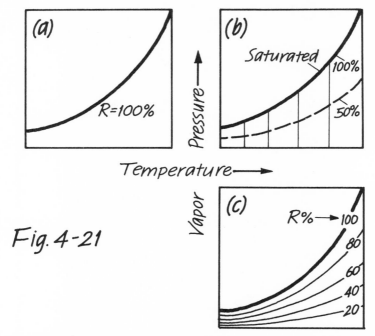

Fig. 4-21

As in **Fig. 4-21c**, we could draw lines for each value of (R), and then we would have a complete **TRe diagram**. But **Fig. 4-22**, which contains an even more complete TRe diagram[5], shows how to convert from one humidity variable to another, for example the **dewpoint temperature** mentioned in **Chapter 3**.

You recognize the **(T-e_s)** curve passing through points **B**, **C**, and **D** and labelled R = 100%. Point **A** is called the **point of state** and gives the combination of T, R, and e — the "state" — of the air sample under consideration. In particular (approximately): T = 27.5 °C, R = 50% (0.5), e_s = 37 mb, and e = 18.5 millibars[G].

(G) There are about 1000 millibars (mb) at sea level pressure. More on that in **Chapter 9**.

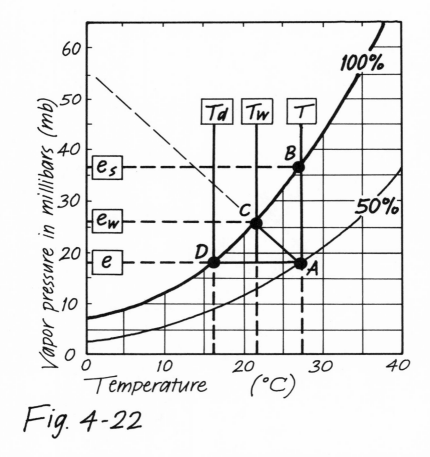

Fig. 4-22

Notice that (0.5)x(37) = 18.5. Several other moisture variables can be defined in terms of the diagram.

The **dewpoint temperature** (T_d) is the *temperature at which the ambient vapor pressure saturates the air.* That is, if the air sample is cooled with no change in moisture content, (Point **A** to Point **D**), it will reach saturation — **dew** will form — at the dewpoint. For the sample in **Fig. 4-22**, T_d = 17°C and e = 18.5 mb as before. At saturation (T = T_d).

Now you can see why, in **Chapter 3**, I said that T_d will stay the same, as long as the moisture content remains the same, *even though T and R range all over the place.*

The **wet bulb temperature** (T_w) is the *temperature at which evaporative cooling saturates the air.* That is, if the air sample is cooled by allowing water to evaporate into it, (Point **A** to Point **C**), it will reach saturation — **dew** will form — at the wet bulb, with a higher value of the vapor pressure than before cooling began. For the sample in **Fig. 4-22**, T_w = 22 °C and e_w = 26 mb, higher than before. At saturation (T = T_d).

The saturation deficit (e_s - e) is *the amount of vapor pressure by which the sample falls short of saturation* (Point **B** minus Point **A**). Often in ecological literature this variable is used to represent the *evaporation rate*. It is NOT a completely reliable measure of the evaporation rate, as discussed in Lowry (1969, Chapter 5).

Plant ecologists and forest fire fighters often use the **wet bulb depression** (T - T_w), and pilots often use the **dewpoint depression** (T - T_d) as measures of ambient humidity. Also, there are several other *ratios*, besides the relative humidity, expressing moisture content. The following table describes them. We will say no more about them other than to define them in case you encounter them in your reading.

Measure	is the ratio of the —	to the —
Relative humidity	ambient vapor pressure	saturation vapor pressure
Mixing ratio	mass of water vapor	mass of dry air
Specific humidity	mass of water vapor	mass of moist air
Absolute humidity	mass of water vapor	volume of moist air

As a means for you to check your understanding of the TRe diagram, the entries in the following table are correct statements about several air samples. The first two things to check are (a) T > T_w > T_d, and (b) R = (e / e_s) for each sample.

Sample	T	T_w	T_d	e_s	e_w	e	R
A	30.	21.	14.5	42.	26.	15.	36.
B	20.	17.5	14.5	23.	20.	15.	65.
C	25.	17.	5.	32.	19.	8.	25.
D	18.	17.8	17.5	21.	20.5	20.	95.

What is all this good for, other than understanding what certain moisture variables mean? For one thing, it gets you right into the simple, direct mathematical variables that govern the movement of water in the environment: evaporation rates, plant transpiration rates, sweat rates having to do with human comfort, and so on. Ecologically, very important stuff.

How can you use the TRe diagram to understand evaporation rates, for example? First, here is the primary rule governing the movement of *water vapor* in the microenvironment:

*water vapor moves from higher to lower **vapor pressure**, and at a rate proportional to the difference in **vapor pressure** along the path.*

If that sounds like the rule about heat flowing along a temperature gradient, don't be surprised. They are both based on the same physical principles. Heat follows the gradient of temperature, and vapor follows the gradient of vapor pressure.

For example, if Sample B, in the table just above, were located at, say, 5cm above a grassy surface, while Sample C were located at 2m above the same place, then we would conclude that *evaporation* was taking place — vapor movement from 5cm to 2m above the turf. The vapor pressure difference between 5cm and 2m is (15-8)mb, and the vapor is flowing from the higher (15) to the lower (8) vapor pressure. The **vapor pressure gradient** is (15-8)mb ÷ (200-5)cm = 0.036mb/cm. There would be flow of vapor from A to C, from D to B, and so on. You get the idea.

Soil moisture

In the soil, the movement of *liquid* water is probably more important than the movement of *vapor* in determining microclimate, but the movement of vapor may be more important in determining the kinds of plants that live in a particular microenvironment. We need to study the movement of both.

Retention and movement of moisture in soil is greatly influenced by the **soil texture**. **Fig. 4-23** represents the proportions of the three basic sizes of *soil grains* — the mineral particles — found in soils of various textures: the smallest is clay (relative size 1x); then silt (10x); and sand (200x). Loam is a mixture of the three sizes. The dark square in the diagram represents a soil with about 10% sand (10% of the *perpendicular distance* from the right edge to the lower left vertex of the triangle), 30% silt, and 60% clay.

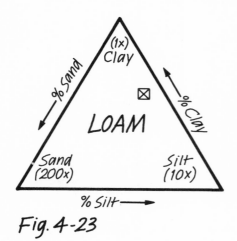

Fig. 4-23

Soil is a mixture of air, water, and a combination of mineral and organic solids.

Fig. 4-24 sets out some of the terminology we need to discuss soil moisture. Of course, the division of the contents into three "pure" layers is artifical — just for clarity of discussion.

The **porosity** (P) is the fraction available for air and water — the fraction *not* occupied by particles.

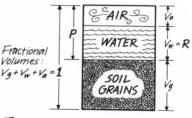

Fig. 4-24

The **water ratio** (R) is the fraction occupied by liquid water (maximum possible value of R is P). The **fractional volumes** (v) permit mathematical discussion of several other ideas. As shown in the figure, the following mathematical statements are true.

$$v_g + v_w + v_a = 1; \quad v_w = R; \text{ and } v_w + v_a = P$$

Earlier we defined and discussed the *heat capacity* (C) of soil and of air and water. Air and water in **Table 4-3** are pure substances, but "soil" as we have just seen, can mean any of a large number of mixtures. Mathematically, the *heat capacity of soil* is a **weighted average (or mean)** of the heat capacities of its constituents[P_11]:

$$(C)_{soil} = \frac{v_g (C)_g + v_w (C)_w + v_a (C)_a}{v_g + v_w + v_a}$$

Back to soil moisture. If you took a microscopic look at two soil samples — silt and sand, for instance — side by side, they would look something like **Fig. 4-25.** The grains are surrounded by thin, liquid water **films**, which in turn surround air-filled **pores** (or voids). What is more, *the air in the pores is saturated* with vapor, because the films provide a source of liquid water for evaporation into the pores until they are saturated.

There are several kinds of force at work on a liquid water molecule in a film.

(a) *gravitational*, directed downward;
(b) *hydrostatic*, due to the pressure of the air in the pore;
(c) *electrochemical*, binding the water chemically to the grain;
(d) *surface tension*, drawing the water by "capillarity" into narrow passageways; and
(e) *osmotic*, a physical force making water move from "pure" water to "salty" water, in this case right next to the mineral grain.

The sum of the last three force is called the **matric potential** (M), while the sum of the last four — everything except gravity — is called the **soil moisture tension** (S), or "suction tension." Tension is expressed in units of *pressure* — negative pressure (suction) — such as millibars (mb). What is more (here it comes again):

*liquid soil moisture moves from lower to higher **tension**, and at a rate proportional to the difference in **tension** along the path.*

Sound familiar? Good. Same physical principle as with heat and water vapor.

Component	Movement follows the gradient of. . .	
Heat	Temperature	higher to lower
Water vapor	Vapor pressure	higher to lower
Liquid soil moisture	Soil moisture tension	lower to higher

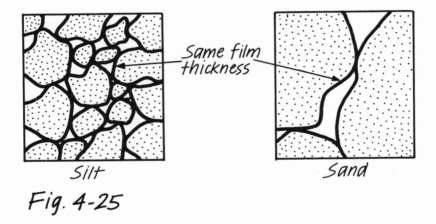

Same film thickness

Silt

Sand

Fig. 4-25

Any mass, in this case a molecule of liquid soil moisture, will move as a result of the *sum of the forces acting on it*. Thus, for example, if the sum of the forces binding the water to the soil grain is stronger than the sum of any other forces, it will stay next to the grain — it won't leave the soil pore. What is more, *the molecules closest to the grain, within the film, are bound most strongly to the grain. Thus, **the thinner the film, the more energy it takes to remove the next molecule from the film**.* Suction, you see, is a measure of the relative amount of energy it will take to remove the next molecule of liquid water from the soil.

Each soil, with its particular mixture of chemicals and grain sizes — and thus of pore sizes — has its own particular relationship between the water ratio (R) and the tension (S). Graphically, such a relationship is represented by the soils's **characteristic curve**. **Fig. 4-26** shows these curves (typical values) for clay, loam, and sand[H].

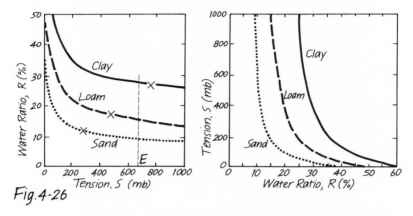

Fig.4-26

From these characteristic curves and the ideas in **Fig. 4-25** we can deduce several ideas about liquid soil moisture and what it takes, for example, for plant roots to remove it.

1) A fine-grained soil, such as clay, has a larger porosity than sandy soil[I], so it has a larger value of (R) at saturation (zero tension).

2) At a given value of (R), clay has a larger value of (S) than does sand[I].

3) Since the larger pores in sandy soil lose a greater fraction of their saturation content than do the smaller pores of clay *before a given film thickness occurs,* (R) is less for sand than for clay *at the same value of (S).*

(H) Both halves of **Fig. 4-26** contain exactly the same information with the coordinates reversed. The reason both are here is that you will probably run into both in your reading, so you shouldn't get confused.

(I) To understand these two ideas, think about the soils this way. If you start with a large chunk of mineral (porosity equal zero), and then start to break it up with a hammer, into smaller and smaller pieces, stirring the pieces after each hammer blow, you notice these things. 1) The more you break it up, the more surface area — called the **specific surface** — is exposed on the faces of the pieces (the grains). 2) When you stir the pieces, they don't come back together as they were before being broken; so that the greater the specific surface, the more opportunities for pores to be "built into" the soil, and the larger is the porosity. 3) A given volume of water spead over all the surface area within a soil will have a thinner film (larger S) in the soil with the larger number of small grains.

4) Just because there is a lot of liquid soil moisture — a large value of (R) — doesn't mean the water is "available" to plant roots — within the energy capability of the roots to extract it. This is particularly true in fine-grained soils such as clay.

We've just said that liquid soil moisture moves from lower tension to higher tension, *regardless of the value of the **water ratio***. In analogy, that's like the other true statements that (a) water vapor moves from higher to lower vapor pressure, *regardless of the value of the **relative humidity***, and (b) heat moves from higher to lower temperature, *regardless of the value of the amount of **heat in storage***.

In **Fig. 4-27** another little "experiment" will demonstrate the meaning of what we have just said about the movement of liquid soil moisture. There are three "soil cores", side by side, touching, on a table top. Since they are all side by side, gravity is acting equally on all the water in the cores. Any movement, therefore, will be because of differences in *tension* (no gravity).

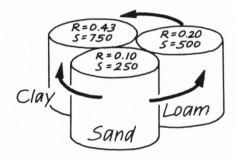

Fig. 4-27

You will notice that the values of (R) and (S) attached to each core are the same as the values marked by the X's on the characteristic curves in the left panel of **Fig. 4-26**. Water will flow from the clay into both the loam and the sand, and from the loam into the sand, until movement ceases when they all have the same value of tension (S). This occurs at the value (**E**) in **Fig. 4-26** after the values of (R) have increased in sand and decreased in clay and loam.

Clearly, these simplified discussions only state the principles at work in real soil. Under field conditions, the fact that rain and snow come irregularly, temperature rises and falls, wind blows and calms, and soil is far from a homogeneous mass below the surface — all these result in a very complex picture of soil moisture movement and distribution. The simple word picture in **Chapter 3** hides the complexity of what happens. To some extent the layering of moisture in the soil reflects the time history of the arrival of the various rains at the surface. Even in the complexity, certain general understandings are possible, and that is what I've tried to offer here.

Two more ideas about soil moisture before we go on. First, in **Fig. 4-28** the effects of *salinity* on soil moisture properties is suggested. Briefly, saline soil acts like clay — the more salt, the more like clay. That is shown in the left panel. To the right are results of growing small plants in soils of various combinations of wetness (R) and salinity, both of which affect the tension (S). The results prove that *the five forces on liquid soil water are additive.* It doesn't matter to the plants what combination of water and salt produce the tension. They react to the tension — the sum of the forces.

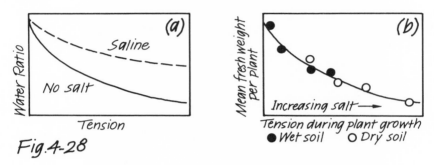

Fig.4-28

The second idea about soil moisture concerns the "crust" that forms on the dry soil surface. Experience tells us its *strength* comes as the drying proceeds and the chemical bonding in the material changes — like the crust on baking bread. But the crust also acts as a *vapor barrier*, reducing the evaporation rate. Negative feedback — more evaporation yields less evaporation. It works like this.

In discussing **Fig. 4-25** we noted that the air in the soil pores is always saturated with water vapor. That is, 100% relative humidity. That being the case, the vapor pressure in the pores is always the s.v.p., and is related to the temperature of the soil according to the s.v.p. curve in the TRe diagram. As (T) increases, so does (e_s). In *very dry* soil, vapor movement is more important (especially to plants) than liquid movement. And since vapor always moves from higher to lower (e), it also moves from *warmer to cooler* soil. Here's what results:

a) during midday, when evaporation would be greatest, vapor in very dry soil moves downward, back into the soil; and

b) at night, when evaporation would be least, vapor moves to the surface *and condenses there as dew.*

Motion and wind

In **Chapter 3** we described in some detail the responses of windstreams to various underlying surfaces, obstacles, and terrain features. But we didn't inquire as to why the wind blows in the first place. In our discussions the wind was just "something that was coming from someplace else." It is worth taking a few moments to see *why fluids move* — why, in our particular case, the fluid called air moves and is called "wind."

Fig. 4-29 shows another of our small, imaginary experiments. We begin with a container filled with fluid (it doesn't matter which fluid — the principles are the same). The fluid is *at rest*, and the pressures[J] are **the same at A and B, and at C and D**. Because the pressures are the same, the *lines of equal pressure* connecting A with B and C with D are *horizontal*. If the fluid in the container is a liquid, there is a definite "top" somewhere above C-D; but if it is a gas, there is no definite "top", as we will see in **Chapter 9**. To go on, we don't need to know which — only that C-D is somewhere below the top.

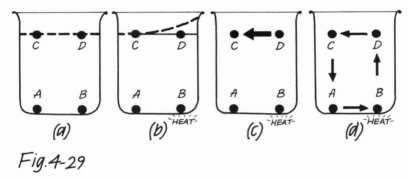

Fig.4-29

In **Fig. 4-29b**, we *heat the fluid column above B*. Because materials expand when they are heated, the column of fluid above B — we'll call it the "column B-D" — expands vertically. Though exaggerated, the sketch shows the line connecting C and D becomes "tilted" and is no longer horizontal. That is, the line of equal pressure (it's called an **isobar**) through C is now above D, which means that the *pressure at D has increased* — it now has more fluid above it, so the pressure has increased, because of the definition of pressure in the footnote. Because the little experiment is imaginary, we can say that no fluid motion has occurred

---

(J) In addition to pressure being the force of molecules striking a surface, as we discussed earlier, it is, in a case like this when the fluid has a "free surface" on the top, a measure of *the weight of the mass above the place* where the pressure is measured.

other than the expansion of B-D. We know also that (a) *the pressures at A and B are still equal:* A = B, but (b) *the pressure at D is greater than that at C: D>C*.

In **Fig. 4-29c**, we see the first motion: the movement in response to the increasing pressure difference between D and C. The fluid is trying to equalize pressures — make all the lines of equal pressure (they are actually *surfaces*) be horizontal. But notice, the minute fluid moves from above B to above A, we get a new situation in which (a) *the pressure at A is now greater than that at B:* A>B, and (b) *the pressure at D is greater than that at C:* D>C.

In **Fig. 4-29d**, we see the result: a **circulation cell** forms in which the fluid moves as shown. The fluid is trying to make the isobars horizontal, or trying to redistribute the heat equally, however you want to think of it. As long as the temperature at B is higher than that at A — it could just as well happen because of *cooling at A* — the circulation cell continues. Because the fluid is trying to equalize the heat content everywhere, there must be some external energy source that maintains the **differential heating** of the surface. Thus, we can say that *differential surface heating is the basic cause of motion*.

Because differential heating occurs on many space and time scales on Earth, so the resulting circulation cells occur on many space and time scales. We have already examined the set of **local circulations**: across shorelines and on sun-heated slopes. Circulation cells occur between the sunny side and the shaded side of an urban canyon, between a warm continent and a cold ocean, and even between the tropics and the polar regions. But (very important) *circulation cells form only when circulations on larger scales are "quiet."* On a stormy day, no cell forms in an urban canyon, or across a shoreline. That is because larger scale circulations do, in fact, distribute the heat and there is no redistribution job left for the smaller cell to do.

Notice that "many space and time scales" involves *diurnal* changes in differential heating in the case of the local circulations, and *annual* (seasonal) changes in the case of the continent-ocean cells, called **monsoons**. More on the tropics-to-poles cells in **Chapter 10**.

In **Fig. 4-30** we can see the "North American Monsoon" at work. In the high-sun season — summer — the flow is onto the warmer land mass, while the reverse occurs in winter. The Asiatic Monsoon — it brings the rains to India — is more dramatic and better known, but any mid-latitude land mass has one. The arrival of summer rains in Arizona is called the "monsoon" there.

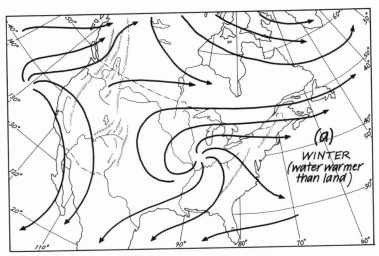

STREAMLINES : Everywhere parallel to the wind.

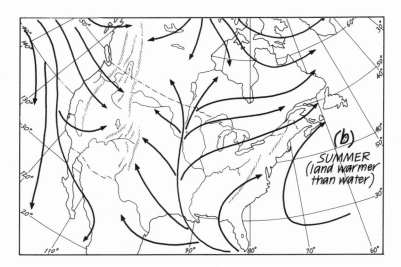

Fig. 4-30

Now is a good time to discuss again the difference between *streamlines and trajectories*, as we did when considering turbulent motion. The lines in **Fig. 4-30** are streamlines — everywhere parallel to the wind. But what wind? *The seasonal average wind.* So that the figure doesn't say the wind is always blowing as shown — in fact, maybe *not even most of the time*[K]. But the large scale, seasonally-changing monsoonal circulation is real and is shown clearly.

If we released a swarm of balloons in Illinois (the "source" of all the mid-continent arrows), on each of several days of winter, let's say, each balloon would follow a *trajectory* — probably no two alike. If you did that every day for a winter, and "averaged" the trajectories, they would probably look like **Fig. 4-30b**. The same for summer, with the results in **Fig. 4-30a**. So what? For one thing, trajectories tell us about *where pollution goes* on a particular day. Seasonal average streamlines tell about what happens, on the average, between regions.

Clouds

We know from the last chapter that clouds are swarms of small water droplets — visible tracers of wind flow. They go where the air goes, responding to all the things wind responds to. I want to tell you now about (a) how clouds form in the first place, (b) our system for naming them, (c) what they tell you about — how they help "diagnose" — the processes at work around them, and (d) why some clouds rain and some don't.

Water droplets don't "just form" for no reason. They are an end result of a complex sequence: (saturation) >> (nucleation) >> (condensation). As we'll see presently, the sequence goes on from there in the clouds that produce rain:

(saturation) >> (nucleation) >> (condensation) >> (aggregation) >> (precipitation).

Nucleation is the process by which water vapor molecules and certain small particles (condensation nuclei) — clay, salt crystals, etc. — in the air "select" each other, so that the vapor finds a site on the surface of the particle, and there changes state — gas to liquid — **condensation**. Nucleation is not completely understood, and only a small fraction of all particles ever act as nucleation sites. "Cloud seeding" involves introduction of nuclei into clouds to alter the otherwise natural course of changes already taking

---

(K) **Chapter 8** contains a lot more about how to manipulate weather data, and what some of the statistical relationships among them are.

place at this crucial point in the chain of events (see the footnote at the end of this section). Nucleation is a subject far beyond our needs to understand.

Be that as it may, for nucleation and condensation to occur, first we need **saturation**. In studying humidity we learned about saturation — the condition in which air molecules, at a certain temperature, are accompanied by water vapor molecules, and the number of vapor molecules is the *largest possible number at that temperature*. If any more vapor molecules enter the volume, some will have to condense[L]. As the temperature rises, that maximum number also rises, according to **Figs. 4-21** and **4-22**.

The question now arises, how does the air become saturated? **Fig. 4-31** describes the various possible pathways leading to saturation, and so on to cloud formation. Broadly speaking, there are two routes: (1) add enough vapor molecules for the present temperature, or (2) lower the temperature enough for the present number of vapor molecules.

Adding moisture can take place from below — evaporation from an underlying water surface — or from above — rain falling from a cloud above and evaporating as it falls. Lowering the temperature can take place in quiet air — by radiation to a clear night sky — or in moving air — by moving up a slope or rising on a local

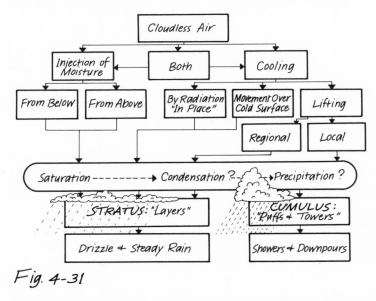

Fig. 4-31

(L) Another topic beyond the scope of this book is the fact that *in very special circumstances*, a few more molecules can be present as vapor, a condition called **supersaturation**.

updraft, or by flowing over a cold surface. Air in a frost pocket (or valley) lies quietly and cools by radiation (see **Fig. 3-26**). Cumulus and stratus clouds form while rising in ways shown in **Figs. 3-29** and **3-30**.

Now that we know how clouds are formed, how are they named? The standard system is given, in a brief form, in the following table. The original system was invented by Luke Howard, an amateur British scientist, in the early 19th Century. Many of his friends were landscape painters, and they expressed a need for understanding and naming clouds they saw. Many details have been added since Howard, as more scientific knowledge of clouds has accumulated[6].

From the table you can see that the three things to know about the cloud in naming it are (a) the height of its base above ground level, (b) whether it is of the cumulus ("Puffs and Towers") or the stratus ("Layers") type, and (c) whether or not it is raining or snowing. High clouds (above 20,000 feet) are made up mostly of ice crystals and are of the "cirrus" type, while middle clouds (6,000 to 20,000 feet) are of the "altus" type. A precipitating cloud has "nimbus" in its name.

NAMING OF CLOUDS

	LAYERS		PUFFS & TOWERS	
RAINING OR SNOWING ? → YES	NO	YES	NO	
Altitude of cloud BASE: 20,000' – 6km.	X	Cirro- stratus (Cs)	Cirrus uncinus (Ci)	Cirrus (Ci); Cirro-cumulus (Cc)
6,000' – 2km.	X	Alto- stratus (As)	X	Alto- cumulus (Ac)
Surface	Nimbo- stratus (Ns)	Stratus (St) Fog (K)	X	Cumulus (Cu)

→ Towers reaching through all layers : Cumulo-nimbus (Cb)

NOTE: "X" means "no special name for this type".

Two names seem to be "exceptions that prove the rules." First, there are *strato-cumulus*, which are formed by processes and have an appearance midway between those suggested in **Fig. 3-30**. Second, *cumulonimbus* have bases below 6000 feet but reach as high as most any clouds reach — usually well above 20,000 feet. They occupy all of the "altitude" categories and are raining; in fact, they are sometimes called **thunderstorms**, though the two names do not refer to exactly the same phenomena.

The classifications with more detail consist mainly of additonal modifiers to names shown here. For example, *alto-cumulus castellatus* are usually shaped like turrets on a castle; and *alto-cumulus lenticularis* are shaped like giant lenses (see **Fig. 4-34**).

While less true of stratus clouds, *cumulus clouds have clear-cut life cycles* of their own. The cycle is suggested in **Fig. 4-32**. An individual cumulus cloud is born on a local updraft, and then *tries* to grow vertically and horizontally. Whether or not it can grow depends upon the local supply of moisture and on the maintenance of the updraft. In dry air and/or air with an inversion "ceiling" (see **Chapter 11**) it cannot grow beyond infancy, and dies. The updraft *begins* over a "hot spot" on the ground, or where two airstreams collide near the surface, or just because an airstream is forced to rise over denser air or a terrain feature. In the absence of terrain forcing further rise, the updraft is *sustained* mainly by the release of the *latent heat of condensation* as droplets form. The extra heat adds buoyancy to the rising air.

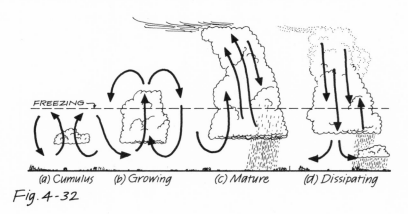

(a) Cumulus   (b) Growing   (c) Mature   (d) Dissipating

Fig. 4-32

With both moisture and a sustained updraft a cloud can grow to maturity, though it may not go beyond the intermediate "growing" or "towering" stage. It may produce rain without growing above the *freezing level* in the tropics, but elsewhere the rise above the freezing level is nearly a necessity. Growth above the freezing level does not guarantee rain, as we shall see presently; but it pretty much assures at least some of it.

Why is it that some clouds rain and others do not? In addition to requiring the moisture supply and the sustained updraft, the production of precipitation requires *the proper mixture of sizes of condensation nuclei*. Reaching the stature of a *fully mature* cumulus cloud requires abundant moisture, a strong and sustained updraft, and a particular range of mid-cloud temperatures. For a mature cloud to become a *thunderstorm*, or even a **hail storm**, requires an even more particular set of circumstances.

In **Fig. 4-33a** we see that the kind — the shape — of an ice crystal forming in a cloud below freezing depends on the temperature at the time it forms. In particular, the **dendritic crystals** form in the range of about -13 to -18 °C. What's special about these spidery crystals we will see in a moment. In **Fig. 4-33b** we see that (a) large drops fall faster than smaller droplets and (b) larger drops absorb smaller drops that they overtake, in the process of **coalescence**. This process is a principal way in which droplets grow into drops big enough to fall out of the bottom of a cloud.

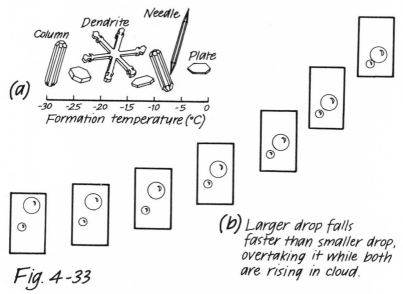

(a)

Formation temperature (°C)

(b) Larger drop falls faster than smaller drop, overtaking it while both are rising in cloud.

Fig. 4-33

It takes a lot of time for larger drops to collect a lot of smaller droplets — millions of them — and *the sustained updraft provides the time* for that to happen. In **Fig. 4-33b** you notice that while the larger and smaller drops are both falling with respect to the air they are embedded in, and the larger drop is falling with repect to the smaller one, they are both rising with respect to the level of the cloud bottom (or the earth's surface) as long as the air in the core of the cloud is rising. Thus, the growing drops and the smaller ones on which they feed are all moving upward and to lower temperatures.

Now, what's special about the spidery, dendritic crystals? When they begin to form, they immediately begin collecting both other ice crystals — by simple mechanical aggregation — and by making liquid droplets freeze on them. In both processes their large surface areas make these processes *extremely efficient*, and so a cloud whose top reaches the temperature range of -13 to -18 °C begins to grow almost explosively[M]. When the largest drops or snow flakes have grown too large to be held up by the updraft, or when the updraft weakens, precipitation falls from the cloud base. And to reach the surface, drops and flakes must not evaporate into the sub-cloud air. If the air there is too dry, and they do not reach the ground, they form a visual "veil" beneath the cloud, called **virga**.

The downrush of large drops in itself helps weaken the updraft by dragging on it, so that the cloud — now dissipating — contains more downdraft than updraft. When these downdrafts reach the ground, they push out in all directions away from the base of the cloud, producing sometimes dangerous winds variously called **gust fronts, microbursts,** and **wind shear**.

So, in the end, you see, for a cumulus cloud to reach full maturity, and for rain to reach the ground, requires just the right combination of several kinds of things, not least of which is the sustained updraft, enhanced by the latent heat of condensation, and the *time* it provides[N].

The extra set of special conditions that turns a thunderstorm into a hail storm is really beyond this discussion, except to say that large raindrops are sometimes "recycled" above the freezing level, where they have ice added to their skins, fall again, are recycled again, and eventually become too large to stay in the cloud and fall out as *hailstones*. Schaefer and Day (1981) give the details of these events effectively and in layman's language.

Returning to the subject of clouds that do not rain, **Fig. 4-34** shows a few of the interesting forms they can take. When, for reasons we need not understand here, airstreams develop a long train

---

(M) The process of droplets freezing on ice crystals is the second principle way in which flakes (and drops when they melt) grow big enough to fall out of the bottom of a cloud. In fact, the droplets and the crystals don't even have to encounter one another. The droplets actually evaporate and then freeze on the surface of nearby crystals in a process discovered by two Norwegian scientists, who gave their names to the **Bergeron-Findeisen** process.

(N) "Cloud seeding", or "weather modification", usually consists of one of two kinds of intervention in a natural cloud. Ice crystals (or mineral crystals that look and act like ice crystals) when placed in a cold, quiet, supercooled stratus cloud — fog — make the droplets follow the **Bergeron-Findeisen** process. The millions of droplets then become a relatively few large ice crystals, and the fog is *dissipated*. Artificial introduction of mineral crystals into a growing cumulus cloud sometimes *enhances* the processes already taking place there in such a way as to "increase the yield" of precipitation. When this works and when it does not is still, after four decades, a subject of great debate.

of wave forms, one of the results is often a series of **wave clouds**, one atop each wave crest. The more humid the airstream, the larger the clouds. *The air is moving through the sequence of clouds even though the cloud forms are standing still.*

When the wave crest is actually produced by a peak in the terrain, the resulting cloud is called a **cap cloud** — *alto-cumulus lenticularis.* The line of wave crests leans upwind above the peak,

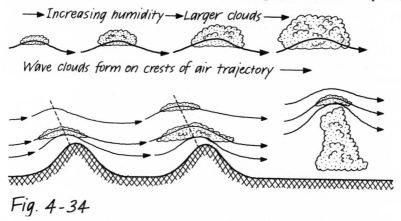

→ *Increasing humidity→Larger clouds* →

Wave clouds form on crests of air trajectory →

Fig. 4-34

and if there are several humid air layers alternating with drier ones, a stack of cap clouds forms like pancakes. Cap clouds sometimes form atop large, mature cumulus clouds.

Transfer processes in plant canopies

Having seen how to describe canopy structure, and something about the nature of various profiles within a canopy, we can now use the ideas from our discussion of soil heat transfer to examine transfer of heat, and other atmospheric properties, within a canopy. **Fig. 4-35** is a reminder of the ideas we developed earlier about gradients and curvature and their rela-

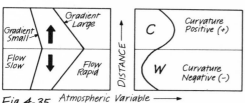

Fig. 4-35 Atmospheric Variable ——→

tionships to the directions and rates of transfer. These ideas, developed for soil heat flow, are equally valid for *transfer of any form of mass or energy within a medium,* such as the air of a plant canopy.

In **Fig. 4-36** are the same three profiles of midday temperature found in **Fig. 3-35b**. Here I have added an analysis of the profiles, using the rules about direction of flow and warming and cooling. The part of the profile formerly called "cooling" is now marked "source of heat" — the layers from which the heat is flowing to other layers of the canopy.

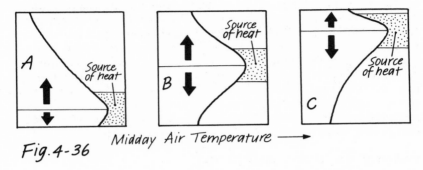

Fig. 4-36 *Midday Air Temperature ⟶*

The concept of a **source** — *from which* flow takes place — and its opposite concept of a **sink** — *to which* flow takes place — are very important in ecology. As we saw in **Chapter 3**, it is the combination of larger insolation and restricted ventilation that makes the mid-canopy warmest at midday. Thus, although *positive curvature* signified "cooling" in soil, here in the canopy air it marks the source of heat for the rest of the canopy. This layer is not cooling, but is acting as the point of maximum introduction of heat energy — the source — for the canopy. The air above the canopy and the soil beneath are the sinks for the heat at this midday hour. At night, the opposites are true — sources and sinks are reversed.

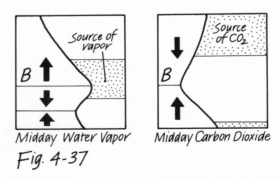

Midday Water Vapor   *Midday Carbon Dioxide*

Fig. 4-37

Fig. 4-37 takes the analysis one step farther by examination of two forms of *mass transfer*, as opposed to the *heat transfer* we just looked at. The masses being transferred are *water* and *carbon dioxide*, again at midday, in Canopy **B**: the coniferous forest.

Water vapor is being introduced in mid-canopy by *transpiration* from leaves in the warmest layers of foliage. There is also a weak source of vapor from *evaporation* at the ground line.

The carbon dioxide is entering into photosynthesis, which *uses* CO_2, in the warmest, most rapidly transpiring foliage of the mid-canopy. Here, the atmosphere above, and the soil beneath, are both sources of CO_2. Details of these interactions among temperature, photosynthesis, and transpiration will be discussed in **Chapter 6**. Information about profiles within canopies becomes a basis for diagnosis, or inference, about the transfer processes taking place there. Soil heat flow and storage are dynamic and ever changing, as we have seen. The addition to the system of solar radiation by day and longwave radiation at night, and ventilation by wind at all times, makes the flow and storage processes even more dynamic and complex in a canopy. The same concepts and ideas we used in soil permit us to analyze what is going on in a canopy.

Putting the parts together: the Energy Balance Concept

Reviewing briefly the concept discussed in **Chapter 2**, and stated in the following equation:

[SUM OF INFLOWS] = [SUM OF OUTFLOWS] + [CHANGE OF STORAGE]

we need to recall that (a) this is a special form of the Law of Conservation, (b) it is a statement of the balance of flows and storage in a defined system during a specified time period, (c) it may be used to describe balances of mass or energy, and (d) it is a statement about *rates* — of flow and of change —which is to say it is about either *[(Energy or mass)/(Area)x(Time)]* or *[(Energy or mass) / (Time)]* . It is the **Energy Balance Concept** when applied to energy flows and storage, and it represents perhaps the most important analytical tool to be described in this book.

Fig. 4-38 is a reminder of the physical form of the concept in the simplified form of **Fig. 2-1**.

Having just finished a discussion, in the last secton, involving flows of solar radiation, the kinetic energy of wind, heat, water vapor, and carbon dioxide within a canopy, you may be wondering about a way to put all those ideas together in a manner that makes the

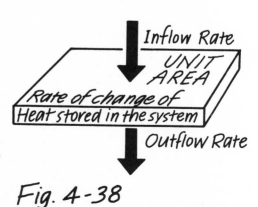

Fig. 4-38

dynamics of the natural system become clear. The Energy Balance concept, together with its twin the Mass Balance concept, is that way.

Earlier in this chapter we set out the several kinds of energy, of which heat is one kind. We need to note, also, the idea of *energy (or mass) transfer* — the rate of movement of energy from one place to another. The following table shows the kinds of energy and energy transfer we will deal with in the analysis of the energy balance of a microenvironmental system, and the set of symbols for the different pathways.

No doubt there seems to you a bit of overlap between the kinds of energy and the kinds of transfer. Let me explain. As we noted earlier, the term "radiation" refers, in the language of physicists, to both the process and to what is transferred. *Sensible heat* is the kind that can be measured with a thermometer, to distinguish it from *latent heat* which cannot.

Kind of Heat Energy	Kind of Energy Transfer			
	Radiant	**Conductive**	**Convective**	**Latent**
Radiant				
Shortwave (Solar)	S	-	-	-
Longwave (Terrestrial)	L	-	-	-
Sensible	-	B	H	-
Latent	-	-	E	E

Note: The flow of heat into and out of the ground is called **B** in this book, after the German "boden" (soil) in the original energy balance studies. In other references it is often called **G**, but the meaning is the same.

Sensible heat can be transferred by *conduction* in the soil and by *convection* in the air. Conduction involves slow and inefficient heat transfer when individual molecules collide with each other and share their internal energy. Convection involves fast and efficient transfer when whole swarms of molecules mix with other swarms, giving the mixed swarm a temperature (heat content) partway between those of the individual swarms before mixing.

Latent heat is transferred when evaporation (or condensation) takes place at the surface of the system — the last column in the table — and also when the wind carries the vapor molecules aloft and away from (or to) the surface, by convection. In most analyses, these two transfers — first by change of state, and then by convection — are combined into one, with the symbol $E^{(O)}$.

In **Fig. 4-39** we see the result of applying these terms of the energy balance concept to a unit area of the earth's surface. We have a system with multiple inflows and outflows, like that in **Fig. 2-3**. Note in particular that we have chosen the system so that it is very thin, and its mass is so small as to be negligible. In this way we can ignore changes in heat storage in the little system we are analyzing.

On the left are the nighttime and daytime *budgets* of radiant energy alone, while on the right are the nighttime and daytime *balances* of all heat energy transfers[P]. A **budget** involves some of the pathways within the **balance**. Here, the *radiation budget* includes only radiation transfers, **S** and **L**, whereas the energy balance includes the radiation budget in addition to the transfers by conduction and convection: **B**, **H**, and **E**.

The terms of a budget do not ordinarily add up to zero — they don't have to balance. The terms of a balance MUST add up to zero — they MUST balance, hence the name. When we say "add up", of course, we are talking about algebraic addition. In our system of analysis, we say that a flow term is *positive (+) when it is toward the system and negative (-) when it is departing from the system.* More on these differences in the next section.

At night, the radiation budget (upper left in Fig. **4-39**) has only two **streams** — pathways — of energy flow: the *incoming* longwave, L_i, and the *outgoing* longwave, L_o. L_o is the longwave radiant energy in **Fig. 4-12**, while L_i is the longwave energy in **Fig. 4-13**. Since this system is a part of the earth's surface, radiant transfers do not occur on the lower surface of the system.

During the day (lower left), the energy of sunlight is added to the radiation budget: S_i and S_o. The terms of the radiation budget are usually added together and given a single symbol, **R**. In formal mathematics[P12], this is:

$$R = S_i - S_o + L_i - L_o$$

To the right in **Fig. 4-39**, the conductive and convective terms are added, so that when *changes in storage are ignored*:

(O) In many references, the symbol **E** refers only to the evaporation (or condensation) rate — the rate at which water changes phase — and the term **L** (latent heat per gram of water) is added — so that the symbol **LE** refers to the transfer of the energy associated with the change of state.

(P) Only in recent years have writers on this subject ceased using terms like "heat balance", "heat budget", and "energy balance" interchangeably. You may encounter all of these in your reading. They are meant to mean the same unless the writer says he means them to be different.

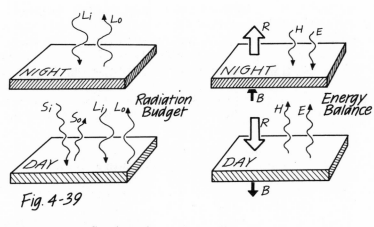

Fig. 4-39

(Inflow) - (Outflow) = 0

At night this is:

$$-R + B + H + E = 0$$

while by day the directions
are reversed, and:

$$+R - B - H - E = 0$$

and both of these amount to:

$$R = B + H + E$$

In using this valuable tool of analysis later in the book, we will
encounter the fact that the little system in **Fig. 4-39** — the unit
area of the earth's surface — is different from other systems we
will need to analyze. It is different in the sense that other systems
require different combinations of *streams* — flow arrows — for
their analysis. For instance, when we later analyze the energy
balance of an *individual leaf* — to represent vegetation in general
— we have to look at it in the manner of **Fig. 4-40**. Here (a) there
are two active surfaces — top (**t**) and bottom (**b**) — for **R, H**, and
E, and (b) no soil flow term, **B**. More on this in **Chapter 6**.

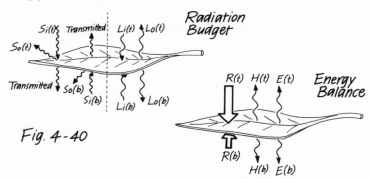

Fig. 4-40

Analysis of the energy balance of the plant canopy we were discussing in the previous section — as opposed to the balance of an

individual leaf in the canopy — usually takes the form of **Fig. 4-41a**. Here the differences from **Fig. 4-39** consist of (a) the need to include changes in storage, Δ, and (b) something called advection, **A**, which is the horizontal flows of **H** and **E** through the vertical walls of the system. Physically, you can think of advection as just *horizontal convection*. Otherwise, all the terms mean the same thing as they did in **Fig. 4-39**.

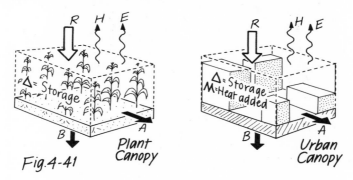

Fig. 4-41   Plant Canopy   Urban Canopy

Fig. 4-41b shows that we can treat the analysis of the energy balance of an urban canyon just as we do in a plant canopy. In fact, the term **urban canopy** is quite standard in research of this kind. The difference between the plant canopy and the urban canopy is the need to include a term representing the introduction of heat, due to things like automobiles and furnaces, in the urban setting. This "source" has the symbol $+M$, and it is always positive.

The energy balance systems of **Figs. 4-39, 4-40**, and **4-41** all have one thing in common: they consist of a unit of area on the earth's surface *chosen to represent a typical unit, representative of the other units of area surrounding them*. This choice is just to keep the size of the numbers small in actual numerical solutions of the balance equations. If you analyze just a unit of area, you can always expand your results to a larger part of the earth's surface by multiplying your results by the larger area involved.

For some analyses it is more meaningful to study the balance of the system as a whole, rather than of a typical, representative unit of area. **Fig. 4-42** shows two examples where this is so. "Typical unit of area" has little meaning on a person, or an animal, or a building, but our method of energy balance analysis is still quite valuable.

In this section I have given you a lot of complex material in a small space. Actually, if you stop and think about it, the complexity has come in the explanations of *how many different ways you*

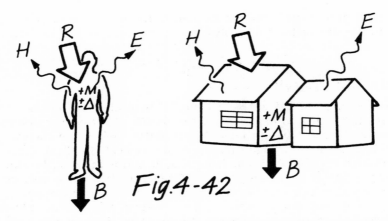

Fig. 4-42

can use one basic idea: the energy balance (EB). It seems more complex than it really is, and by the time we have used the EB to study each of these different systems, one at a time, it will all become clear to you[Q].

In the next section we will look at actual numerical studies of EBs for several types of surface. From them you can begin to see how to pick up subtle differences, representing distinct changes in the microenvironments they represent. You will begin to view the information in a numerical EB as a means for *diagnosing* the energetic effects of physical changes and differences in the microenvironment. You will begin to see, truly, "how the microenvironment works."

In **Chapter 5** we will study individual people, and in **Chapter 6** individual leaves and animals. These chapters will give you enough information about "how to read EBs", you can study those you encounter in the literature with great understanding. Though it is not my intention to offer you a complete set of EBs, in **Appendix A** I have included several specialized examples, including some

---

[Q] Let's put all the EB equations for midday conditions (R positive) side-by-side to summarize what we have just said.

(a) The surface of **Fig. 4-39**: $+ R - B - H - E = 0$

(b) The leaf of **Fig. 4-40**: $+ R_t + R_b - H_t - H_b - E_t - E_b = 0$

(c) The plant canopy of **Fig. 4-41a**: $+ R - B - H - E \pm A \pm \Delta = 0$

(d) The urban canopy of **Fig. 4-41b**: $+ R - B - H - E \pm A \pm \Delta + M = 0$

In these first four equations, the individual flows are (Energy)/(Time(Area)).

(e) The person in **Fig. 4-42**: $+ R - B - H - E \pm \Delta + M = 0$

In this last equation, the individual flows are (Energy)/(Time) for each area to be discussed in **Chapter 5**. The symbol ($\pm$) means "plus or minus"; that is, it is possible for this term, unlike most of the others, to be either.

for urban sites, beyond the few needed in the text to show you how to use this method of analysis.

The Energy Balance Concept: analyzing the microclimate

One of the major advantages of the EB method of analysis is that *it enables you to organize all the details about energy flows in one understandable package.* In particular, look at the energy flows in **Fig. 4-39** through the course of a typical cloudless, summer diurnal cycle — 24 hours beginning at midnight. First, in **Fig. 4-43a** the five terms of the radiation budget for this grassy surface are shown changing hour by hour. Values plotted above the *zero line* (+) represent energy arriving at the system, while values plotted below (-) are departing energy streams. This sketch is idealized — not actual measurements — but it makes clear the following important features of the radiation component of microclimate:

(a) at each moment, the values add up in such a way that the **net radiation** equation holds true: $R = S_i - S_o + L_i - L_o$;

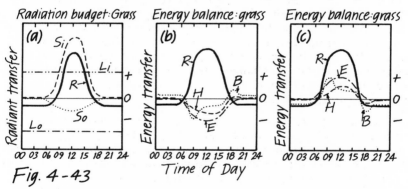

Fig. 4-43

(b) the shortwave terms, $+S_i$ (the insolation) and $-S_o$, follow the course of the sun through the cloudless sky, both being zero at night;

(c) as noted earlier in our discussion of *albedo*, $-S_o$ is the reflected part of $+S_i$, so that $aS_i = S_o$;

(d) $+L_i$ is less than $-L_o$ (see **Figs. 4-12** and **4-13**), partly because it does not arrive in all wavelengths and partly because the *absolute* — Kelvin — temperature of the sky is slightly lower than that of the ground surface;

(e) L_i and L_o change little during the cycle — quite different from S_i and S_o — because the Kelvin temperature changes very little;

(f) the *net radiation*, **R**, is small and negative at night, but much larger by day, peaking near noon;

(g) the times of *zero net radiation* are just after sunrise and just before sunset; and

(h) the daily net value of **R** is positive — more radiant energy arrives than leaves — so that the grassy surface has an excess of radiant energy through the course of the daily cycle.

The curve for net radiation, **R**, in **Fig. 4-43a** is transferred directly to **Figs. 4-43b** and **4-43c**. Also in those figures are the conductive and convective components of the equation: **R = B + H + E**. It is clear, by comparison of **Fig. 4-43a** and **Fig. 4-43c** that *the daily excess of radiant energy is all used up by the heating of soil and air, and the evaporation of water at the surface.*

The only difference between **Fig. 4-43b** and **Figs. 4-43a** and **43c** is that in the middle panel *all components are shown as positive at midday.* I include both versions of the presentation because you will encounter both in your reading. They both mean the same things. The main reasons for ever using the format of **Fig. 4-43b** are to make clearer that **R = B+H+E**, and to conserve printing space.

Figs. 4-43a and **4-44a** are the same — the radiation budget of a grassy surface. In **Fig. 4-44a** the positive daytime area and the negative nighttime area for the net radiation are highlighted to make clear the fact that the net for the 24 hour period is positive.

Fig. 4-44 provides you with the first opportunity to glimpse the workings of the physical microclimate system by comparing the behavior of the energy balance at three very different surfaces: (i) grass, (ii) an irrigated (wetted) crop, and (iii) a bare, dry surface. In all three parts of the figure the incoming streams — S_i and L_i — are the same, so as to be certain any differences in **R** are due to

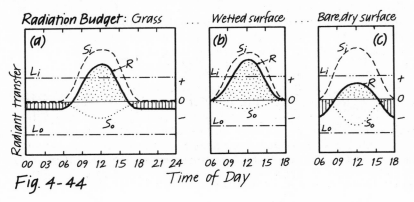

Fig. 4-44

the differences in the surfaces themselves. Here are the kinds of things you should note in making your comparisons with the grass:

(a) the darkened surface (smaller albedo = smaller S_o) of the wetted field, and the fact that it is cooled by evaporation (smaller L_o) yields a larger net radiation through the day, and almost a zero value for **R** at night;

(b) the opposite effects of the large albedo (reflection) of the hot, dry surface — larger S_o and larger L_o — produce a larger net loss both at night and for the whole day.

Taking the net radiation curves for the three surfaces *directly to Fig. 4-45*, and adding the conductive and convective components (**B, H,** and **E**) to get the *energy balances* of the surfaces, we see the results of having different kinds of surface and different amounts of net radiant energy to distribute among **B, H,** and **E**. By *midday* comparison with the grass surface:

(a) *the wetted crop* put all the extra radiant energy, and more, into evaporation: **E** is much larger, while **B** and **H** are smaller; and

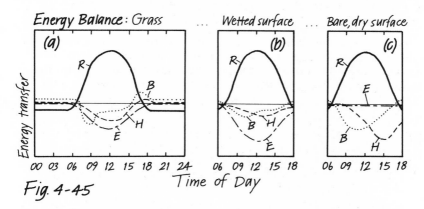

Fig. 4-45

Energy Balance: Grass ... Wetted surface ... Bare, dry surface

Time of Day

(b) *The dry surface*, with no water to evaporate (**E** = 0), divides the smaller **R** about equally between **H** and **B**, with **B** being larger in the forenoon and **H** larger in the afternoon.

Some "rules of thumb" — clearly seen in **Fig. 4-45** — about what is called the **partitioning of R among B, H, and E**: (i) E gets first call on the energy of **R** and is very large whenever there is abundant water available to be evaporated, (ii) the presence of wind enhances the *convective* terms **E** and **H**[R], and (iii) the soil heat flow term, **B**, dependent on the relatively inefficient process

---

[R] See the remark, now more meaningful, about wind effects in **Fig. 3-7a.**

of *conduction*, can be thought of as taking the usually rather small "leftovers" from **E** and **H**.

The last exercise in "reading the microclimate" from an energy balance diagram is in **Fig. 4-46**. Here we have some results of actual measurements[7] rather than idealized curves. The first two panels are energy *balance* diagrams using the format in which all midday values are plotted as positive. **Fig. 4-46a** is very similar to the *balance* for the dry surface above, and it shows clearly the effect of rising wind in midday: **B** and **H** reverse magnitudes when the wind comes up on the dry lake bed.

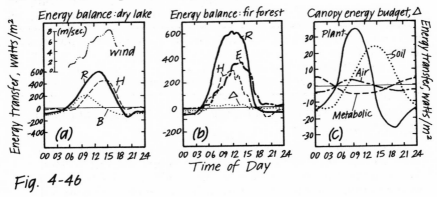

Fig. 4-46

Fig. 4-46b is the energy *balance* for the canopy of a fir forest. At first glance it appears very much like the grass surface above. A closer look, however, reveals that there is no soil heat flow term, **B**, but instead a term describing the heat stored in the materials of the canopy, Δ. This canopy storage term was introduced in **Fig. 4-41**. Notice the very small midday magnitude of Δ, indicating that the sun's energy is dissipated almost entirely by transpiration from the trees and by the warming of winds passing through the treetops. Very little energy penetrates into the canopy.

Another way in which to present the quantitative contents of an energy balance is as a *table*. In the following table, the balances for the dry lake and the fir forest are condensed in a form used in **Appendix A** to summarize energy balances for many different surfaces.

Finally, **Fig. 4-46c** is the energy *budget* for the canopy and soil of a fir forest. Recall we said a *budget is part of a balance*. Here we have the budget, Δ, which is part of the balance of **Fig. 4-46b**. The alternate format used is that of positive terms above the zero line, and negative terms below. Here, a positive term represents *heat energy being stored in the air, soil, or plant materials of the canopy itself*, such as the large mid-morning heat

Summaries of energy balances in **Figs. 4-46a** and **4-46b**: percent of maximum $R^{(m)}$

Surface	Ref	R*	Time	R	B	H	E	Δ
Dry lake bed	a	600.	Morning	63.	-33.	-30.	0.	-
			Afternoon	75.	-3.	-72.	0.	-
			Night	-13.	13.	0.	0.	-
Fir forest	a	600.	Morning	83.	§	-35.	-45.	§
			Afternoon	75.	§	-23.	-50.	§
			Night	-10.	§	10.	-5.	§

m Values of the midday net radiation, **R***, are in watts per square meter — W/m².
a = Oke (1978)
§ — These terms are not separated in the published balance.

storage in the warming plant parts and the large mid-afternoon heat storage in the soil[S].

This is a *budget*, so the four terms *do not have to add up to zero*. In fact, they add up to Δ, which you see in **Fig. 4-46b**, and in the actual numerical values of the coordinate scales, is *a very small amount of energy* compared, for example, with the solar energy arriving at the top of the canopy at midday.

The "metabolic" term requires a few last comments. It represents the net effects of plant metabolism — photosynthesis and respiration — which we will discuss in **Chapter 6**. It has a negative value at midday *when heat energy is being removed from the canopy to form chemical energy in plant tissue*. As you will see in **Chapter 6**, most of this energy is returned to the canopy as heat (positive value for "metabolic") at night when photosynthesis is absent. At its largest, the metabolic term is at most a few percent of the value of **R**.

At this point you may be thinking that the analysis of the physical microenvironment, even using the method I have told you is advantageous and revealing, *is pretty complicated*. Your thought would be most appropriate: the working of the physical microenvironment — and its results called **microclimate** —are complex. The point is that *you are capable of understanding the complexity if you need to*. The next section should help.

(S) In case you hadn't noticed, the soil term, **B**, is absent in **Fig. 4-46b** but present in **Fig. 4-46c**. That's just because the original sketches for the two different forests in these figures[7] were that way. To reduce confusion, just think of the contents of **Fig. 4-46c** as being (Δ + **B**).

The Energy Balance Concept: managing the microclimate

Another great advantage of the EB method of analysis is that it gives you a way to *anticipate — predict — the effects of alterations made to a microenvironment.* In fact, the discussion in this section of the ways in which one can alter the EB by altering various components of the physical microenvironment will make clearer exactly how the individual processes of radiation (**R**), conduction (**B**), convection (**H**), and evaporation (**E**) operate.

We will look first at several of the possible means for modifying the net radiation budget, **R**. Take the terms of the budget — $R = + S_i - S_o + L_i - L_o$ — one at a time.

(1) *Incoming shortwave* ($+ S_i$)**.** This is the insolation, and the amount of energy involved is discussed in detail earlier in this chapter. One may *increase* it by arranging for reflections from nearby surfaces onto the system being modified. One may *decrease* it by shading.

(2) *Outgoing (reflected) shortwave* (-S_o = -aS_i)**.** Changing the albedo (see **Table 4-3**) by altering surface materials provides the principal means for modifying this component. In turn these means can be achieved by (i) materials of different color, (ii) changing the wetness of existing materials, or (iii) changing the roughness of the surface. **Fig. 4-47** shows that **internal reflection and absorption** at roughened and wetted surfaces causes reduced reflection. At each point where a light ray strikes the surface, part is absorbed, and the remainder proceeds as reflected

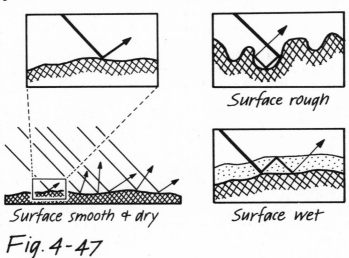

Surface rough —

Surface smooth & dry

Surface wet

Fig. 4-47

energy. The wet film on the surface acts like a "duct" in its effect on the light rays. In both the roughened and wetted cases shown, only one fourth as much energy leaves the surface as first reached it.

(3) *Incoming longwave* ($+L_i$). The sky radiates toward the surface, in a manner suggested by **Fig. 4-13**. The amount of energy in that stream depends mainly on the temperature of the air aloft and on the amount of cloudiness (and particulate pollution) there. *Cold, clean, clear air makes* ($+L_i$) *have very small values* — a small radiant energy contribution to the surface. Placing a material — such as lathe or fibre screen — above the surface causes ($+L_i$) to increase because the material then acts more like a warm black body — like a thin cloud — as in **Fig. 4-12**.

(4) *Outgoing Longwave* ($-L_o$). This term is controlled almost exclusively by the surface temperature itself — the higher the temperature, the larger is ($-L_o$). In a sense, then, this term is the **dependent variable** in the balance, depending on the effects of the other terms.

Having looked term-by-term at the net radiation, **R**, look now at what controls the magnitudes of **B**, **H**, and **E**.

(5) *Soil heat flow* (**B**). **Table 4-4** suggests that control of the magnitude of **B** lies mainly in the thermal properties of the surface material — soil or water. Those properties, in turn, depend on the particular combination of mineral and organic solids, air, and water in the soil, as discussed in the section on soil moisture. The number of these possible combinations is large, and the amount of heat involved in **B** is usually small, so that manipulation of soil structure to alter surface temperatures becomes of marginal utility[8]. For large changes in **B**, warming or cooling the soil artificially is possible by introduction of electrical or ducted fluid systems beneath the surface. More discussion of such strategies as these is in **Chapter 8**.

(6) *Sensible convective heat flow* (**H**). The magnitude of this term depends upon three things: (i) the temperature difference between the surface (source by day) and the air stream (sink by day), (ii) the surface roughness, and (iii) the wind speed. Usually in a natural setting the temperature of the airstream is not very subject to modification, but its speed is. One can either provide **wind sheltering** (to reduce the speed) or **wind channeling** (to increase it). Artifical ventilation by wind machines is also possible. Rough surfaces cause more turbulent motion, and thereby enhance the turbulent transfer processes, as discussed in connection with boundary layers in **Chapter 3**.

(7) *Evaporative (convective) heat flow* (**E**). In a manner similar to **H**, the magnitude of this term depends upon three things: (i) the vapor pressure difference between the surface (source by day) and the air stream (sink by day), (ii) the surface roughness, and (iii) the wind speed. Beyond the remarks just made about **H**, wetting the surface is the major control on **E**.

Even with an understanding of both the individual components and what controls them, and the EB method of analysis, you can see that *"microclimatic management" is as much an art as a science.* Understanding the science can come through study such as is found in this book. The art comes, as you can guess, with your experience and consultation with those who have experience. *The art and the science support each other:* useful application usually requires both.

Table 4-5 summarizes the ideas in this section, in particular the *interactive* nature of the system. In the words of an ecological precept: *you can't change only one thing.* The table covers the most likely linkages in a system such as the grass turf. More discussion of these ideas, in a more "practical" vein, is in **Chapter 8**.

Mass balances

Because both energy and mass are conserved in the changes and flows that take place within the physical microenvironment, we can study flows of mass by using the same kinds of thinking, with the same kinds of caution and for the same benefits, that we used for energy. Though heat energy is about the only kind of energy studied this way by microclimatologists[T], several kinds of mass are studied in analyses of **mass balance**.

Fig. 4-48 suggests, in the format of **Fig. 4-41**, how a **water balance** is analyzed in canopy systems[9]. The principal inflow is *precipitation* (**p**), while the outflows are *evaporation* (**e**) and (liquid) *drainage* (**d**). Flows that can be either inflows or outflows, depending on circumstances, are (liquid) *runoff* (**r**) and (gaseous) *advection* (**a**). In the urban canopy, the addition of water by *combustion* processes — as from automobile exhausts and industrial outlets — is called (**m**). In both canopies, any change in the total amount of water in the system is (∂).

The balance equations for water, in the form given above for heat energy, are:

the plant canopy: $+ \mathbf{p} - \mathbf{e} - \mathbf{d} \pm \mathbf{a} \pm \mathbf{r} \pm \partial = 0$ and

the urban canopy: $+ \mathbf{p} - \mathbf{e} - \mathbf{d} \pm \mathbf{a} \pm \mathbf{r} \pm \partial + \mathbf{m} = 0$

(T) In **Chapter 9** you will have a look at the ecological study of energy in the form of chemical bonds in animal and plant tissue.

Table 4-5 Summary of means for altering components of the microenvironmental energy balance

Alteration	S_i	S_o	L_i	L_o^a	B	H	E
Reflectors	+	+	o	+	+	+	+
Shading	-	-	+	-	-	-	-
Decreased albedo							
darker colored material	o	-	o	+	+	+	+
roughened surface	o	-	o	+	+	+	+
wetted surface	o	-	o	-	?	-	+
Overhead screen	-	-	+	-	-	-	-
Altered soil structure							
increased heat capacity	o	o	o	-	+	-	?
increased admittance	o	o	o	?	+	?	?
Supplementary soil heat	o	o	o	+	+	+	+
Increased ventilation	o	o	o	-	-	+	+
Increased surface temperature[a]	o	o	o	+	+	+	+

(o) = no change; (+) = increase in magnitude (not necessarily direction); (-) = decrease in magnitude, not necessarily direction; and (?) means the change could be any of the others listed, depending on other information.

a Of course, changed surface temperature is usually the objective of the alterations, so it must be considered to be a dependent variable.

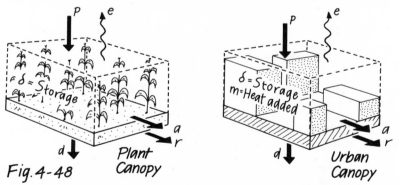

Fig. 4-48 Plant Canopy / Urban Canopy

Care must be taken in studying a mass balance to specify what mass is being balanced. For example, rather than balancing *carbon dioxide*, which can be altered significantly in chemical changes, for most ecological purposes it has been more meaningful to balance *carbon*, which, of course, is a component of carbon dioxide[10].

Notes

(1) In addition to the references in **Note 2** of **Chapter 3**, consider the following sources of data on observed insolation: Kondratyev (1969), Lowry (1980), Miller, (1981), and Givoni (1969).

(2) These topics are discussed clearly and concisely in Oke (1978, Chapter 1) and Lowry (1969, Chapter 3).

(3) See Oke (1978, Chapter 2) and Lowry (1969, Chapter 4).

(4) Though he also uses considerable mathematics, Byers (1974, Chapters 6 and 15) explains these matters clearly and carefully.

(5) For still more detailed versions of the TRe diagram, see Lowry (1969, page 71), Platt and Griffiths (1964), or McGuinness et al (1980, page 95).

(6) About as good a discussion as you can find on cloud naming and cloud processes is in Schaefer and Day (1981).

(7) **Fig. 4-46a** is after Fig. 3.1 of Oke (1978) from observations by Vehrencamp. **Fig. 4-46b** is after Fig. 4.21 of Oke (1978) from observations by McNaughton and Black. **Fig. 4-46c** is after Fig. 167 of Geiger (1965) and the accompanying table, from observations by Baumgartner.

(8) For more detail refer to Lowry (1969), Oke (1978), and Rosenberg *et al* (1983).

(9) For a complete study of water balance methods and results, see Miller (1977).

(10) For analyses of such mass balances as those for water, carbon, oxygen, and nitrogen, see Scientific American (1970).

Peptalks

(P₉) Also, this chapter will include the mathematics of several concepts, only for those who want the math. My experience tells me there are many who can make use of the formulas, for instance, behind the sun path diagrams. As noted earlier, using the math keeps you from having to get along with only the graphs prepared by someone else, though many designers have quite successfully done just that.

(P₁₀) Too much physics? The little imaginary experiment is just a way of helping explain these concepts. But maybe you aren't even sure what "pressure" is. It is the *force per unit*

area exerted when the moving molecules of gas under the bell jar strike the face of the manometer. Pressure is exerted by any group of molecules — they don't have to be gases — but here we are talking about gases. If you still need to know something about physics here, I guess consulting a book with a title like "General College Physics" would help.

(P₁₁) Though the weighted average of soil heat capacity is important, it is the general concept of weighted average I want you to focus on here. It is a kind of average in which *the more important components get more weight* — more importance — *in the average*. It will come up many times in ecological discussions, so now is a good time to know about it. It is (sorry about that) another example of a concept that almost requires math to discuss, so here goes. Each component is given a **weight** (w) to be attached to the value (**V**) of the component in the average. If we have three components, as in the discussion of soil, the weighted average looks like this:

$$(\mathbf{V})_{ave} = \frac{w_1 \, (\mathbf{V})_1 + w_2 \, (\mathbf{V})_2 + w_3 \, (\mathbf{V})_3}{w_1 + w_2 + w_3}$$

The ordinary, everyday "average" is an **arithmetic average** is which *all the weights are equal*.

(P₁₂) This is just another one of those times when mathematical symbols allow us to make complex statements simply. These equations are just to allow you to check your understanding of the ideas involved in **Fig. 4-39.**

5. How are we doing here?
Humans and the atmosphere

We have just examined the concept of the energy balance, and noted that the ideas about **(Rate of Inflow) = (Rate of Outflow) + (Rate of Change in Storage)** are applicable in studying the energy economy of any number of systems. In particular, we have examined the connections between the diurnal and seasonal energy balances of various natural *surfaces*, including plant canopies, and the microclimates associated with those surfaces.

In this Chapter we will examine the connections between the energy balance of a human being and the levels of comfort associated with the various microclimates to which the person is exposed.

In consideration of a *surface* we spoke briefly about "managing" the microclimate to alter the temperature of the surface system itself. In that management problem, the surface is mostly an inanimate system responding *passively* to natural changes in the energy environment, and to the changes the manager is able to impose on the environment and on the system.

In considering the energy balance of a person, it quickly becomes apparent that the management problem is quite different. "Fine tuning" the energy balance is much more important, and the reactions of the person/system are quite *active*, since the system's temperature, unlike that of the surface, must be kept within a very narrow range of values.

Man's energy balance: how it works

Fig. 4-42 shows the energy balance system for a human being. **Fig. 5-1** shows the same in different form. Here, the *heat load* on the person is associated with the incoming radiation (the "radiant heat load") and with the heat energy released by metabolic processes ($+M$) within the system itself. The rate of metabolic heat production in human beings is nearly constant (unless he is exercising).

The *dissipation of the heat load* is associated with the radiative, conductive, and convective heat energy flows from the system to the environment. In **Fig. 4-42** the symbol ($\pm \Delta$) was included to indicate that changes in storage within the human body are possible. In **Fig. 5-1** that symbol is missing, to indicate that, while possible, *changes in heat storage within the human body must be avoided*.

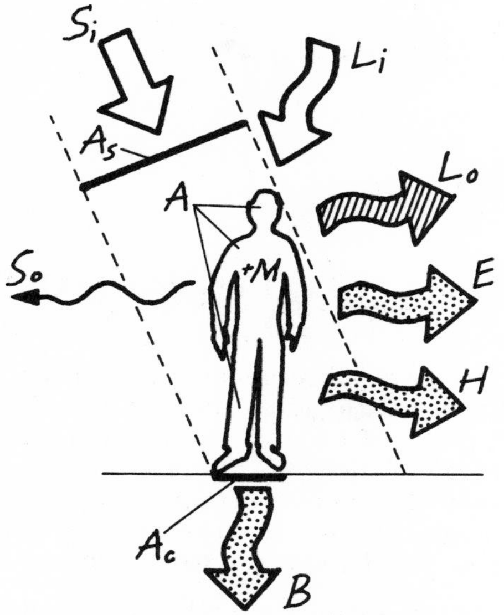

The other major differences between the two figures are the representations of the *radiation budget* and the introduction of *areas* (**A**) in **Fig. 5-1**.

Fig. 5-1

In **Chapter 4** we saw that the radiation budget is the net effect of the four radiant streams of energy, and that the net effect is summarized in the net radiation (**R**):

$$R = S_i - S_o + L_i - L_o$$

In particular, this net radiation appears in **Fig. 4-42**, while all four radiation streams appear in **Fig. 5-1**, in addition to the conductive (**B**) and convective (**H,E**) streams.

Discussion of the use of the various *areas* involved in the analysis will make clear how **Fig. 4-42** is a very simplified version of **Fig. 5-1**. **Fig. 5-2** shows how the *shadow area* (A_s) of the stand-

Lowest sun

Highest sun

Fig. 5-2 Shadow area, A_s

ing person changes as the sun rises higher in the sky. In **Fig. 5-1** this area is shown between the sun and the person, but in fact it is exactly the same as the area of the shadow cast by the person *on a surface perpendicular to the sun's rays.*

Because, as in **Chapter 4**, the physical units of (S_{in}) are (energy) $\div$ [(area)x(time)], multiplication of (S_{in}) by the area on which the direct solar radiation shines — the same as the area of the shadow cast by the person — yields the total amount of direct solar energy arriving on the person's body:

Total sunshine arriving = (Energy per area per time) x (Area)
= (Energy per time) = (S_{in})x(A_s)

Fig. 5-3 shows the results of the fact that, as time goes on after sunrise, (S_{in}) increases while (A_s) decreases. The product $(S_{in}A_s)$ does not just increase until noon and then decrease again. Instead, depending on the solar geometry — the angle at which the sunshine arrives on the person — the product is usually greatest in mid-morning and mid-afternoon, decreasing at midday. As we shall see in **Chapter 6**, this midday decrease does not occur for four-legged animals. Thus, the midday escape from the largest radiant heat load is one of the results of walking erect.

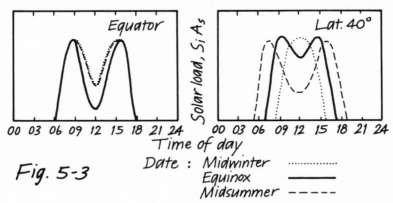

Fig. 5-3

Date : Midwinter ....
Equinox ————
Midsummer – – – –

Since the fraction of $(S_{in}A_s)$ that is *reflected* is the albedo (**a**), then the total amount of sunshine actually *absorbed* by the person is $(1-a)(S_{in}A_s)^{(P_{13})}$.

A moment's thought will tell you that it is only the incoming solar radiation — the sunshine — that is associated with the shadow area (A_s). The conductive loss to the ground (B) is associated with the *contact area* (A_c), while all the other terms are associated with area (A), which is the total surface area of the person, (A_t), minus the contact area: $A = (A_t - A_c)$. Areas (A) and (A_c) are shown in **Fig. 5-1**. Another moment's thought will tell you that, contrary to being a nuisance, this careful consideration of areas permits one to take account, in studying a person's energy management and comfort, of the role of *body size and posture* in the study.

In summary of the energy balance for a person, recall that the basic framework is:

(Rate of Inflow) =
    (Rate of Outflow) + (Rate of Change in Storage),

and that this word equation may be written, *for midday conditions*, as a mathematical equation like this:

(Rate of Inflow) $(1-a)(S_{in})(A_s) + (L_{in})(A) + (M)$

=   =

(Rate of Outflow) $(L_{out} + E + H)(A) + (B)(A_c)$

+   +

(Rate of Change in Storage) $(\pm\Delta)$

Again, writing such an equation gives us the advantage of seeing, in a very condensed form, the relationships among the various properties and processes that go into the energy balance. Organizing the various terms in the equation in the following way often helps to see the energy balance as an interaction of environment and system:

Environmental variables	(S_{in}) and (L_{in})
System variables	(1-a), (A_s), (A_c), and (A)
Environment/system variables	(E),(H), and (B)
Quasi-constant dependent variables	$(+M)$, (L_{out}) and $(\pm\Delta)$

Every term in the equation has a place in that listing. The *environmental variables* are those that would be present whether or not the person was present. Here the environmental variables are the two incoming radiation streams.

The *system variables* are those that are controlled by the person. Here they are the "color" of the clothing (a), and the various areas that depend on the size and posture of the person.

The *environment/system variables* depend on conditions in both the environment and in the system. The evaporative term (E)

depends on the vapor pressure difference between the person and the environment, while the (**H**) and (**B**) terms depend on the temperature difference between the person and the environment.

The *dependent variables* are the outcomes of the interactions of all the other terms. For human beings the metabolic rate (+**M**) changes relatively little, except during exercise. (L_{out}) depends on the body's surface temperature[A], while the body does or does not change in heat content. ($\pm \Delta$), is determined by whether or not all the other terms add to zero.

Writing out and understanding all the mathematical terms will now permit us a very efficient discussion of the way in which a person fits himself, in relation to energy flows, into the microenvironment.

Man's energy balance: the central problem

Man's central problem in managing his own energy balance is to *keep the value of (Δ) equal to zero.* That is, his problem is to match energy inflow and energy outflow — match dissipation with heat load. (Δ) must remain equal to zero because the body's core temperature, (T_c), must remain constantly near $37\,°C = 98.6\,°F$. This requirement for a constant core temperature puts Man in the class of animals called *homeotherms*.

Unlike the surfaces and canopies discussed in **Chapter 4**, homeotherms have both a large internal, metabolic source of heat (**M**) and a need to keep their core temperatures within a narrow range. This necessary energy relationship of animals with the environment has resulted, through evolution, in several general principles which apply, in particular, to Man:

1) the core temperature must be near a value that permits rapid biochemical reactions associated with fast, precise brain and muscle responses — responses related to the roles of both predator and prey;

2) the appropriate core temperature is larger, most of the time, than the air temperature in which the animal functions;

3) a side benefit of the metabolic energy conversion to brain and muscle function is that the heat released helps keep the core temperature above air temperature;

4) the narrower the range within which the core temperature must remain, the greater is the number of special internal mechanisms for "fine tuning" the energy balance — keeping (Δ) at zero — and the greater is the amount of biochemical energy required to maintain and operate those mechanisms; and

---

(A) See the second law of radiation in **Chapter 4**.

5) the relationships of body size and shape to metabolic rate in homeotherms — to be explained in **Chapter 6** — have been set by the fact that homeothermic evolution has taken place almost exclusively in "tropical" temperatures near 28 °C. In that sense, then, *Man* — among other homeotherms — *is inescapably a tropical animal*.

Man's energy balance: a simple mathematical model

While the last equation is a convenient way to list most of the relevant factors in an organized way, there is another equation that is a better basis for discussing the nature of Man's problem of energy management:

$$\textbf{Rate of Heat Outflow} = \frac{(\textbf{T}_c - \textbf{T}_e)}{\textbf{Resistance}}$$

It probably doesn't look like the same equation to you, but it tells the same story as the one in the previous section, only in a different form. The *body core temperature*, $(\textbf{T}_c)$, has already been mentioned. It is the requirement that Man brings to the problem. Combining all the environmental factors — temperature, sunshine, humidity, and wind — in a certain way permits us to express the net result of all these factors as one temperature, the *environmental temperature*, $(\textbf{T}_e)$. It is the requirement that the environment brings to the problem. The second principle in the list of the last section is that $(\textbf{T}_c - \textbf{T}_e)$ is a positive number — that is, $(\textbf{T}_c)$ is larger than $(\textbf{T}_e)$ — so heat must be flowing from the body to the environment whenever the body is in *thermal equilibrium* with its environment — $(\Delta) = 0$. In this simple mathematical model, the **resistance** is the value of the net effect of all the body's thermal feedback regulators needed to "tune" the outward flow of heat from the body. Being in the denominator, the resistance becomes larger when the heat loss rate needs to be smaller, and vice versa[B].

Man brings $(\textbf{T}_c)$ to the problem, the environment brings $(\textbf{T}_e)$ to the problem, and Man's manipulation of the **resistance** provides the solution to the problem.

Thermoregulation

We will discuss how various factors combine into $(\textbf{T}_e)$ in the next section. In this section we will examine the various methods available to a person for *thermoregulation* — regulating the resistance.

(B) Many readers will recognize that the **resistance** is inversely proportional to the **thermal conductivity** described in **Table 4-4**. It was mentioned back in **Chapter 2** that we can view a human being as an energy flow system, with inflow, outflow, and feedback. Here, many pages later, we have begun to discuss Man's energy balance in just that way.

Begin by noting that the means for thermoregulation in Man may be viewed according to this scheme, in which "coarse" and "fine" are analogous to tuning a radio:

	Coarse Tuning	**Fine Tuning**
Voluntary	Changing location	Buttoning or unbuttoning clothing
	Changing clothes	Small changes in posture
		Ingesting cold or warm food & drink
Involuntary	(*Hibernation* in some homeotherms, similar to *diapause* in insects)	Changes in heart & breathing rates
		Changes in sweat rate, shivering
		Dilation or contraction of peripheral blood capillaries

The table by no means lists all the means for thermoregulation, but it gives enough examples to make a brief discussion meaningful. For one thing, it suggests the difference between *voluntary* and *involuntary* thermoregulation. The first is based on conscious decision by the person, while the latter comes as a result of "internal programming" of the body — we don't really *decide* when the feedback begins or ends, nor how much is required.

The phrase "changing location" includes a wide variety of actions, from moving out of the sun into the shade, or out of the wind into calm air — that is, "indoors" — to burrowing underground or moving to the tropics.

The involuntary *cardiovascular* mechanisms — increased heart and breathing rates, and dilated capillaries — pump more heat-bearing blood to the skin surface faster and expel heat and evaporated water more rapidly from the lungs. Increased sweat rate is a temporary "emergency" mechanism to increase evaporative cooling, while shivering is a temporary emergency mechanism to increase $(+M)$.

In a form similar to **Table 4-5**, we may summarize partially the means for human thermoregulation in **Table 5-1**.

To conclude, thermoregulation for human beings involves regulating heat energy inflow and outflow to equal each other, thereby maintaining an equilibrium in which $(\Delta) = 0$. *Regulating the inflow* (heat load) involves manipulation of the values of (a), (S_{in}), (L_{in}), (A_s), or to a lesser extent (A), while allowing $(+M)$ to remain constant. *Regulating the outflow* (dissipation) involves manipulation of the values of (T_e) — explained in the next section — (**A**), (**A**$_c$), (**B**) and the involuntary physiological mechanisms involved with (**E**) and (**H**).

Table 5-1 Summary of means for thermoregulation in Man

Means	HEAT LOAD		DISSIPATION		
	$(1-a)(S_{in})(A_s)$	$(L_{in})(A)$	$(E)(A)$	$(H)(A)$	$(B)(A_c)$
Changing location					
sun to shade	-	0	0	0	0
windy to calm	0	0	-	-	0
wet to dry air	0	0	+	0	0
warm to cold	0	-	0	+	+
Changing clothes					
light to heavy	0	0	-	-	-
light to dark	+	0	0	0	0
Changing posture					
curling into a ball	-	-	-	-	0
lying on cold soil	0	0	0	0	+
Drinking cold water	0	0	0	-	0
Increasing heart rate	0	0	0	+	+
Increasing sweat rate	0	0	+	0	0
Dilation of capillaries	0	0	0	+	0

0 = no change; (+) = increase in magnitude; (-) = decrease in magnitude.
NOTE: In this table, the responses indicated are the *initial* responses. In actuality, *secondary* responses may well follow. For example, drinking cold water may later reduce the sweat rate and (**E**)(**A**) when thermal stress is reduced.

In **Chapter 6** we will see that small homeotherms, birds, and cold-blooded animals have different and additional means for thermoregulation. Even plants have rudimentary feedback for temperature control, mostly as a means for controlling water loss.

Human comfort

In the context of the human energy balance, we are "comfortable" when the heat load and the dissipation rate are equal — thermal equilibrium — *without resort to any of the feedback mechanisms.* Use of these mechanisms requires energy, so we can say also that *comfort accompanies minimal energy use in maintaining thermal equilibrium*.

We are "uncomfortable" when feedback control is necessary and *very* uncomfortable when emergency mechanisms, such as sweating or shivering, are required. More than that, emergency mechanisms must be used for only short periods of time. Discomfort arises in an environment where (Heat Load) exceeds (Dissipation Rate) — a *hyperthermic environment*. When (Dissipation Rate) exceeds (Heat Load) the environment is *hypothermic*.

Hyperthermia can arise either because the heat load is too large or because the dissipation rate is too small, or some combination of those conditions. The reverse holds for hypothermia. The point here is that feedback can operate on (Heat Load), (Dissipation), or both.

A sunny, hot, humid, calm environment increases (Heat Load) and reduces (Dissipation), and is hyperthermic. A sunless, cold, dry, windy environment is hypothermic. In these two statements — opposite sides of one coin — lies the basis for the concept of the *environmental temperature*, (T_e), sometimes called the *equivalent temperature*, in which the four environmental factors are combined into one number, expressed as a temperature.

When expressing comfort numerically, it is clumsy and difficult to comprehend if the scale is physical, for example in units of energy flow. It is much more meaningful when the scale expresses a *departure from a "standard"*, comfortable condition. An alternative approach expresses an *equivalent sense of comfort*. The most common method for expressing comfort illustrates the point:

a) specify the standard person — for example, a sedentary, middle-aged adult in good health, and wearing light street clothes;

b) specify the standard environment *except for the air temperature* — for example, no sunshine, calm wind, and 100% relative humidity; and then

c) express the existing environment *as equivalent, in feelings of comfort, to the standard environment with air temperature equal to (T_e)*.

Let's put numbers on that explanation. Given the standard person, indoors (no sun and calm air) in 100% relative humidity and 90 °F air temperature, (T_e) would be 90 °F. In another environment with a gentle wind, a lower relative humidity, and a higher air temperature, (T_e) could be 90 °F, too. That would be so because *the person would feel equally comfortable* in both environments, since the wind and greater evaporative cooling in the second case has offset the effect of the higher air temperature. In this context,

(T_e) is usually called the *effective temperature*, but the meaning is the same.

Fig. 5-4a shows Steadman's system[1] for estimating (T_e) for hyperthermic and hypothermic — "stressful" — environments. **Fig. 5-4b** shows a similar system employed by the U.S. National Weather Service. The slight differences between them are based on the differences, described above, between the choices of "standard person" and "standard environment." To understand the meaning of one of these two systems is to understand both of them, and most of the several other systems ordinarily encountered.

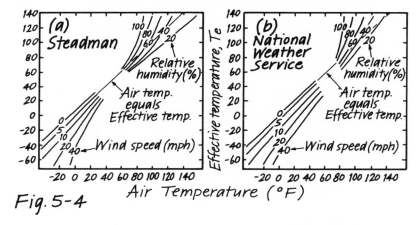

Fig. 5-4 Air Temperature (°F)

Begin with a discussion of the lower left corner of Steadman's chart, where the hypothermic environments associated with *wind chill* appear. We enter the chart with values of air temperature (T_a) and wind speed, and exit with a value of (T_e), in this case often called a "wind chill index." In this system it is assumed there is no excess of solar radiation. Further, because cold air cannot hold very much moisture even at saturation, relative humidity plays no part in obtaining (T_e). It is clear from the chart that:

(i) for a given air temperature, (T_e) decreases as wind speed increases;

(ii) for a given wind speed, (T_e) decreases as air temperature decreases;

(iii) (T_e) does *not* decrease *linearly* as wind speed increases[C]; and

(iv) (T_e) equals or exceeds (T_a) only in very light wind.

(C) This *non-linear* relationship of wind speed and temperature difference is mathematically equivalent to the following:

(T_a - T_e) is proportional to (wind speed)^X, where X is a number less than one. The physical basis for the relationship lies in the change in thickness of the person's microscale boundary layer as wind speed changes (see **Chapter 3**).

In the upper right corner of Steadman's chart, the hyperthermal environments appear. We enter the chart with values of air temperature (T_a) and relative humidity (RH), and exit with a value of (T_e), in this case variously called a "humiture index," or a "sultriness index." In this system it is assumed there is no excess of solar radiation. Further, hot air is very near to skin temperature, so that (H) is nearly zero[D], and wind speed plays no part in obtaining (T_e) beyond the assumption that a very light wind is blowing. It is clear from the chart that:

(i) for a given air temperature, (T_e) increases as (RH) increases,
(ii) for a given value of (RH), (T_e) increases as air temperature increases,
(iii) $(T_e) = (T_a)$ with high (RH) only in cool air, and
(iv) $(T_e) = (T_a)$ with low (RH) only in very warm air.

The differences between **Fig. 5-4a** and **5-4b** are small in cold environments but greater in hot environments. As noted above, the differences are because of different assumptions about the standard person and the standard environment used for comparisons, but understanding one system means understanding the other as well. The point here is primarily that there are different systems for expressing "comfort" by means of a value of (T_e), so that you must be clear which system is used in what you read and present to others.

Fig. 5-5 shows another method for displaying (T_e) in hyperthermic conditions. On the *TRe diagram* described in **Chapter 4**, lines of equal (T_e) are straight and conform completely to the four points made just above concerning the relationships among (T_a), (T_e), and (RH) in hot environments. Knowing that the core temperature (T_c) must be near 37 °C, or 98.6 °F, we can see that all the combinations of (T_a) and (RH) to the right of the line $(T_e) = 100$ in **Fig. 5-5** will *reverse the flow of heat*, and increase the body's heat storage since $(T_c - T_e)$ will be negative. In this zone of the chart lies heat death from hyperthermia.

Although the subject of regional, continental, and global scale climates will be considered primarily in **Chapter 10**, take a moment to see the subject of human thermal comfort on scales larger than local process. **Fig. 5-6** suggests how comfort, in the context of human bioclimate, changes across the globe. The terms "hot", "mild", "cool", and "cold" are themselves anthropocentric terms — human-centered — as suggested in **Table 3-1**.

(D) Recall from the discussion of **Fig. 4-15** that heat flows from a source (in this case the warm skin) to the sink (in this case the air) at a rate proportional to the difference in temperature between them. That holds true for both conduction and convection, so the flow from warm skin to warm air may be quite small. It is related to the fact (have you noticed?) that you cannot *feel* a wind blowing when it is at skin temperature.

For all its complexity and sense of reality, **Fig. 5-6** has major shortcomings for any planner or designer. Aside from the fact that categories such as those mapped do not actually exhibit sharp regional boundaries, there are two other major problems with the

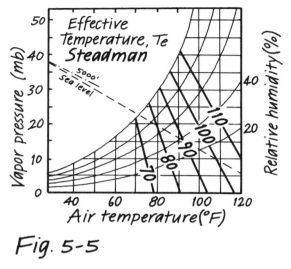

Fig. 5-5

information in **Fig. 5-6**. First, the map is *static* — it is the same from one year to the next — because it is based on "average" conditions. Second, the map does not take account of the well-recognized fact that human beings become *acclimated* — they "get used to" stress after a period of exposure, both physiologically and behaviorally.

The problem of a static map is solved simply by making more maps, designed to show variability. Dealing with the problem of humans' ability to acclimate is a bit more complicated. In fact, the static map is presented here mainly as a means for introducing the idea of the *Weather Stress Index*[2], *which takes acclimation into account.*

Fig. 5-7 shows two frequency distributions of observed (T_e) for, let us say, midday conditions during July. The first panel is for a station whose July climate is cooler and more variable — a wider variety of observed temperatures — than the climate at the station in the right panel. More particularly, an observed value of (T_e) = 28 °C is relatively rare during July at the cooler station in **Fig. 5-7a**, while it is commonplace in the July climate of **Fig. 5-7b**.

Placing a numerical statement of rarity on (T_e) = 28 °C at each station is the basis for the *Weather Stress Index* (WSI). If that value exceeds 90% of all observed values in **Fig. 5-7a**, we say the WSI is

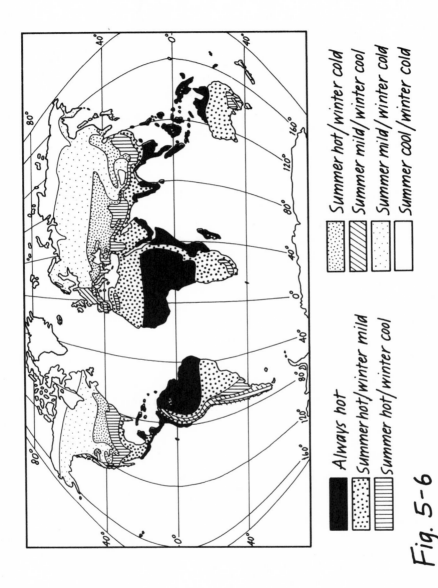

Fig. 5-6

Always hot
Summer hot/winter mild
Summer hot/winter cool

Summer hot/winter cold
Summer mild/winter cool
Summer mild/winter cold
Summer cool/winter cold

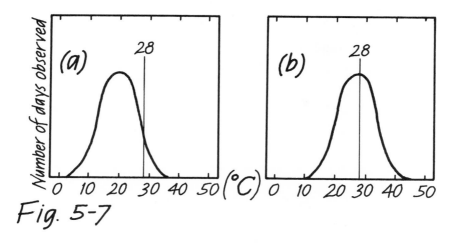

Fig. 5-7

90 on a July day when $(T_e) = 28\,°C$ is observed at that station. At the station in **Fig. 5-7b** a July day when $(T_e) = 28\,°C$ would have a WSI of about 50. Thus, *the WSI is a statement of what exists relative to what one is accustomed to* at the time and place in question.

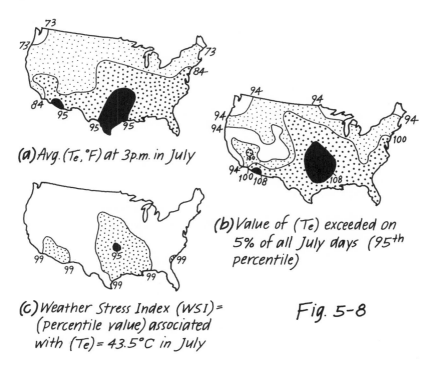

(a) Avg. $(T_e, °F)$ at 3 p.m. in July

(b) Value of (T_e) exceeded on 5% of all July days (95th percentile)

(c) Weather Stress Index (WSI) = (percentile value) associated with $(T_e) = 43.5°C$ in July

Fig. 5-8

Mapping the idea of the WSI is instructive. **Fig. 5-8** shows several results from Kalkstein and Valimont (1986). **Fig. 5-8a** shows a map of the average value of (T_e) — the "central" value in a distribution such as those in **Fig. 5-7**, only in °F. It is another static map, and I am presenting it mainly for a comparison with **Figs. 5-8b** and **5-8c**.

Fig. 5-8b gives the temperatures that are "rare" in different parts of the United States in July — "rare" meaning "exceeded once in twenty July days" in this map. We can see, for example, that the average temperatures of (T_e) = 95 °F that seem so hot in **Fig. 5-8a** are rare only in the far West and far North of the 48 states in July. Finally, in **Fig. 5-8c** we see another aspect of "rarity." Here the map tells us that a really hot day — (T_e) = 111 °F = 43.5 °C — is rare everywhere in July, but that it is least rare in the mid-plains, where it is *exceeded* (Whew!) 5% of the time in July. The idea of the WSI is relatively new. It ought to be more widespread, and of greater utility to planners and designers, in the future.

Man's technological specialties: clothing and architecture

The biological species of Earth are often viewed ecologically as being various combinations of generalists and specialists. Human beings are primates with specialties of *"time awareness"* — seeing backward and forward in time, with "management" as a by-product — and *advanced information transfer*, to include the ability to transmit accumulated experience and knowledge both between individuals and between times.

In the context of this book on atmospheric ecology, *Homo sapiens'* greatest specialty — his greatest leverage on the energetic environment — is probably the power of his technological mastery of clothing and architecture. While that mastery may not give Man advantage in the "climatic homeland" — the tropical regions where we find today's representation of the climates of most of evolutionary time — it has allowed Man to expand his home range far beyond the tropics, into the deserts, the polar regions, and even to the Moon, as no other single species has done.

That expansion, while a marvel when compared with other species, is achieved at a cost — often enormous — not borne by the species that have "stayed put" in one narrow range of environments. As noted earlier, the feedback mechanisms that combine to represent the *resistance* to heat exchange between Man's body core and his physical environment require energy to construct, maintain, and operate. Those are the costs of expansion. *Clothing and architecture may be viewed and discussed as technological exten-sions of the resistance to heat exchange between Man's body core and his physical environment,* as summarized in the equation:

Rate of Heat Outflow $= \dfrac{(T_c - T_e)}{\textbf{Resistance}}$

Viewed in that way, clothes may be considered to be extensions of the body, and buildings may be considered as extensions of clothing, with many similarities in the sense of the energy balance of the homeothermic human being.

Clothing increases both the amount of thermal resistance availa-ble to the body and the ability to fine-tune its value at any moment:

1) with color change, the albedo (**a**) can change;
2) with the trapped, still air spaces of insulation (see **Table 4-4**) the thermal conductivity, and thus (**H**), can be reduced;
3) with "vapor barrier" material in the clothing, the evaporative cooling and water loss, (**E**), can also be reduced;
4) with the still air inside and in front of a "parka hood" both (**H**) and (**E**) can be reduced;
5) in hot, humid conditions thin clothing may provide minimal insulation, with wind action almost directly on the skin, to permit large values of both (**H**) and (**E**); and
6) acting as wicking, thin clothing may enhance evaporative cooling and an increase in (**E**).

Buildings — architecture — are, at least among modern peoples, *permanent* while clothing is *intermittent*. Clothing may easily be put on and taken off, but it is "inanimate" with no controls of its own over the resistance it provides. A building, on the other hand, is THERE once it's built. Rather than putting it on and taking it off, one goes into it and comes out of it. It can be "animated" — made automatically responsive to thermal stress — by installations of thermostatted heating and cooling systems. In the sense of the energy balance, this heating and cooling are part of the "meta-bolic" term (**M**), with cooling now representing a *negative (M)*, absent in the body or in clothing.

A carefully designed building may change its responses to seasonal changes in energy flow; but more often it is either a response to the most stressful conditions of one season, or else an all-season optimization, probably sub-optimal under the most stressful, short-term conditions encountered. As an example of one-season optimization, nearer the equator, it may minimize its "shadow area" (A_s) and its absorptivity (large albedo, (**a**)) in the hottest, high-sun season — minimizing the heat load $(1\text{-}a)(S_{in})(A_s)$

— while presenting a larger profile (**A**) to cold season winds from a different direction, increasing convective heat loss, (**H**)(**A**), during the low-sun season.

A building may be seen, mathematically and energetically, as *a body with a person at its core — with a requirement for a nearly constant temperature in the living space that is its core.* A building may be considered as to its "coarse tuning" and its "fine tuning" characteristics, in the same sense as the human body was considered in **Table 5-1**[3].

Energy implications of architecture

Clothing and buildings may be viewed as extensions of the concept of *resistance*. In particular, the energy balance of the interior space may be reduced to the equation

$$\textbf{Rate of Heat Outflow} = \frac{(\textbf{T}_{inside} - \textbf{T}_{outside})}{\textbf{Resistance}}$$

in which the core of human space, with its temperature (T_{inside}), is surrounded by a "shell" of **resistance** to heat flow, separating the core from the environment, with its temperature ($T_{outside}$). The *optimal core temperature for comfort*, (T_{inside})*, is usually near 21°C = 70°F in modern western societies, but any of a considerable number of values may be found in the communities of Man, depending on geography, climate, and culture.

The exterior temperature ($T_{outside}$) is equivalent to (T_e), as discussed earlier. With a constant resistance, then, the heat flow between the core space and the environment is proportional to the temperature difference ($T_{inside} - T_{outside}$). Said another way, *to maintain the core temperature at its optimal value, together with a constant resistance, by using imported "metabolic" energy, the metabolic energy must be consumed at a rate proportional to* (T_{inside})* - ($T_{outside}$). In this statement, the only variable is the fluctuating external temperature. The statement is the basis for the twin concepts of **Heating Degree Days (HDD)** and **Cooling Degree Days (CDD)**.

Fig. 5-9 represents these twin concepts graphically. The annual temperature record shows the variability of ($T_{outside}$) through the year, while (T_{inside})* = 70°F — the thermostat setting. Each day in the heating season, the *average daily temperature* is approximated by:

$$(T_{average}) = (1/2) (T_M + T_m)$$

in which T_M and T_m are the *maximum* and *minimum temperatures* recorded at a great many weather stations.

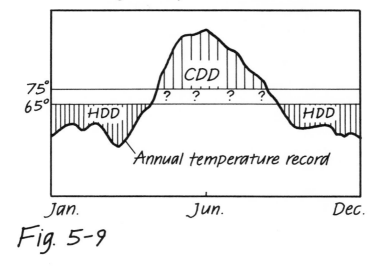

Fig. 5-9

On each day, according to the *Degree Day* (**DD**) concept, the energy requirement to maintain the constant core temperature is proportional to the number of (**DD**) on that day:

$$\mathbf{DD} = (T_{inside})^* - (T_{average}).$$

Adding each day's contribution into a sum[E] gives a *sum of degree days* $\Sigma(\mathbf{DD})$ for the period — for example the month of February in a given year — proportional to the energy requirement to maintain the constant core temperature.

In the heating season $(T_{inside})^*$ is larger than $(T_{average})$, so that $(T_{inside})^* - (T_{average})$ is a positive number, and we speak of (**HDD**) and $\Sigma(\mathbf{HDD})$ being proportional to the energy for heating.

In the cooling season — assuming the use of air conditioning — $(T_{inside})^*$ is smaller than $(T_{average})$, so that $(T_{inside})^* - (T_{average})$ is a negative number, and we speak of (**CDD**) and $\Sigma(\mathbf{CDD})$ being proportional to the energy for cooling[E].

In **Fig. 5-9** there seem to be two values for $(T_{inside})^*$ — 65 °F and 75 °F — whose average value is 70 °F. Actual experience with modern buildings shows that "the heating thermostat clicks on" when $(T_{average})$ falls below 65 °F, and "the cooling thermostat

(E) Degree Days are never negative numbers, even when it is very warm outside. When accumulating (**HDD**), ignore all negative temperature differences, and when accumulating (**CDD**) ignore all positive differences. When accumulating (**DD**) of both kinds, simply accumulate all differences as if they were positive numbers.

clicks on" when ($T_{average}$) rises above 75 °F. The reason the small area in **Fig. 5-9** between 65 °F and 75 °F is filled with (? ? ?) is that in *published climatological data representing $\Sigma(CDD)$*, the "threshold temperature value" is sometimes 65 °F and sometimes 75 °F. Most of the time it is 65 °F, but when you use tabulations of that kind of weather data, you should be certain which value is used.

The Degree Day concept is mere theory unless it can be shown that energy use really is proportional to $\Sigma(\mathbf{DD})$, either heating or cooling. **Fig. 5-10** gives an example of such evidence for the whole city of St. Louis, Missouri[4].

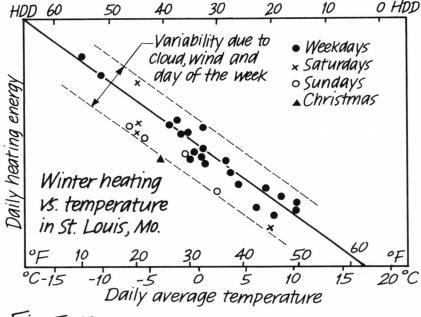

Fig. 5-10

Clearly, there is a general relationship between falling ($T_{average}$) and energy demand for heating in St. Louis. As mentioned in discussion of (T_e), there is variability due to cloud and wind. The figure shows, also, that there is variability around the general relationship due to *variable per capita demand*, as represented by data from

  (i) *Weekdays* (Residential + Commercial + Industrial),
 (ii) *Saturdays and Sundays* (Residential + Commercial), and
(iii) *Christmas* (Residential).

Of course, things aren't that simple, and even on the different kinds of demand days there is variability in cloud and wind, but the general connection among all these variables is clear in **Fig. 5-10**.

Fig. 5-10 represents the Temperature-Energy relationship for a large urbanized area. What about the relationship for an individual building? **Fig. 5-11a** suggests the answer for heating individual buildings, and **Fig. 5-11c** for cooling. Within these panels you can see several kinds of variability. First, there is the *variability of the "threshold temperature"* at which energy demand begins to increase. Second, there is the *variability of the "resistance",* represented here by the slopes of the lines for individual buildings. Thermally efficient buildings have large resistances and small slopes, and *vice versa* for inefficient buildings.

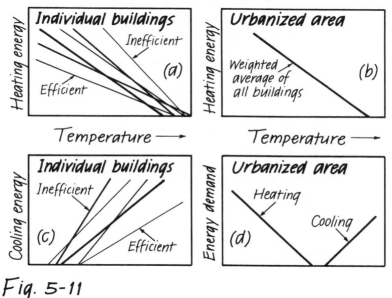

Fig. 5-11

Fig. 5-11b is similar to **Fig. 5-10**, representing an urbanized area as a weighted average of all the buildings, weighted according to energy demand for each building. **Fig. 5-11d** shows the relationship of environmental temperature to the combined demands of heating and cooling for an urbanized area. Demand is zero — along the bottom of the figure — between 65° and 75°F.

"Efficiency" itself has several components. The *first* component has to do with the materials of the structure, through which the heat must move — good or bad insulation. The *second* component has to do with the systems used for heating and cooling. For example, electrical heating is more efficient, in this sense, than burning fuel in the building. Evaporative cooling — in hot, dry climates — is much more efficient than the cooling systems using refrigerants and pumps, which use energy to move heat from inside to outside the building in hot, humid climates.

Maps, tables, and charts of "normal" (long-term average) $\Sigma(\mathbf{HDD})$ and $\Sigma(\mathbf{CDD})$ for many localities are widely available[5], but one must be careful about using them in the context of planning and design. Here are examples of the reasons that care must be taken:

1) In mountainous terrain, modest differences in elevation and exposure, between the climatic data station and the site being planned, can make major differences in microclimate, and therefore between designed and observed energy demand;

2) the "base period" on which the averages of $\Sigma(\mathbf{HDD})$ and $\Sigma(\mathbf{CDD})$ are calculated may be, climatologically, quite different from the present or future for which planning and design are intended; and

3) the variability in resistance — building materials and energy conversion technology — the slope of the Temperature-Energy relationship — cannot be included in the tabulations and mappings of $\Sigma(\mathbf{HDD})$ and $\Sigma(\mathbf{CDD})$.

All the same, maps of $\Sigma(\mathbf{HDD})$ and $\Sigma(\mathbf{CDD})$ separately are reasonably common. **Fig. 5-12** contains information on [$\Sigma(\mathbf{HDD})$ + $\Sigma(\mathbf{CDD})$] for the eastern two thirds of the United States, a much less common kind of map. In the context of *regional planning*, or of a *national energy and housing policy*, this kind of map — allowing for the reasons for care just listed — is more directly revealing of the impacts of climate on architecture.

The result of adding heating and cooling demands together for an entire year — $[\Sigma(\mathbf{HDD}) + \Sigma(\mathbf{CDD})]$ — discloses what intuition may well have told you already: *in a climate where the outside temperature is never very far from* $(\mathbf{T}_{inside})^*$ *regardless of season, the annual demand is minimal*. In the eastern United States — the mountainous West being ignored for reasons just explained — that area of minimal annual demand seems to be in the heart of the "Sun Belt", to which domestic migration has been maximal in recent years.

Another way to suggest the implications of human migrations interacting with climatic change is in **Fig. 5-13**[6], where historical trends of *energy demand per capita* are shown for (a) the real, increasing, migrating population of the country and (b) an imaginary, static, constant population of the size and distribution observed in 1980. The graph shows that, *within the same variable* *climate, and assuming constant* technology, the migration of population more and more to the Sun Belt has reduced the national *per capita* demand for energy.

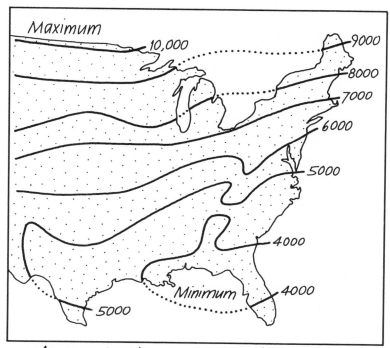

Average heating + cooling requirements
(Fahrenheit Degree Days)

Fig. 5-12

In another example of how mathematical symbols permit condensation of complex ideas, the following explains the variables graphed in **Fig. 5-13**. Using a large, fast computer with *annual* values of $\Sigma(\mathbf{HDD})$ and $\Sigma(\mathbf{CDD})$ and population census values for individual states, the researchers calculated several variables:

$$\Sigma_{State} = \left\{ \; [\Sigma(HDD + CDD)] \times (\text{State Population}) \; \right\}$$

$$\Sigma^{*}_{State} = \left\{ \; [\Sigma(HDD + CDD)] \times (\text{1980 State Population}) \; \right\}$$

$$D_1 = \frac{\text{Sum of } \Sigma_{State} \text{ from 48 states}}{\text{National Population}}$$

$$D_2 = \frac{\text{Sum of } \Sigma^*{}_{\text{State}} \text{ from 48 states}}{1980 \text{ National Population}}$$

1) for each year with a population value, the population *for each state* and the $\Sigma(\mathbf{DD})$ for heating and cooling were multiplied together to get (Σ_{State}); then

2) the values for all 48 states were added, and the sum divided by the national population for that year, to get D_1; then

3) Step (1) was repeated using the 1980 population and its distribution, to get ($\Sigma^*{}_{\text{State}}$); then

4) Step (2) was repeated using the 1980 population and its distribution, to get D_2.

D_1 (migrating) and $_D2$ (static) are plotted in **Fig. 5-13**.

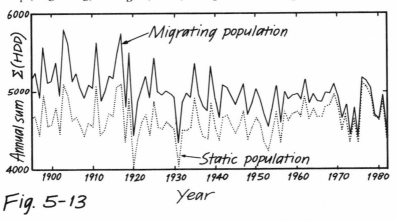

Fig. 5-13

Architecture and the Urban Energy Balance

The discussions of **Figs. 5-10** and **5-11** have made clear that the linkages between climate and architecture are complex, and that the linkages for an entire urbanized area are systematically related to the linkages for individual buildings. Here is one view of the set of linkages[7].

The familiar old city at the heart of most European urban areas may have been planned in the sense that it contained a cultural center such as a fort or a cathedral, and that its site was selected for its commercial and military characteristics. But it was not planned in the modern sense in which a network of streets and other service arteries is designed all at once. The differences between old and new produce impacts worthy of comment concerning weather and climate in small-scale environments. Likewise, as we will explore in the following sections, these cities have effects on the atmosphere.

One difference between old and new is in the *general morphology of the street plans*. The typical old city has narrow and winding streets that follow what were originally footpaths and animal trails, in turn probably dictated by topography. Furthermore, it often tends to have a concentric circular pattern, centered on the cultural hub. Such an urban area is unlikely to have windflow channeling in long, straight urban canyons, as newer parts of the city are. As a result of this more effective trapping of small pockets of airspace, humidity is usually higher among the buildings of an older city where vapor from fountains and kitchens enters the atmosphere.

With narrow streets and only small courtyards, extra trapping of sunlight by multiple reflection (see **Fig. 4-47**) is less likely, since these spaces — including the upper walls — are more often in shadow than if streets were wider and courtyards larger. *The net results for temperature are a reduced range between day and night and between summer and winter.* These results may be seen as advantages or disadvantages, depending on the overall climate. Maintenance of higher humidity and lower midday summer temperature would be a distinct advantage in a dry, subtropical locality, but probably not in higher latitudes.

Another difference between the older and newer parts of cities is in the *homogeneity of land use*. Old cities are very limited in extent, as a rule, reflecting the compactness of the original urban settlements. With communication limited mainly to those within a short walking distance and with the need for proximity of the agricultural fields to the military defenses, land in the settlements was likely to be used exclusively for streets and roofed areas. Parks, ponds, and green areas were a waste of urban space. This is still evident in most old city sections. The point of this discussion is the *higher population density* and the *greater expanse of unrelieved urban surface* — granitelike materials and the absence of open water — in older urban areas.

Buildings in older cities have thick walls, small windows, a small but usually uniform number of stories, with a resulting elevated, nearly uniform roof surface, and few large courtyards or wide streets. The relative absence of unroofed ground is probably due to the high premium on protected space in the city.

The generally uniform morphology of the older buildings is probably due to the limitations of the contemporary technology. Walls are both membranes and supporting members for upper stories and roofs — the more stories, the thicker the ground-level walls. Skeletal and wall materials, and the means for hoisting them, probably limited the number of stories in the older buildings to no more than half a dozen, quite unlike the modern materials and technology. The need for a maximum amount of bearing capacity for each running foot of outer wall made small windows the rule.

As noted, the air among closely packed buildings is likely to be more humid, less windy, and with less temperature variability. These are the same characteristics of the air found in a *cave* — the "structure" with the maximum *resistance* — and in the structures of pueblo and saharan peoples. When walls are thick and windows small and tight, relatively small amounts of fuel are needed in cold weather. The volume of inner space per unit of exterior surface is largest in a cubical building — even larger in a sphere — where families share interior walls not exposed to heat loss and gain, and a single roof stands between the sky and several levels of occupants.

Older urban buildings are, in the context of energy efficiency, probably better reponses to climate — by accident rather than by design — than newer ones. In the same context of energy flow and the urban energy balance, different architectural styles produce recognizable effects on the rate at which energy is released, by radiation and convection, to the atmosphere. Broadly speaking, the city's *shapes and materials* affect the way in which the solar heat load is partitioned among (L_{out}), (**H**), and (**E**), but the "*anthropogenic heat*" — (**M**) — affects the total amount of heat dissipated. These differences will be discussed in more detail in the following sections of this chapter.

Table 5-2 presents data that describe the populations, areas, and estimated metabolic, or anthropogenic, energy flows — (**M**) — in a

Table 5-2 Populations and Anthropogenic Energy Flows in Selected Urban Areas

City	Population* (Millions)	Area* (km²)	Population density (10³/km²)	Energy (M) per Capita (kW/Capita)	Energy (M) per unit area* (W/m²)
Hong Kong	4.4	92.	47.8	0.06	3.
Manhattan	1.7	59.	28.8	5.5	159.
Sheffield	0.5	49.	10.2	2.0	20.
West Berlin	2.3	233.	9.87	2.1	21.
Moscow	6.4	878.	7.29	17.4	127.
Sydney (center city)	0.1	24.	4.17	13.7	57.
Budapest	2.0	525.	3.81	11.3	43.
St. Louis	0.75	250.	3.00	5.3	16.
Cincinnati (summer)	0.54	200.	2.70	9.6	26.
Brussels	1.0	400.	2.50	11.2	28.
Los Angeles	7.0	3500.	2.00	10.5	21.
Chicago	3.5	1800.	1.94	27.2	53.
Fairbanks	0.03	19.	0.80	22.8	19.

* Data from Landsberg, *The Urban Climate*, (1981, page 76)

selected group of urban areas. The population and areal data are combined in the *population density*, while these two kinds of data are further combined to express *energy release rates per unit area and per capita*.

The cities in **Table 5-2** are listed in order of decreasing population density. Because of the complex linkages of architecture, land use patterns, and climate none of the other variables in the table is simply correlated with the density. **Fig. 5-14** begins to make some sense of the nature of these linkages by showing that, in this sample, *energy release rates per unit area exhibit only limited variability*, compared with the variability of other columns in the table, even over a large range of populations and areas. That is, as population densities increase — people live and work closer together — energy use per capita tends to decrease, one compensating for the other. This accords with the idea suggested above, that increasing the amount of human space above each unit of land area tends to reduce the energy used to make that space habitable, and thus the amount of energy released to the atmosphere.

Mathematically, this point is made by noting that the slanting line in the figure expresses the following:

$$(\text{Population density}) \times (\text{Energy release per capita}) =$$

$$(\text{Energy release per unit area}) = K$$

where **K** acts very much like a constant, varying much less around its average value[F], near 30 W/m^2, than the values in any other column of **Table 5-2**. To put this average value in a larger perspective, we note that a typical *midday* value for the *net radiation*, (**R**), in a midlatitude city near Equinox is 500 W/m^2 and a typical value averaged over 24 hours is 100 W/m^2. Obviously, the various cities do not all plot exactly on the line. Generally speaking, the more industrialized and the less tropical is a city, the farther it plots *above and to the right* of the line. Hong Kong is densely populated, but tropical and without heavy industry, so we find it well below the line. Small and non-industrial Fairbanks is more northerly than large and highly industrialized Chicago, but both have similarly high energy use per capita.

It is notable that the two largest urbanized areas represented — Manhattan and Moscow — have distinctly larger, but still such similar, values of per area energy consumption (about 150 kW per m^2) despite what is generally thought of as their great differences in architecture. Greater New York — not just Manhattan — would

---

(F) The values for Manhattan and Moscow were not used in arriving at this average. In fact, the values for Manhattan are not those from the original Landsberg table, since errors in that table were discovered after its publication[8].

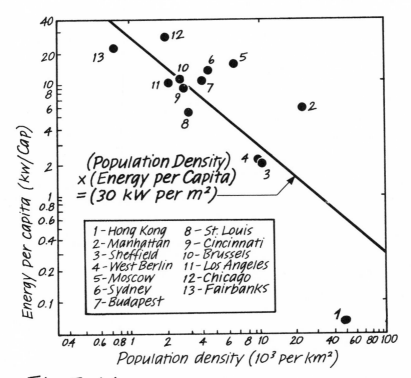

Fig. 5-14

appear much more similar to Moscow, of course, but these data show dramatically some of the major energy interactions of modern urban areas and the atmosphere.

Before leaving the subject of energy, architecture, and the atmosphere take a moment to consider how relationships such as those in **Fig. 5-10** might be expected to change as the urban landscape changes, or how the relationship for an individual building (see **Figs. 5-11a and 5-11c**) changes as the building itself is caused to change.

Fig. 5-15 makes use of a technique called *double mass analysis* to examine several scenarios concerning such changes. To the left in the diagram, before a change takes place, accumulating annual values[G] of $\Sigma(DD)$ — a measure of climate — and of actual energy use remain on the same plotted straight line, because the two are in a very nearly constant proportion to one another.

(G) In this case, the "time unit" in **Fig. 5-15** is one year.

After a change in the city or the building, in which the proportion changes, the subsequent plottings will be along a line that departs from the original, straight line. In *Scenario A*, there is no change, and the slope of the line remains the same. In *Scenario B*, the energy use increases gradually through time after the change. That could be because of gradually increasing population or industrialization, in the case of a city, or of incremental installation of more energy using devices, in the case of the individual building.

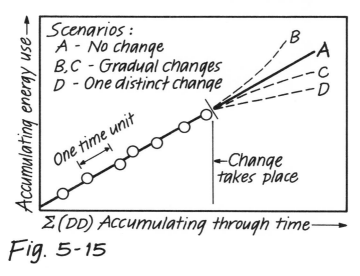

Fig. 5-15

In *Scenarios C and D* the energy demand decreases — gradually in the case of C and instantly in the case of D. For a city, Scenario C might represent, for example, an increase in energy efficient housing without an increase in industrial activity. For an individual building, Scenario D would result from a "retrofitting", or "weatherization", with insulation and similar modernization.

Urban climates: what and why

The term "urban climate" has been at least marginally familiar to designers and planners for more than a decade, but only much more recently has the meaning of the term been dissected[9]. Surely "urban climate" implies not just the climate observed in the city, and how it might differ from that nearby, but also *the differences between what is observed in the city and what would have been observed there if the city were not present on the landscape*. Thus, it is really the **urban effects on local climate — the changes caused by designers and planners — that we are considering. And since cities have atmospheric effects outside their**

borders, we must be concerned also about *the differences between what is observed in the city's environs and what would have been observed there if the city were not present on the landscape —* **urban effects on regional climate**.

If this dissection sounds a bit artificial or pedantic, consider that, while it is easy to observe the climate in a city, inferring the *effects of the city on the climate* requires knowing what the pre-urban climate was like at that place under the same regional weather conditions — a much more difficult task. While I recognize it is of more true importance to specialists in climatology than to designers and planners, I intend to show presently that the importance of this dissection of meaning can have an impact on the work of designers and planners as well.

Fig. 5-16 introduces several terms needed for this discussion, as well as depicting the city and its environs as a particular example of the kind of "patchiness" of landscape mentioned in consideration of **Fig. 3-18**. Clearly, the city deforms the airflow, so we must not be surprised that it changes other things about weather and climate as well.

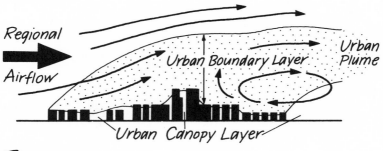

Fig. 5-16

As discussed in both **Chapters 3** and **4**, atmospheric behavior within a *canopy* needs to be considered as somewhat separate from atmospheric behavior in general. In fact, that's what we were doing in the last section when we considered urban architecture and morphology in relation to the microclimates of semi-isolated urban spaces. Thus, the term **urban canopy layer** in **Fig. 5-16** formalizes the need for this separation from the **urban boundary layer**.

We mentioned in the last section that three major differences between an urban "patch" on the landscape and a rural patch are the city's *shapes, materials, and rate of release of anthropogenic heat*. Let us add two items to that list to form the basic *causes of urban effects on the atmosphere*.

Compared with the countryside that existed before the city at that place:

1) cities have more complex shapes, often with as much vertical surface as horizontal surface, that respond differently to both sun and wind;

2) cities have distinctly different surface materials, with greater ability both to conduct heat quickly from the surface and to store the heat once it is conducted away (see **Table 4-4**);

3) cities remove liquid water (and snow) quickly from the surface, thereby reducing the opportunities for evaporative cooling of urban air;

4) cities have many additional sources of heat energy (**M**) — space heating, air conditioners, automotive traffic, industrial processes — to add to the heat load that must be dissipated; and

5) cities have dirtier, more turbid air, thereby (a) responding to the flow of radiant energy much as if they had a permanent layer of thin cloud above them, and (b) providing clouds with a different mixture of condensation nuclei (refer to the discussion of **Fig. 4-33**).

Fig. 5-17 addresses the first idea on the list: city shapes responding to sunshine. The cityscape allows many more opportunities for internal reflection and absorption of sunlight, thereby *changing — usually reducing — the albedo* (fractional reflection upward to the sky). This is particularly true in early morning and late afternoon when the sun can strike vertical surfaces that would simply not be there unless the city were there. As suggested in the figure, it is not just the heights of the buildings and the fact that they are often of lighter colored and more reflective materials, but also their spacing — the widths of canyons — that act to alter the albedo. The effects of city shapes on windflow were considered in **Chapter 3**.

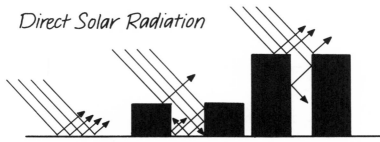

Direct Solar Radiation

Internal reflection & absorption patterns complicate the urban albedo

Fig. 5-17

Urban climates: temperature

Having established what we mean by urban climates, urban effects on climate, and the causes of the effects, look at some of the results. Then, following the pattern established early on in the book, we will inquire as to why the results are as they are. Begin with temperature effects, mainly because these effects were among the first ones measured[H] and are the most widely recognized among non-specialists. **Fig. 5-18** introduces several ideas about urban effects on temperature. Using data from both fixed stations and mobile instruments — first bicycles, then autos, and now aircraft and even satellites — one may map the field of temperature in and around a city. Most mappings have been done for nighttime hours, as will be discussed presently.

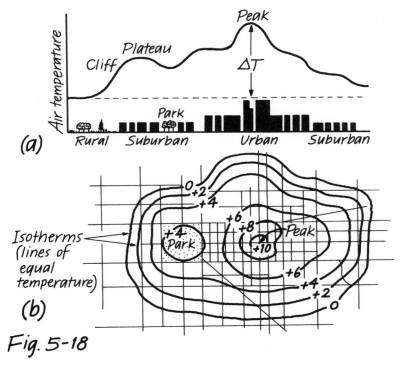

Fig. 5-18

(H) Luke Howard (see Landsberg, 1981) first noted and reported temperature differences between London and its environs in the early 1800s. He is the same Luke Howard who invented our system for naming clouds (**Chapter 4**). Before that time, urban effects on the atmosphere were recognized mainly as the effects on air quality.

Without exception, the map looks like the topographic map of an isolated island, with steep cliffs at the margins and at least one distinct "peak" inland. The name early given to this configuration was the **urban heat island (UHI)**. An example of it is in **Fig. 5-18b**. Temperature maps on some nights in some cities are considerably more complex than this simple example, but the major features of the map —cliffs, plateau, and peak — appear consistently.

Fig. 5-18a contains the added information that the temperature difference between the rural surroundings and the highest peak is often used to express the "strength of the UHI" and is called (ΔT), or alternatively $\Delta T_{(u-r)}$. The largest observed values for (ΔT) for any one city occur under nighttime conditions of clear sky, calm winds, and dry air. Within a typical diurnal cycle with clear sky, calm winds, and dry air (ΔT) usually changes as in **Fig. 5-19**.

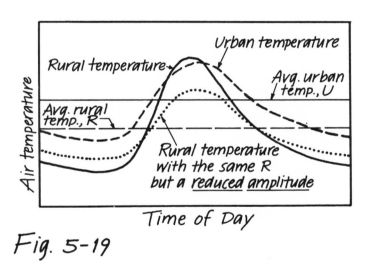

Fig. 5-19

As climatologists became more aware of the observed consistency of many temperature patterns, they began to look into the reasons for them. **Fig. 5-16** suggests at once that regional wind speed and city size ought to be important variables in determing the intensity of the heat island: $\Delta T_{(u-r)}$. With special observations, under a variety of weather conditions in a few urban areas[10], the picture began to emerge that $\Delta T_{(u-r)}$ responds to wind speed much as in **Fig. 5-20a**: increasing wind speed "wipes out" the heat island and reduces ΔT. This is in accord with the principle noted in **Chapter 4** that "circulation cells form only when circulations on larger scales are quiet."

Fig. 5-20a tells the story of temperature and wind for one size of city, but what about the story when varying city size — population, for example — is included? **Fig. 5-20b**, based on a classic figure from Oke, shows clearly that *in light winds and clear skies* ΔT_{max} is related to city population in a strikingly consistent way. Combining data on $\Delta T_{(u-r)}$, wind speed, and population for a sample of Canadian urban areas has led Oke to a useful summary. The summary may be expressed in the form of the simple equation[1] in **Fig. 5-20a**:

$$\Delta T_{(u-r)} = (1/4) \text{ x (Fourth root of Population)} \div \text{(Square root of wind speed)}$$

The fact that North American and European cities seem, in **Fig. 5-20b**, to act differently in apparently similar conditions suggests there is some combination of architectural and energy-use variables at work, as in the discussions of the previous section. This suggestion is made even more vivid when data from a mixture of Australasian and Third World urban areas (not shown)[11] are added to **Fig. 5-20b**. When they are plotted together with the North American and European cities they form no consistent pattern as do the two groups in the figure.

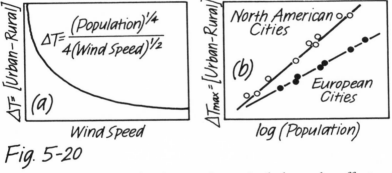

Fig. 5-20

Arrival at questions like the one just raised about the effects of architecture and energy use on otherwise simple appearing relationships led urban climatologists, as it leads us now, out of the area of *what* and into the area of *why*.

Leaving aside for the moment the question of the effects of architecture and energy use, examine the causes of the relationships in **Figs. 5-18b** and **5-19** concerning the general warmth of

(I) Other studies, more detailed than we need here, show that two additional variables seem to be related to the magnitude of ΔT: the rate of heat release (**M**) and the way air temperature changes with altitude — the air temperature profile — in the upwind rural area. This last variable is related to the concept of the thermal stability of an air layer, to be discussed in connection with the dispersal of air pollutants in **Chapter 11**.

cities compared with their environs, and the fact that the warmth is greatest under nighttime conditions of clear sky, calm winds, and dry air.

Fig. 5-21 harks back to the energy flow diagrams of **Chapter 2** and includes ideas from our previous discussion of the urban energy balance. In the upper part of the figure, the energy balance of a rural surface is suggested and the urban balance is in the lower part. In this presentation, *the daytime storage of heat in the system, followed by nighttime release of the same energy*, is a feature not given great attention up to now in this book.

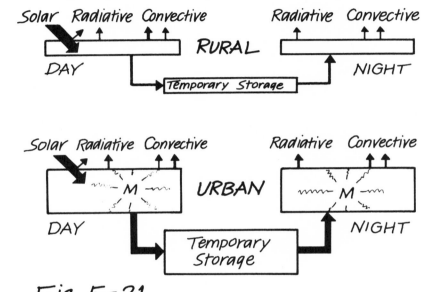

Fig. 5-21

With the same insolation on both systems, but with an added component of the heat load — (**M**) — and a decreased albedo in the urban system, the figure suggests the remark made in the last section; namely, the city's shapes and materials affect the way in which the solar heat load is partitioned, but the anthropogenic heat (**M**) affects the total amount of heat dissipated. These factors result in the outcomes shown in **Fig. 5-19**.

In **Fig. 5-19** the diurnal temperature pattern in the urban site may be seen as the result of two kinds of change from the diurnal temperature pattern in the rural site. Though the two changes take place simultaneously, imagine them as separate. *First*, the "rural" curve has its maximum decreased and its minumum increased — the amplitude is reduced with no change in the *average* temperature — yielding the intermediate curve. *Second*, the whole curve with reduced amplitude is then warmed by about the same amount night and day, yielding the "urban" curve.

Given the list of five kinds of difference between a rural "patch" and an urban "patch" and the two kinds of change just noted, we can link these five differences and two changes like this:

1) The reduction in amplitude results from two of the differences: (a) the thin cloud of pollutants, and (b) the more rapid storage of heat in the city's granite-like materials by day followed by its release at night.

2) The general warming at all hours results from three of the differences: (a) the addition of (**M**) in the city, (b) the reduced evaporative cooling in the city, and (c) the reduced convective dissipation rate in the city due to the lower wind speeds there.

With these remarks about the effects of the urban landscape on the energy balance, it seems a good place to pause and remark about the apparent contradiction between the facts that (a) (Δ**T**) is least at midday regardless of season (**Fig. 5-19**) and that (b) people often experience distinct hyperthermic feelings in an urban canyon at midday in summer. Common experience is that being in the countryside at midday in summer is distinctly cooler than being downtown.

The contradiction may be resolved by noting again that *human comfort is strongly related to the relationship of heat load to dissipation rate*, and that standing in an urban canyon exposes a person to an *extra heat load* — longwave radiation from the warm walls of surrounding tall buildings — and a *reduced convective dissipation rate* because of the reduced wind speed. The convective dissipation both by sensible heat loss (**H**) and evaporative cooling (**E**) are reduced in light winds. The result of the extra heat load and the reduced convective dissipation rate, in summer at least, is a feeling of overheating.

We have explained the extra urban warmth, and the fact that it is greatest at night, with calm winds and clear skies. It remains to explain why the differences in urban architecture and energy use among cultures produces the differences it does in the relationships of (Δ**T**), wind speed, and city population. The small dip in the temperature pattern in the urban park of **Fig. 5-18** is the clue to the answer.

Oke (1982) discovered, through a combination of laboratory experiments and analysis of temperature and wind observations from a wide variety of the world's cities, that in fact *the urban heat island is a very local affair.* The fraction of the cold night sky one can see from the bottom of an urban canyon — and thus the fraction blocked out by warm urban walls — mostly determines the radiative cooling rate in that location. The urban park acts like a local rural patch, and so it cools faster than the surrounding canyons. Further, deep canyons — typical only of large cities — cool most slowly. The net result is that after sunset the cooling in open areas runs ahead of the cooling in the deep canyons, and the UHI results.

So, it turns out that *the shapes and locations of the buildings, rather than the wind travel distance over the city or the population, are most important in determining the strength and configuration of the urban heat island.* **Fig. 5-22** shows the details of the relationships among the **sky view factor,** Ψ (the fraction of overlying hemisphere occupied by the sky), the height (**H**) and width (**W**) of the canyon[J].

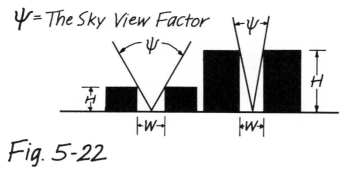

Ψ = The Sky View Factor

Fig. 5-22

The proof of the pudding about the sky view factor is in **Fig. 5-23**, where data on $(\Delta T)_{max}$ from North American, European, and Australasian urban areas behave in the same way[K] with respect to the value of Ψ in center city, whereas they behaved quite differently with respect to population (**Fig. 5-20b**).

When all is said, then, most of the urban effects on temperature *in the Urban Canopy Layer (UCL) where people live* — (**Fig. 5-16**)

(J) For the mathematically adept reader, these relationships are summed up for us by Oke (1981) as follows. Consider a point in the exact middle of an urban street, with walls the same on both sides, and the street "infinitely long" in both directions from the point. The *wall view factor* (the fraction of overlying hemisphere occupied by the walls) is given by:

$\Psi_w = (1/2)(\sin^2\Theta + \cos\Theta - 1) \div (\cos\Theta)$ in which $[(2H/W)] = \tan\Theta$. Finally, the wall view and sky view factors are related by $\Psi = (1 - 2\Psi_w)$.

(K) **Fig. 5-23** imitates Fig. 7 in Oke (1981), who reports that the mathematical formula for the line in the figure may be taken as either

$(\Delta T)_{max} = 15.27 - 13.88(\Psi)$ or $(\Delta T)_{max} = 7.45 + 3.97 \ln(H/W)$

— are determined by the work of designers and planners. In the former view, in which the UHI was the product of larger scale, aggregated effects of the urban fabric — in the *Urban Boundary Layer (UBL)* — one could assume only master planners who long ago laid out the city were responsible for the UHI.

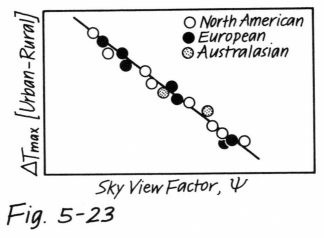

Fig. 5-23

It wasn't until I became aware of the truth contained in **Fig. 5-23** that I connected two other facts I had known for some time.

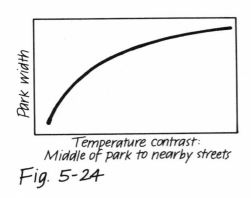

Temperature contrast:
Middle of park to nearby streets

Fig. 5-24

First, the two lines in **Fig. 5-20b** converge at a population of about 100 people — a very local scale. Second, East European scientists learned some years ago[L] that urban parks produce local temperature contrasts directly related to their size, as suggested in **Fig. 5-24**. The fact that all these kinds of data and ideas converge to one coherent whole, and that predictions about urban temperature effects are now so accurate, makes us rather certain we truly understand the temperature climates of cities.

(L) I became aware of this when a Hungarian climatologist, Ferenc Probald, showed me a Russian article during a Hungarian-American conference in Budapest. It is a shame such information doesn't always slip across borders.

Urban climates: moisture

In **Chapter 3** we went to some lengths to show that relative humidity (**RH**), being responsive to changes in both temperature and vapor content, is not very reliable as a measure of vapor content by itself. We made the point that the dewpoint temperature (T_d), and less often the vapor pressure (**e**) and the mixing ratio, is used by climatologists when comparing vapor content between air samples. Because urban landscapes have differences in both heat content (temperature) and vapor content (humidity) as compared with their rural suroundings, it should not be surprising that studies of urban moisture seldom show consistent patterns of relative humidity, in either space or time.

On the one hand, urban areas, compared with nearby rural areas, have fewer "natural" sources of water vapor — plants and open water — and better removal of the temporary sources arriving as precipitation (rain and snow). On the other hand, cities have their special "manmade" vapor sources from automobiles and industrial activities, not to mention the burning of fuels for space heating. A reasonable statement about sources, then, is that *the amount of vapor entering the urban air is different from that in the countryside not so much in the amount entering over a period of time as it is in the times and places it enters — the source distribution.*

As we noted in an earlier section, humidity tends to be higher in the smaller pockets of trapped airspace found in areas most typical of the centers of older cities. Thus, ventilation, as well as source distribution, plays a role in determining urban humidity patterns, and these patterns are not very consistent from one city to another, or from one kind of day or time of year to another. Based on studies to date and for the reasons suggested here, *it is difficult to generalize about the humidity components of urban climates* except to say that they differ by relatively small amounts from rural environs.

Humidity and vapor content, however they are measured, are continuous — they are always present in both time and space. It is quite different for *precipitation amounts*[M], the other principal variable related to urban moisture. Rain and snow are continuous in neither time nor space. As one result, their true distributions are very much more difficult than humidity to measure and characterize from the measurements taken. In addition, measurable differ-

---

(M) The following discussion is about precipitation *amounts*, as distinct from how often rain falls (frequencies) or the chemical quality of the rainwater, each the subject of many studies.

ences in humidity and vapor content exist primarily under calm conditions, whereas measurable differences in precipitation occur primarily under stormy — and thus windy — conditions.

Because of these differences in continuity between humidity and precipitation, humidity — like temperature — may be measured by "continuous" methods employing both networks of fixed stations with continuously recording instruments, and "traverses" of mobile instruments. Precipitation, however, must be measured mostly by networks, with little chance of being sure how much has fallen, and when, between points of the network. The use of radar has overcome this difficulty somewhat, but the great expense of such techniques has left us with radar observations from only a very few cities for rather short periods of time. In addition, radar observations give very imprecise information about amounts, especially when amounts are small.

The reason I am mentioning all these differences and difficulties, between continuous temperature on the one hand and discontinuous precipitation amounts on the other, is to be able to explain why *climatologists have been able to reach generalized and firm conclusions about urban effects on temperature — the kinds in the previous section — while they still debate, and have yet to reach generalized and firm conclusions about, urban effects on precipitation amounts.*

As I have done in the case of urban effects on temperature, I will discuss the kinds of spatial patterns of precipitation amounts one observes around cities, the ideas that have been used to explain the observations, the difficulties with these explanations, and something of the doubts that arise from these difficulties.

Networks of rain gages, beginning in the 1920s in Germany, have produced studies suggesting that if urban areas have effects on precipitation amounts, the effects usually result in:

1) slightly greater amounts within the city under relatively unstormy, drizzly conditions not producing very much rain anywhere; and

2) distinctly greater amounts "downwind" from the city center, under stormy conditions, along the customary tracks followed by rainstorms in the region.

Studies from both smaller and larger cities in several parts of the industrialized world have all suggested these same conclusions. **Fig. 5-25** is a sketch showing essential details of a famous study of summer rainfall amounts in and east of Chicago. The strange shape of the network map results from the following components: a circle with a 30-mile radius centered on the central Chicago

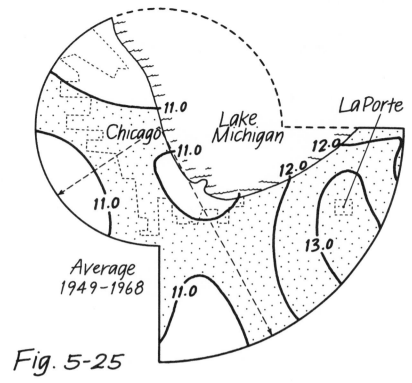

Fig. 5-25

lakefront, and the southeastern quadrant of a circle with a 60-mile radius centered on the same place, with parts of both cut out — unobserved — over Lake Michigan.

Within this network, average summer rainfall totals are expressed by **isohyets** — lines connecting points with equal totals of rainfall amount. At Chicago, summer totals are typically near 11 inches[N] over the several decades for which the map shows the averages.

Much has been made of the small region of distinctly larger totals centered in the southeast — downwind — around LaPorte, Indiana. This region has come to be known among climatologists

---

(N) "Eleven inches of precipitation" means that a rain gage has caught an amount of water whose depth in the gage is 11 inches over the time period specified, in this case an average summer. An unspoken assumption is that, *if held where it fell*, the water would have a depth of 11 inches on the ground in all nearby surroundings. More modern studies express these depths in millimeters, but that should not trouble you if you encounter such a result. Just use this: 1 inch = 25.4 millimeters. More troublesome are the facts that (a) the presence of the gage itself may affect the amount it catches, and (b) measurements only at fixed points can obscure major variability between the points, due to such things as the effects of terrain features and open water in rivers and lakes.

as the *LaPorte anomaly*. It is quite distinct, and undeniably present on maps drawn for different groups of years and by different analysts. So, even though the interpolations required for drawing isohyets are subjective — the patterns can differ substantially depending on who draws them — few argue that the LaPorte anomaly is present. Agreeing on the explanation of it is quite another matter[12]. Before mentioning the kinds of explanations offered for the urban effects such patterns seem to disclose, and the difficulties that arise for these explanations, let me show you several more results — all of summer precipitation totals around major cities in the American midwest. **Fig. 5-26** shows the map of summer

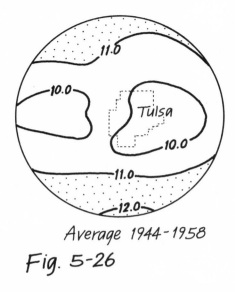

Average 1944-1958

Fig. 5-26

isohyets around Tulsa, Oklahoma, presented within another circle with a 30-mile radius. Again, the typical summer total is 11 inches, averaged over a decade and a half, and there is a pattern of isohyets that suggests a "plume" of slightly more rainfall downwind of the city center.

Fig. 5-27 shows both the relationship of St. Louis, Missouri, to major rivers nearby and the patterns of summer totals of precipitation for two groups of years. Again, the maps are on circles with 30-mile radii, and the summer totals are in the neighborhood of 11 inches. On both St. Louis maps, the largest amounts shown are in clusters to the east and northeast — downwind — of the city center. The map for 1971-1975 shows more detail than the earlier map because the network of raingages was more dense.

There are many other maps of precipitation amounts, for other times and other places — both urban and rural — available in technical reports. These four — one each for Chicago and Tulsa, and two for St. Louis[13] — will help make the points in the following discussion of the explanations and their difficulties.

Why do these effects appear — the small increases in the city on drizzly days, and the larger increases downwind on stormier days? As to why clouds rain, we have already established that the basic requirements (**Chapter 4**) are a vapor supply, a sustained updraft,

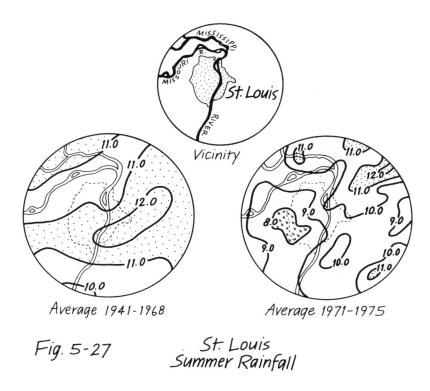

Average 1941-1968   Average 1971-1975

Fig. 5-27   St. Louis
Summer Rainfall

and the right combination of sizes in the swarm of nuclei in the cloudy air. How do cities change these three things?

It is in doubt just how cities affect the supply of water vapor, but it is clear that, *at some places and some times,* the vapor concentration is distinctly higher because the city is there. Those places are near industrial areas, and those times are usually on humid summer afternoons.

Cities produce local updrafts — and also downdrafts — because of their rougher shapes (mechanical turbulence) and their warmer surfaces (thermal turbulence). (See the footnote in **Chapter 3**).

Finally, cities add to the passing air swarms of dust and soot particles that may, if they are the right sizes and of the right materials, act as condensation nuclei (see **Chapter 4**). So, on the face of it, we should expect more precipitation in and *downwind* from cities, since any cloud passing or initiated over the city requires time — and therefore distance — for the rain processes and cycles to be completed and the rain to fall to the ground.

It would seem that rainfall maps and maps of the UHI are essentially the same kind of tool of analysis applied to two different climatological elements: temperature and precipitation. Since the two kinds of analysis have produced recognizable — and to some extent explanable — patterns, one may reasonably conclude that in both kinds of map, urban effects have been revealed.

What is more, other kinds of analysis have been applied to both temperature and precipitation amounts, with explanable results[O]. One such analysis is shown in **Fig. 5-28a**, in which the *ratio* of the summer precipitation total downwind of central St. Louis to the total in the central city appears to increase through two decades. These results appear similar to those for trends in city center temperatures[14]. The easy explanation is that urbanization

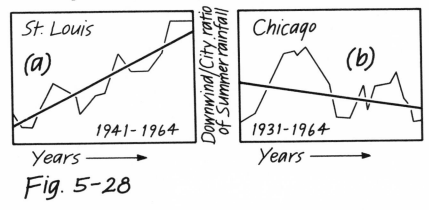

Fig. 5-28

and industrialization in Greater St. Louis increased — with increasing effects on rainfall — through these years.

Finally, though mostly beyond the scope of our immediate interest, researchers have published reports that measures of "storminess" other than just precipitation amounts show map patterns similar to those just presented. *Thunder (and therefore lightning) and hail* are reported to be affected by the passage of storms over urban and industrial areas (e.g. Changnon *et al*, 1977).

We have explored the observed patterns of precipitation amount and the proposed explanations of the patterns. Now we will consider the difficulties with these analyses and explanations. *Then I*

(O) Other kinds of analysis applied to both temperature and precipitation but not discussed here are (a) statistical differences between weekday and weekend patterns, (b) statistical differences due to differences in urban population, and (c) observed differences in short-term, small-scale local behavior of temperature and of clouds in particular kinds of weather. The interested reader may learn more about these studies in Oke (1982) — temperature — and Changnon (1977) — precipitation.

will try to show why all these detailed considerations are impor-
tant for designers and planners.

In addition to the list of differences between temperature and precipitation amount we have already constructed — those related to space and time continuity, occurrence in calm and windy weather, and systems for measurement — the following list further suggests that the firm conclusions regarding temperature cannot so easily be translated into firm conclusions regarding precipitation amount.

1) Studies of one city through time[15] have shown increases in the UHI but no changes in precipitation patterns. This difficulty harks back to the comment made at the beginning of an earlier section, that true urban effects on climate are those that show up *only after a city is established.*

2) Precipitation-producing processes are more sensitive to altitude variations of the underlying terrain than are temperature-determining processes. Thus, we cannot avoid asking what effects of terrain on updrafts — mixed in with effects of urbanization — might appear in rainfall maps[16]. Likewise we must ask about effects of other landscape features on the precipitation-producing processes: effects of major rivers and lakes on the moisture supply, and effects of wind patterns on updrafts at major shorelines (**Figs. 3-24** and **5-25**).

3) The increasing trend in **Fig. 5-28a** at St. Louis tends to lose its impact when one notes the decreasing trend in **Fig. 5-28b** for Chicago[17].

4) The changes in summer rainfall patterns in recent years — e.g. differences between the two maps of **Fig. 5-27** and the changes in **Fig. 5-28a** — could be explained by changes in the proportions of the different kinds of rainstorm as well as by changes in urbanization and industrialization. The argument for a change in proportions could be supported by the fact that rainfall amounts for 1971-1975 in **Fig. 5-27** are *less* than those in 1941-1968, both in the city center and downwind, even though Greater St. Louis certainly did not become *less urbanized and industrialized* during the same period.

5) Researchers who have argued for the "enhancement" of precipitation amounts downwind from cities have seldom, if ever, considered that cities might *suppress* rainfall amounts. This suggests that these researchers might be "seeing only what they are looking for" in the results and observations they have analyzed.

Urban climates: so what?

Now I will try to show why all these detailed considerations are important for designers and planners. Obviously, very few people choose to live in a city or in the countryside primarily because of small differences in temperature or precipitation. Some might make the choice on the basis that the snow-free season, and thus the gardening season, might be slightly longer in the city; or that the mid-summer comfort level might be greater in the country; or that heating bills are a bit smaller in the city. But mostly the decision has to do with factors other than weather and climate.

That being the case, why should designers and planners be interested in the details discussed in the last three sections? More particularly, why should they be concerned about the assertion that the firm conclusions regarding temperature cannot so easily be translated into firm conclusions regarding precipitation amount?

In my opinion, designers and planners need to be interested, on one level, simply because they need to be aware of *all the effects their work has on the landscape* — both local and regional. That is part of being a professional whose work effects people other than his clients.

Also, again in my opinion and on another level, designers and planners need to be aware that *the conventional wisdom presented to them by scientists is sometimes deserving of further scrutiny*. To make the point, especially about doubts concerning precipitation amounts, consider the following conundrum (Lowry, 1979).

On the night of 13 July 1977 Greater New York City suffered a staggering electrical blackout. It is claimed that the cause was directly linked to the fact that "a severe summer thunderstorm had just swept across . . . the vicinity of the Indian Point No. 3 nuclear power plant . . . flashes of lightning (then knocking out) two 345-kilovolt lines." Nineteen minutes later two more lines were knocked out by lightning strikes, and then three minutes after that, another line[18].

Assume for the moment that the claims are correct about urban areas enhancing lightning storms — that *the city itself was the cause of its own miseries*: the huge economic and social impacts. If the claims are correct, one might expect massive legal and insurance claims against urban designers, regional planners, and who knows who else, on the grounds that the lightning damage was their work rather than "an act of God", as lawyers so often claim.

There is reason, therefore, for designers and planners to hold serious doubts concerning the claims about urban areas enhancing lightning storms, or even changing rainfall patterns. Even without the claims about lightning, hail can damage both crops and sensitive structures, and increased rain and runoff can overload carefully designed drainage systems.

The real point, of course, is the one made above: that *professional designers and planners need to be aware that the conventional wisdom presented by scientists is not always beyond further scrutiny*. I choose the story about urban effects on storminess to make this point, because it is the story I am most familiar with. And I give the long list of footnotes and references, because the debate is still in progress and I must document my version of the arguments.

Notes

(1) Professor Steadman calls (T_e) the "apparent temperature", and his standard person is outdoors, walking slowly, and dressed in a light woolen suit. The results of his elaborate research, which includes both theoretical calculations and experiments with human subjects, are brought together in Steadman (1984).

(2) See Kalkstein and Valimont (1986, 1987).

(3) Rather than trying to present great technical detail about the interactions of buildings and climate, I refer you to Olgyay (1963) and Givoni (1969).

(4) After Fig. 4.9 in Landsberg (1981), who used data from Turner.

(5) These materials are available in publications of the American Society of Heating, Refrigeration, and Air Conditioning Engineers (ASHRAE). They are also available in various textbooks such as McGuinness *et al*, (1980, Fig. 5.3 and Appendix C) and Ahrens (1982, pages 130-131).

(6) From the *Network Newsletter, Environmental and Societal Impacts Group (ESIG),* Volume 1 No. 4, Summer 1986. Box 3000, Boulder, CO 80307.

(7) This view is adapted from Lowry (1979).

(8) In personal correspondence, Professor Oke has detailed the various oversights that have, through several years, led to certain errors in Landsberg's table. Recounting the sequence is beyond our needs here, but the errors are by now well known to specialists.

(9) Perhaps it is permissible for me to suggest that the awareness dates from publication of "The Climate of Cities" (Lowry,

1967) and that the consideration of meaning was first under-taken in a systematic way a decade later in Lowry (1977).

(10) Professor T. R. Oke and his students have led the way in producing both temperature observations and analyses of them. Many of the figures in this section are based on their work. A recent summary of their thinking can be found in Oke (1982b).

(11) The Third World cities referred to in connection with **Fig. 5-20b** are described by Jauregui (1986) but are not included in **Fig. 5-23**, from Oke (1981) and (1982b).

(12) The intricate twistings and turnings of the debate on the LaPorte anomaly can be found mainly in the pages of the *Bulletin of the American Meteorological Society*, beginning with Changnon (1968) and ending (or has it?) with Changnon (1980).

(13) The maps for Chicago and Tulsa, and for the period 1941-1968 at St. Louis, are from Huff and Changnon (1972). The map for 1971-1975 at St. Louis is from Changnon *et al* (1977).

(14) For examples, see Figs. 5.3 and 5.4 in Landsberg (1981).

(15) The only city I am aware of given this before-vs-after exami-nation is Columbia, Maryland, as described in Landsberg (1979).

(16) There are reports in the technical literature showing that (a) one can get rainfall maps such as those in **Fig. 5-27** *even over very flat, rural terrain* (e.g. Changnon, 1962) and that (b) *even modest terrain features in the absence of urbaniza-tion* can produce rainfall maps such as those in **Fig. 5-27**, (e.g. Huff *et al*, 1975).

(17) These two time trends of the "rural/urban ratio" are from Huff and Changnon (1972).

(18) The quotes are from *TIME*, 25 July 1977, as described in Lowry (1979).

Peptalk

(P₁₃) If you are confused about these modifications, from **Chapter 4**, of the mathematical discussion of the energy balance, be content for the moment in knowing that you will encounter these various mathematical forms in what you read. Taking the time now for thinking it through will leave you prepared for that encounter. Also, you should appreciate now that clear thinking requires great care when adapting an analysis of microclimate to an analysis of human comfort. The general ideas of the energy balance are the same in both cases, but the details are not.

6. Who else is here?
Animals and plants and the atmosphere

In this Chapter we will extend our consideration of the energy balance concept and its ability to provide insights into the lives and behavior of living organisms, in particular animals and plants.

In **Chapter 5** we noted how human beings are active participants in the management of their energy balances, and that other kinds of animals have different and additional means for thermoregulation, as compared with human beings. We will examine those means in this Chapter.

In **Chapter 5** we also noted that surfaces, as compared with animals and people, are essentially inanimate, responding passively to changing heat energy loadings. In this Chapter we will also examine the responses of plants to changing heat energy loadings, noting that they respond at a level somewhere between the passivity of inanimate surfaces and the activity of animals and people.

In this Chapter you will encounter considerably greater detail than before on the *insights offered by the use of mathematical models of the energy balances* of animals and plants$^{(P_{14})}$. Before I introduce you to these details, consider what may be the primary advantage — the payoff — for you in understanding the details. I contend that *the payoff comes as insights into **the role of form in shaping a system's response to atmospheric variables***, particulary as to energy management.

Animals and plants compared with human beings

We can make use here of the detailed explanations of the human energy balance in **Chapter 5**, but first we need to take note of the ways in which smaller homeotherms differ from Man and how plants differ from both.

We have already noted that human beings have a relatively constant **resting metabolic rate** (+**M**) — not working and exercising — and a well developed variety of technological controls — clothing and housing — over the resistance to the flow of heat between the environment and themselves. Large homeotherms, such as cattle, also have relatively constant resting metabolic rates, but relatively fewer technological controls.

Smaller homeotherms, on the other hand, have variable metabolic rates. In fact, *the smaller the animal, the larger the range of variability*, and the greater the dependence on that variability as a response to thermal stress. Compared with Man, other homeotherms have a relatively short list of "technological" reponses — mostly haircoats that may be erected (feathers for birds) and shelters of invariant design. They display seasonal variability in haircoats — thickening and thinning, lengthening and shortening, changes of color — but not diurnal variability as in Man.

In summary of the human-animal comparison, animals have much more variable metabolic rate and body size, while Man is more dependent on technology. For both, metabolic heat is a major and essential contribution to the energy balance, and the ability to move to a refuge — a "hiding place" — is a major defense against stress.

Plants, of course, are *immobile* and must withstand whatever stress comes, without the ability to seek a hiding place. Without large body mass and with very small metabolic rates, plants have much less control than animals over their energy balance. The major result is that the everyday temperatures of plant parts range much more widely than those of animals, being never very far from the temperature of the air surounding them.

The animal kingdom

Though we have spoken only of **homeotherms**, in fact animals come in several other kinds. Homeotherms are often called "warm-blooded" animals, while the "cold-blooded" animals — snakes and lizards — are known as **poikilotherms**. Without internal means for thermoregulation — "fine tuning" —they depend mainly on the "coarse tuning" that comes with changing location. They seek sun and shade, or burrow to where temperatures are less extreme, and their body temperatures normally range more widely than those of homeotherms.

Insects are like poikilotherms in the sense that they use location as a principal means for thermoregulation. Adults of many species, however, have a form of "coarse tuning" that consists of rapid metabolic heat production through chemical change. Bees, for example, can "fire up" on a chilly morning and raise their body temperatures to the range needed for the fast brain and muscle responses of foraging. Their temperatures might have reached that range anyway on that morning, but their biochemical response enables them to activate their foraging cycle earlier and thereby have more time at work.

Birds depend for thermoregulation on variable metabolic rates, and the variable insulation in feathers that comes with their ability to "fluff" and install dead air pockets within the feather coat.

In what follows, most attention will be given to homeotherms, leaving the reader to explore the thermoregulation of poikilotherms, insects, and birds elsewhere.

Atmospheric effects on animals

A useful way to provide an overview of the many kinds of animal-atmosphere interaction is presented in **Fig. 6-1**. Some atmospheric effects act directly on the animals of interest, while other effects act indirectly through other animals. Understanding the **direct effects** usually involves an analysis of the animal's energy balance in one form or other.

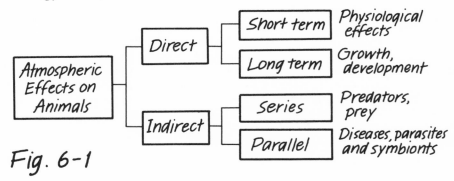

Fig. 6-1

The *short-term direct effects* operate over short time intervals, such as minutes and hours within a day, and involve the details of thermoregulation we have been discussing. The *long-term direct effects* operate over time intervals such as weeks, months, and whole life cycles of individuals. This Chapter is concerned primarily with direct effects.

Understanding the **indirect effects** of the atmosphere on animals usually involves an analysis of the **population dynamics** of various interacting species, each in turn interacting with the atmospheric environment, a subject far beyond the scope of this book. Those indirect effects involve the dynamics of such processes as predator-prey relationships and the impacts of diseases and parasites on animal populations. Two or more species acting in concert to maintain each other — symbiosis — are called **symbionts**.

I have further subdivided indirect effects — perhaps only in the interest of symmetry with the subdivision of direct effects — into those where animal species are involved *linearly* with each other in matters of consumption (predators and their prey interacting within a **food chain**) and those where species are involved *side-by-side*, so to speak.

Using the terminology from systems of electrical resistances, I call the linear relationships **series effects**, the name used when resistances are connected end-to-end. Resistances connected to two common points — side-by-side — are said to be in **parallel**.

As noted, most of what follows is concerned only with direct effects, and most of that on homeotherms. The purpose of the discussion of **Fig. 6-1** is primarily to emphasize the fact that there are a multitude of effects of the atmosphere on animals.

The importance of animal body size

Except for differences due to age, the range of body sizes in Man is not great. Considering the whole variety of homeotherms, however, differences in body size — and therefore *body mass* — are huge. Excluding whales because they are marine animals not exposed to extremes of atmospheric conditions, the range is that from elephants to mice and shrews. The range of sizes, of course, reflects the range of "jobs" the various species do in their ecosystems — the range of **niches**[A]. The linkage between body size and metabolic rate — and so, the energy balance, as noted in **Chapter 5** — is the fact that homeothermic evolution has taken place almost exclusively in tropical temperatures.

Body size has three impacts on an animal's energy balance — the management of its heat energy flows. The first impact is connected with the body's *heat storage* rather than to thermoregulation, while the second impact is connected with the body's *heat production* — the metabolic rate. The third impact is connected to the thickness of the microscale boundary layer around the animal.

The principle involved with the body's *heat storage* is contained in the idea of the **thermal time constant**, a principle that applies equally well to inanimate bodies, such as logs and rocks. That idea

---

(A) Contrary to the notion held by many non-specialsts, the term *ecological niche* refers to the role a species plays — its "occupation" — in a community of many species, and not to the physical location — the *habitat* — of that species.

is illustrated in **Fig. 6-2**, where the temperatures of two bodies — one small, the other large — are traced against the time elapsed since they were placed in a cold environment[B].

As common experience tells us, the bodies cool rapidly at first, then more and more slowly as their temperatures approach that of the surroundings. In the terms of heat transfer in **Chapter 4**, heat flows from the warmer to the colder at a rate proportional to the temperature difference, which in this case is becoming less and less as time goes on. *For any particular temperature difference, the rate of temperature change in the body is smaller for the larger body.* Thus, body size — body mass — plays an important role in the responsiveness of a body's temperature to its surroundings.

From our discussions of heat flow and storage in **Chapter 4**, you have probably already figured out that this difference between a larger and a smaller body is because *each time unit sees the loss of a smaller **percentage** of the total heat content of the larger body*, and since the temperature is a measure of the total heat content, each time unit sees a smaller drop in temperature in the larger body.

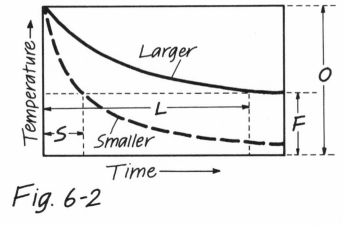

Fig. 6-2

Rather than explain the whole phenomenon in complex terms of body mass, surface area, and temperature differences, we customarily express the differences between bodies in terms of a single variable: **the thermal time constant**. As shown in the figure it is *the time it takes for the temperature difference between the body and its environment to be reduced to a certain fraction* (**F**) *of the original difference* (**O**) — a longer time (**L**) for a larger body, and shorter time (**S**) for a smaller body. The fraction is immaterial: the

(B) The discussion could be related just as well to a body placed in a *warm environment* simply by rotating the figure on a horizontal axis.

two time constants are (approximately) in the same proportion regardless of that fraction.

In the terms of animal thermoregulation, then, we can say from the first principle, concerning body heat storage, that *the larger the body size, the greater its thermal mass, and the longer the animal, upon leaving a "refuge", can stand thermal stress before having to resort to "emergency" controls.*

The principle involved with the body's *heat production* is contained in **Fig. 6-3**, where the body weight of a homeotherm — logarithmic on the lower scale and linear on the upper — is related to its resting metabolic rate (+M) — logarithmic on the righthand scale and linear on the lefthand. The particular mathematical relationship in the figure is for an imaginary animal, but it is the kind invariably used to describe the relationship. Let me explain.

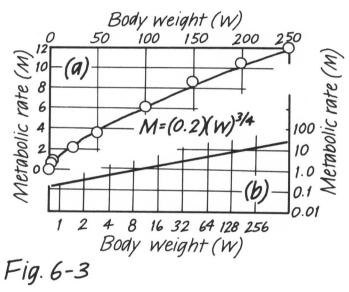

Fig. 6-3

The mathematical equation — the "model" — says that

(Resting Metabolic rate, M)
   is proportional to (**Body Weight, W**)$^{3/4}$

where the exponent — the "power" (3/4) — *less than one* means that as the body weight increases, the metabolic rate increases increasingly less rapidly (P_{15}). Why should that be so?

The answer stems from three observations that agree with intuition:

(a) an animal contains — stores — heat in proportion to its body *volume*; and

(b) an animal exchanges heat with its surroundings in proportion to its body *surface area*, so that

(c) an animal must resupply its body's stored heat, as it is lost to the environment, in propoprtion to its **surface-to-volume ratio**.

Using the simple geometric form of a sphere to complete the argument, we note that a sphere's surface area is proportional to the *square of its radius*, (r²) whereas its volume is proportional to the *cube of its radius*, (r³). Thus the surface-to-volume ratio is proportional to the power (2/3).

The exponent in **Fig. 6-3**, it seems, ought to be (2/3) — a *theoretical* value — rather than (3/4) — an *experimental* value. The difference lies in the facts that (i) an animal's body is not a simple geometric form, but is a complex of several connected sub-volumes, and (ii) some of the surface area of some sub-volumes is pressed next to — "hidden by" — some of the surface area of some other sub-volumes, and is not, therefore, available for heat transfer to the environment.

Fig. 6-4 makes clear that the model equation — especially the (3/4) power — holds true over the whole range of animal body sizes both homeotherms and others[1].

The figure shows *experimental results* confirming that the linkage between the surface-to-volume ratio and the metabolic rate is

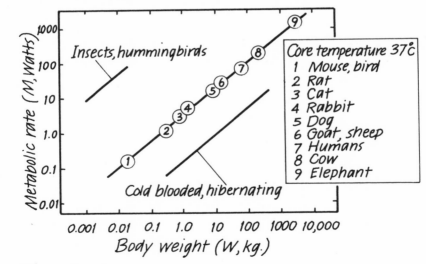

Fig. 6-4

pretty well universal for animals. It shows also that animals with the need to maintain a higher core temperature — insects and humming birds in this example — exhibit a higher proportion than that of homeotherms with core temperature near that of Man: 37 °C. Likewise, those with the need only for a lower core temperature — poikilotherms and hibernating homeotherms — have a lower proportion, but all three categories follow the rule of the power of (3/4).

While the power (3/4) is clearly linked to the size and proportions of a body, the proportion (3.2) is a result of the fact, as noted earlier, that most of animal evolution has taken place under tropical conditions, with (T_c) near 28 °C.

In the terms of animal thermoregulation, then, we can say from the second principle, concerning body heat production, that *while total metabolic energy increases as body size increases, the metabolic energy **per unit of body weight** decreases as body size increases.*

For the animal's behavior, this means larger animals — especially meat eaters[C] — need not spend large amounts of time feeding, while smaller ones — especially those that eat plant materials and have higher core temperatures — feed almost constantly unless they are asleep. Meat eaters consume food containing high *concentrations* of biochemical energy, while plant eaters' food is much less concentrated. Lions can rest and socialize a lot, but humming birds and shrews are always at the job of feeding.

The third principle involved with the thickness of the microscale boundary layer around the animal's body is contained in **Fig. 3-16**, where the cylinder represents the body. The larger the body, the thicker the boundary layer, the greater the boundary layer *resistance*, and the smaller the convective flow of heat to the environment at a given wind speed. This extra thickness is an advantage in a cold environment and the opposite when the environment is tropical.

Animals and the atmosphere: management of heat energy

As noted in earlier discussions, animal evolution in tropical environments has produced core temperatures that permit

---

(C) Another classification of animals is according to what they eat and their position in a food chain. Vegetable eaters are **herbivores**, meat eaters are **carnivores**, those at the end of the food chain are **top carnivores**, those that eat both plants and flesh are **omnivores**, and those that eat dead plant and animal materials are **reducers**. This organization and its energy balance for Earth are discussed in **Chapter 9**.

appropriately rapid biochemical reactions for fast, precise brain and muscle responses. We noted also that the same evolution has produced a close correlation of body size (and shape) with the resting metabolic rate, such as to match the core temperature with the requirement of thermal equilibrium:

(Heat inflow rate) = (Heat outflow rate).

In human beings we associate an environmental temperature in the range of 22-28 °C with *comfort*[D], and comfort, in turn, with the absence of any need to resort to thermoregulatory feedback mechanisms. Whether or not animals can be said to be "comfortable" is debatable. What is not debatable, however, is the fact that *energy used for short-term thermoregulation in an animal is energy that cannot be used instead for other purposes*. It pays in ways other than those pertaining to comfort, that is, for an animal to manage its energy balance so as to spend most of its time near $(T_e = 28 °C)$.

In simplest terms, the idea just introduced says that an animal has levels of priority for allocating its energy, roughly in this *order of decreasing priority:*

a) the basic physiological operations of life, including thermo-regulation;

b) maintenance and repair of tissues, muscles, and various organs:

c) feeding, including foraging, handling of food, the act of eating, and digestion; and

d) growth and devlopment, which includes all aspects of completing its life cycle.

Staying alive and healthy is the first priority; growing gets any leftover energy. In this sense, then, a healthy, well-fed animal whose environmental temperature stays near some particular optimal value, (T^*), has surplus energy to devote to getting on through its life cycle.

We looked into the way a human being manages his heat energy balance in **Chapter 5**, and that discussion gives us a basis for doing the same for other homeotherms. In particular, **Fig. 6-5** is the equivalent, for a four-legged animal (a **quadruped**), of parts of **Figs. 5-1** and **5-3**. In general terms, we see that a quadruped — represented by a cylinder with a horizontal axis — does *not* experience a midday dip in its radiant heat load the way a person (a **biped**) does.

(D) The lower end of this range is near the optimal thermostat setting *for clothed, western man*, mentioned in **Chapter 5**, while the upper end is near the tropical temperatures in which animal evolution took place.

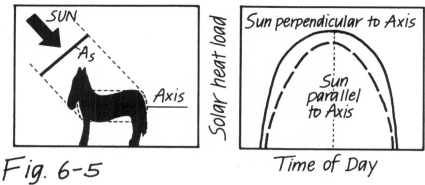

Fig. 6-5

In **Chapter 5** we presented the energy balance for a person in the basic framework of:

(Rate of Inflow) =
    (Rate of Outflow) + **(Rate of Change in Storage)**

noting that this word equation may be written, *for midday conditions*, as a mathematical equation like this $^{(P_{14})}$:

(Rate of Inflow)	$(1-a)(S_{in})(A_s) + (L_{in})(A) + (M)$
=	=
(Rate of Outflow)	$(L_{out} + E + H)(A) + (B)(A_c)$
+	+
(Rate of Change in Storage)	$(\pm \Delta)$

Then we classified each term in the equation like this:

Environmental variables	(S_{in}) and (L_{in})
System variables	$(1-a)$, (A_s), (A_c), and (A)
Environment/system variables	(E),(H), and (B)
Quasi-constant dependent variables	$(+M)$, (L_{out}) and $(\pm \Delta)$

As is often done by researchers who study the energy balance of various animals, we take the mathematical ideas for human beings and modify them in order to emphasize different ideas in considering the energy balance of animals. To begin, we will *ignore the heat flow to the ground or floor* beneath the animal — $(B)(A_c)$ — and *the possibility that heat stored in the body can change* — $(\pm \Delta)$. It's not that heat doesn't flow to the ground, or that heat stored cannot change in animals; it's that we want to pay attention to other problems.

Next we consider both shortwave and longwave radiant heat loads in one term, (R_{in}), so that we have:

(Rate of energy flow) = (R_{in})(A) + (M) = (L_{out} + E + H)(A)

The state of affairs after these modifications is shown in **Fig. 6-6**, except that the areas (**A**) are not shown directly.

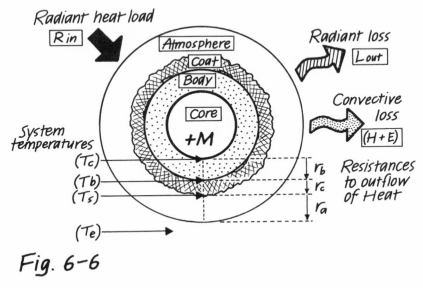

Fig. 6-6

Our last step in the modification[2] is to collect several energy streams into one: (i) the radiant heat load, (ii) the metabolic rate, and (iii) the evaporative cooling are all put into one term, called the **effective heat load**, (Q_{eff}). Since our animal has a haircoat, we can ignore cooling due to the evaporation of sweat on the surface. Thus (Q_{eff}) is, simply put, *the radiant heat load, plus the metabolic heat load, minus the relatively small energy loss due to the evaporation of water in the lungs during breathing.*

In **Fig. 6-6** we are reminded that heat flow outward from the core to the air is retarded by three kinds of *resistance*: that of the body mass itself (r_b), that of the haircoat (r_c), and, that of the microscale bounary layer of air (r_a). These three layers together constitute the body core's *insulation* from the thermal environment. As we saw in the last Chapter, when their sum is large, a large temperature difference can exist between the core and the environment: (T_c - T_e).

The order of *increasing size* for these three resistances is: air, body, and haircoat. As we saw in **Chapter 3**, any body has a microscale boundary layer whose thickness can be "fine tuned" by changes of posture. As we saw in **Chapter 5**, any animal body has cardiovascular resistances that can be "fine tuned" physiologically. The haircoat resistance, being usually the largest, is the major, thermoregulatory, "coarse tuning" control for animals living in non-tropical environments. These animals have exhibited growth of a haircoat, rather than changes of body shape and metabolic rate, as their response to living outside the tropics.

Fig. 6-7 shows the results of experimental measurements made on the coats of a wide variety of animals, expressed as resistances of the haircoat to conductive heat loss. The scatter of the data points expresses the variability of coats due to age, thickness and length[E], even within one species. Notice that a layer of still air always has a larger resistance than a layer of haircoat of equal thickness.

Summarizing the energy balance model we have come up with after modifications[2], when we ignore the areas for simplicity, it looks like this:

(Effective Heat Load) = (Radiant Loss) + (Convective Loss)

$$(R_{in} + M - E) = (Q_{eff}) = (L_{out}) + (H)$$

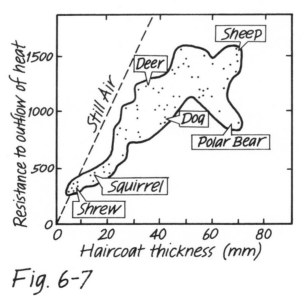

Fig. 6-7

(E) Although the haircoat is primarily a means for coarse tuning the resistance to heat loss, the animal's ability to **erect** the haircoat — increasing thickness and resistance — is a form of fine tuning. Hairless Man still has the vestigial ability to erect his haircoat. It appears when we have "goose bumps" upon being chilled.

Finally, we have come to the reason for all the modifications. We now have a view of the animal's energy balance in which:

Factor	includes information on . . .
(Q_{eff}) -------------------	sunshine, metabolic rate, and evaporative cooling;
(L_{out}) -------------------	the body's surface temperature (see the second radiation law);
(H) ---------------------	the body's size, haircoat, and core temperature, the body's surface temperature; the air temperature; and the wind speed.
Environmental variables ------------	sunshine, wind speed, and air temperature
System variables ----	body size, haircoat, core temperature, and metabolic rate
Environment/system variables ------------	evaporative cooling (depends on vapor pressure difference), convective heat loss rate (depends on the difference between the body's surface temperature and the air temperature)

Notice in the lists in the table that we have specified a distinction between the body's constant core temperature (T_c) — a requirement for the animal — and the body's surface temperature (T_s), which can vary according to circumstances. At the skin-haircoat interface In **Fig. 6-6**, the temperature is (T_b), and beyond the animal's surface is the atmospheric environment, with its equivalent temperature (T_e).

A moment's thought will tell you that *when the animal is in thermal equilibrium, equal rates of heat are flowing through all the resistance layers*. If these flows were not equal, thermal equilibrium (see **Chapter 2**) would not be maintained. With unchanging values for (T_c) and (T_e), the animal's skin and surface temperatures — (T_b) and (T_s), located between the core and the environment — must change so as to make the flows equal. In that sense, then (T_b) and (T_s) *are **dependent variables** representing the animal's thermal response to a changing thermal environment*. This line of thinking can be summarized in a familiar-looking relationship as follows:

**The rate of Outward heat flow
from the Core to the Environment** =

$$\frac{(T_c - T_b)}{r_b} = \frac{(T_b - T_s)}{r_c} = \frac{(T_s - T_e)}{r_a}$$

When we solve the mathematical equations of our model in different combinations, we gain insights such as those in **Figs. 6-8, 6-9,** and **6-10**. For example, if we let all the system variables have constant values in the equations, that is the same as saying "the animal doesn't change." We can then allow the environmental variables to change in various ways, as in **Fig. 6-8**.

In **Fig. 6-8a** we see, not surprisingly, that as the effective heat load $(Q_{eff})^{(F)}$ rises, with a constant wind speed and air temperature, so does the surface temperature.

Fig. 6-8b shows the effect of changing air temperatures on the same process: an unchanging animal system with an unchanging wind speed. The result is that, with a given heat load, the surface temperature (T_s) must rise higher — and so must (T_b) — to maintain the same temperature difference between surface and air, and thereby the same convective dissipation rate, (H).

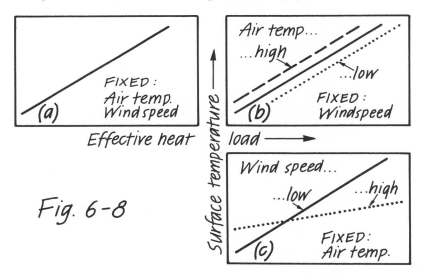

Fig. 6-8

Fig. 6-8c, on the other hand, shows that the animal's surface temperature *is less responsive to increasing sunshine when the wind speed increases*. That is, when the wind speed increases, a given increase in (Q_{eff}) yields a smaller increase in (T_s). It is so because increases in solar energy absorbed are more quickly blown away and do not go to increasing the surface temperature: (H) is larger.

(F) We know that by definition $(R_{in} + M - E) = (Q_{eff})$, so that when $(M-E)$ remains constant in a "constant animal", increasing (Q_{eff}) really means increasing sunshine (R_{in}).

What happens to (**H**) and (**T$_s$**) when the animal and, therefore, its microscale boundary layer, are larger? **Fig. 6-9a** gives the answer: because of the third effect of increased animal size, the boundary layer is thicker and the increased radiant heat load is dissipated less effectively. The system temperature rises more than for a smaller animal.

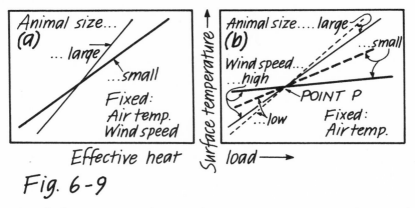

Fig. 6-9

Fig. 6-9b includes both variable wind speed and variable animal size, with one air temperature. It is, therefore, a combination of the information in **Figs. 6-8c** and **6-9a**. In two versions of the same idea, it says (a) a large animal in light winds is extremely responsive to any increase in sunshine, while (b) a small animal in high winds is scarcely responsive at all. **Fig. 6-9b** says also that (c) for a given heat load a smaller animal has a larger change in surface temperature for a given change in wind speed.

The point P in **Fig. 6-9b** is the point where the heat load makes the surface temperature just equal to the air temperature — part of (**T$_e$**) — so that the dissipation is entirely radiative, and there is no convective loss: **H** = 0.

These various statements about the results in **Figs. 6-8** and **6-9** are examples of the design insights obtainable from only one form of calculation with the mathematical model: the form in which the response of (**T$_s$**) to changing (**Q$_{eff}$**) is examined for different combinations of wind speed, air temperature, and body size.

Notice in passing that these results are from *a model that treats the animal as a static system* — no stress responses, for example, in either of the important controls of evaporative cooling or metabolic rate. In the next section we will consider a model that allows for changes in both of these controls.

Before leaving the model in which the animal is static, consider another form of calculation, in which the animal, *including its heat loss rate*, which is to say all the temperatures (T_c), (T_b), and (T_s), is held constant in the presence of varying heat load, air temperature, and wind speed.

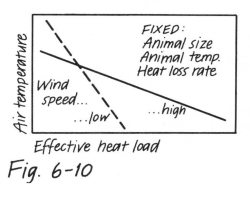

Effective heat load

Fig. 6-10

Fig. 6-10 gives one result. With the animal remaining in thermal equilibrium, with the same rates of heat dissipation by radiation (L_{out}) and convection (H), an increasing effective heat load (Q_{eff}) must be offset by a decreasing air temperature (only one part of T_e), the more so in lighter winds. Clearly, as by now intuition will tell you, stronger winds make a system like this less responsive to changes in its heat load, since the stronger wind makes for more efficient disposal of any additional heat load.

Animals and the atmosphere: physiological responses to temperature

Recall from the discussion of human comfort in **Chapter 5** that the *environmental temperature*, (T_e), is a combination of information on air temperature, humidity, and wind. It is not affected by solar radiation (sunshine) in its usual relationship to human comfort. Recall also, from earlier in this Chapter, that the *effective heat load*, $(Q_{eff}) = (R_{in} + M - E)$, does include information on sunshine.

In the previous section, when (Q_{eff}) varied, it was because the sunshine was changing. In this section we will hold the sunshine constant, so that when (Q_{eff}) varys it is because (M) and (E) are changing. These changes in (M) and (E) are *physiological* in nature, not simply temperature responses, as in the static model of the previous section.

Fig. 6-11 shows how an animal's metabolic and evaporative controls can be expected to respond to a changing thermal environment. The coordinates in both panels are (T_e) — strictly environmental — and (M) — partly physiological. The results shown accord with observed behavior, and any good mathematical model will produce similar results.

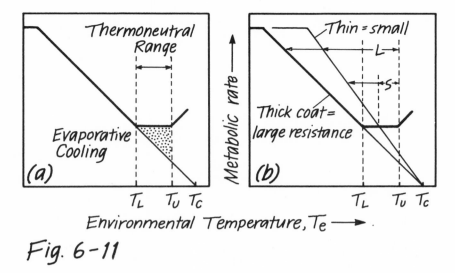

Fig. 6-11

Start at the upper left corner of **Fig. 6-11a** where the homeotherm is exposed to a cold environment, and its *metabolic rate is
maximal* — the largest physiologically possible for the animal —
while its *evaporative cooling rate is minimal* — from slow
breathing only. During the first slight warming, the animal's
response does not change: though warming, the environment is
still so cold it dissipates all the heat the animal can generate.

Soon the environment warms enough that the animal responds
physiologically with a slight reduction in its metabolic rate — the
heavy line begins to slope downward toward the right. This
response continues as the environment warms, because in

$$(Q_{eff}) = (R_{in} + M - E) = (L_{out}) + (H)$$

the animal matches a decreasing convective loss (**H**) — the difference (T_s-T_e) is becoming smaller — with a decreasing metabolic
rate (**M**). Sunshine (**R_{in}**) and water loss (**E**) are constant. At last, the
environmental temperature rises to the **lower critical tempera-
ture,** (**T_L**), *at which time the metabolic rate is at the smallest
value physiologically possible for this animal.* With still a need to
reduce (**Q_{eff}**) as (**H**) gets smaller, the animal responds physiologically with *increased evaporative cooling*[3].

In the **thermoneutral range** where the metabolic rate is constant, the environmental temperature, though now *very warm*, is
still below the body's core temperature, so that heat continues to
flow outward to the environment. When (**T_e**) warms to the **upper
critical temperature,** (**T_U**), the animal reaches its maximum physiological limit of evaporative cooling: it cannot deliver body water

for evaporation any faster. (T_e) is still less than (T_c), and heat loss, though now very small, is still outward to the environment.

The animal's physiological response to continued warming above (T_U) is an increased heart rate — the heart beats faster in an attempt to carry heat more rapidly from the body core to the surface — which shows up in an increasing metabolic rate, since the extra pumping requires extra energy. Eventually (T_e) rises above (T_c), and heat begins to flow *into the body core*. The animal loses all control of its thermoregulatory responses, and suffers hyperthermia, leading to death unless something changes.

Fig. 6-11b suggests the differences from the previous discussion when the animal's haircoat is thicker or thinner. Seen in the context of the familiar relationship

$$\textbf{Rate of Heat Loss} = \frac{(T_c - T_e)}{\textbf{Resistance}}$$

we note that a thicker haircoat — larger resistance — permits a larger temperature difference — colder (T_e) when (T_c) is constant — for a given rate of metabolic heat production. This is what appears in **Fig. 6-11b**. It also appears there that the animal with the thicker coat, other things being equal, has a wider thermoneutral range (L); and the better insulated animal can expose itself to a far colder environment before reaching its maximum rate of metabolic heat production. I trust the relationships of these ideas to design are clear, in particular the relationships between **Figs 5-11** and **6-11**.

This little table will give you some idea about the temperatures we have been discussing. Several things are clear at once. First, the cow's large body size and haircoat — both contribute to a large resistance — yield a wide thermoneutral range. Second, the bird has an elevated core temperature, needed for its greater precision of movement. Then, the bird's relatively wide thermoneutral range is made possible by the good insulative properties of a fluffable feather coat.

Animal	T_L		T_c	
	°C	°F	°C	°F
Bird	20	68	40	104
Cow	-20	-4	37	99
Man	32	90	37	99

For animals, as for buildings, there are various physical forms and principles available to the designer for, let us say, cooling.

Though cattle and horses have haircoats, still, like Man, they can sweat profusely over large fractions of their skin area, making (**E**) very large.

Birds, on the other hand, cannot sweat through dry feathers, so they resort to *panting*, and the passage of large volumes of air over the wet surfaces inside their lungs. Birds, when panting, can also pass even larger volumes of air through "dry sacks" inside their bodies, so that heat is lost without the loss of water. For birds, then, panting has two aspects: evaporative (**E**) and convective (**H**) cooling.

Jackrabbits, in desert heat, pump large volumes of warm blood through their huge, hairless ears so as to "dump" heat by convective loss, making (**H**) huge. Man, without the rabbit's haircoat, can dilate his near-surface capillaries in conjunction with an increased heart rate to increase blood flow and accomplish the same kind of dumping by an increase in (**H**).

Dogs pant rather than sweat, of course, but the cooling takes place mostly in the mouth rather than in the lungs. The variety of designs in thermoregulatory systems is wide amongst the species of Earth — too wide a variety to be discussed here, where all we can do is suggest the variety with a few examples[4].

With these ideas, you can probably watch animals around you, doing everyday things, and see their **behavioral thermoregulation** in action. Notice the places they rest, the things they do, and how they orient themselves to sun and wind to perform the coarse tuning for thermal equilibrium. Birds are especially good subjects for such watching, because they are visible and because, in selecting resting places, they have such a wide choice of combinations of sun and wind. Cattle and horses in fields are also good subjects for observing. With really only one choice of place to be, they resort instead to orientation of their cylindrical bodies to sun and wind. Finally, people at leisure in urban settings are often very revealing of the principles described here.

Animals and the atmosphere: climate space

David Gates has used his mathematical and physical treatment of the animal energy balance to develop the concept of **climate space**, an idea that addresses not so much *what* an animal can do in response to thermal stress, but rather *where* — spatially — he can live without experiencing undue stress. In the treatment of climate space, the spatial dimensions are not "geographical", but "microclimatic" — sunshine, temperature, and wind. The translation of climatic space into geographical space, of course, is accomplished by knowing what microclimates are where on the landscape, the subject of **Chapters 3** and **4**.

Fig. 6-12a sets the stage for our consideration of climate space. The coordinates are $(Q_{eff}) = (R_{in} + M - E)$ — in this case $(M-E)$ doesn't change — and (T_{air}), which is one of the components of (T_e). Begin with the line called "cave or nest." In a cave, nest, or burrow, an animal is surrounded on all sides by walls and air of uniform temperature, and the radiant heat load increases as the wall temperature increases[G]. This curve describes that environment.

On a clear night, the animal outside under the sky would be exposed to a relatively warm hemisphere below — the ground — and a relatively cold hemisphere above — the sky. Hence, for a given air temperature, its radiant heat load would be less than when it was in the cave. The second curve shows that environment.

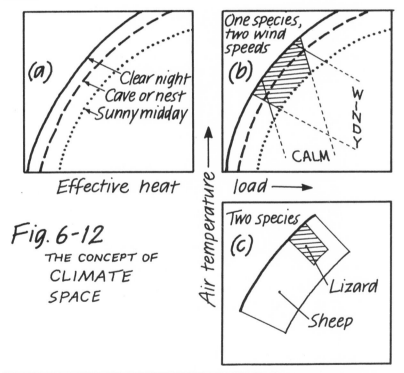

Fig. 6-12

THE CONCEPT OF

CLIMATE

SPACE

(a) Effective heat — Clear night — Cave or nest — Sunny midday

(b) One species, two wind speeds — WINDY — CALM

(c) Two species — Lizard — Sheep

Air temperature → / load →

(G) The line is curved because, as you recall from Footnote C of **Chapter 4**, the radiation from a black body increases as the 4th power of the temperature. The line would be straight if the increase were according to the *first* power.

On a sunny day, the animal outside under the sky would be exposed to not only the longwave radiation encountered at night and in a cave, but also the solar radiation in sunshine. The amount of sunshine absorbed would depend on the color of its coat, so the (Q_{eff}) for a given air temperature would be slightly greater for dark fur, up to a maximum for a completely black haircoat. This environment, with a "medium" fur color, is shown by the last curve in **Fig. 6-12a**.

Atop the information in **Fig. 6-12a** we add the information in **Fig. 6-10** to get **Fig. 6-12b**. The sketch shows two pairs of crossing lines, one at higher temperatures and one at lower temperatures. In each pair, higher wind speed is associated with a more horizontal line, in conjunction with a certain value of (**M-E**) and thermal equilibrium, as discussed in conjunction with **Fig. 6-10**. *The pair of crossing lines at higher temperatures is associated, in the mathematical model, with the smallest value of the animal's variable metabolic rate*, and those at lower temperatures with the largest rate.

As a general rule, higher wind speeds are more stressful in cold environments because they increase (**H**), and calm air is more stressful in hot environments when a larger (**H**) is needed for equilibrium. Following this rule, I have shown the limits to the animal's climate space — the shaded area in **Fig. 6-12b** — associated with higher speeds at lower air temperatrures, and vice versa. These limits are set by the animal and its physiological characteristics, such as its size and range of possible values for (**M-E**). The other limits on the shaded area are those set by the environment: the clear night and the sunny midday limits.

Fig. 6-12c compares the climate spaces of two kinds of animal: the lizard with narrow tolerances, and the sheep with wide tolerances in the microclimates it can inhabit without thermal stress.

Animals and the atmosphere: optima and limits of tolerance

We have been discussing *short-term direct effects* of the changing atmospheric environment on animals, in which the animal responds in a way to keep its energy balance in a state of equilibrium.

As noted in the discussion of **Fig. 6-11**, there is a temperature near (T_L), at which the metabolic rate is minimal and no resort to evaporative cooling is required to maintain a constant (T_c). There, in other words, the animal's energy requirement for thermoregulation is minimal. In addition, as has been noted already, evolution has determined that (T_L) for most animals has a value slightly below the core temperature, which in turn permits biochemical reactions to proceed rapidly.

Fig. 6-13a shows this idea in the form of a **biological response** to environmental temperature. The biological response may be, for purposes of this discussion, any one of many things — such as growth rate, meat production, or motor ability — depending on the subject under consideration. The main point of the figure is that there is an **optimum environmental temperature**, (T^*), surely related to minimal thermoregulation, at which the biological response is at a maximum.

At the **lower and upper threshold environmental temperatures**, (T_t), all available energy is used for immediate needs only, and none is available for the biological response under discussion[H]. Between these threshold temperatures, the biological response increases from zero to the maximum possible value.

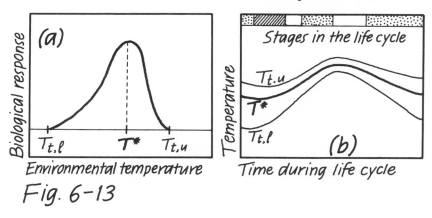

Fig. 6-13

Clearly, the value of the optimum environmental temperature, (T^*), includes the results of minimal thermoregulation and rapid, efficient biochemical reactions. **Fig. 6-13b** suggests that the animal may not necessarily have the same optimum temperature throughout its life cycle, nor that it be equally sensitive, at all stages of that cycle, to departures from optimum. There, the graph shows the relationships of the threshold and optimum temperatures to the age of the animal, more specifically to the *stage of development* of the animal. At midlife, for the hypothetical animal pictured, the

---

(H) These *threshold temperatures* are not to be confused with the *critical temperatures* in **Fig. 6-11**, which are much nearer to the *optimum temperature*. Furthermore, there is another pair of temperatures, farther from the optimum, at and beyond which life ceases and death occurs. These are not shown, but they are called the **lower and upper lethal threshold temperatures**.

value of (**T***) is higher, and the **range of tolerance** is narrower than before or after.

In fact, **Fig. 6-13b** depicts, mainly for illustration of the point, an *extreme, hypothetical case* of the variability of the value of (**T***) and the range of tolerance through a life cycle. Real animals, especialy larger ones, show much less variability. As we shall see presently, however, **Fig. 6-13b** is quite appropriate for the same relationships in the life of a typical plant.

We have noted that the rates of biochemical reactions involved in various biological responses play a role in determining the value of the optimum temperature, (**T***). **Fig. 6-14** suggests that these rates, at any one time, may depend on earlier events in the animal's life. This carry-over of past effects into the present is called **thermal preconditioning**.

Fig. 6-14a, based on actual experimental data, refers to the hatching of the eggs of an insect species that "requires" a chill period during incubation; that is, *the hatching success is always better with than without the chill.* Said another way, the absence of the chill period creates an extremely adverse thermal preconditioning.

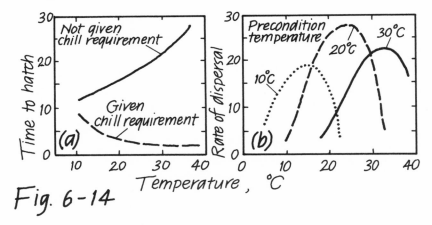

Fig. 6-14

Fig. 6-14b, also based on actual experimental data, refers to the maximum speed with which young fish can swim[5]. That speed depends not only on the temperature of the water at the time of observation, but also on the temperature of the water in which the young fry were held for a period prior to the observation: the *precondition temperature. Clearly what is optimum at any moment can depend strongly on prior events in the life of an organism*, whether the biological response is hatching success or motor ability.

Animals and the atmosphere:
growth and development

Moving from immediate, short-term responses of animals to the atmospheric environment into discussion of responses over longer, connected time periods takes us to discussion of *long-term, direct effects*. The idea of the preconditioning of optima and limits suggests how complex can be the matrix of atmospheric effects on an entire life cycle. **Fig. 6-15** provides us with a schematic structure to the concept of **life cycle** for purposes of discussion.

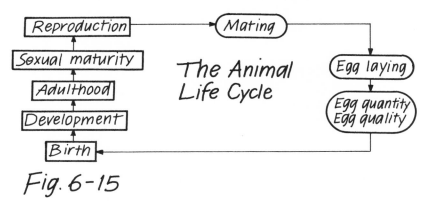

Fig. 6-15

Not only does the atmospheric environment affect the outcome of each stage in the cycle, but different combinations of atmospheric factors are important at each stage. In addition to that, at each stage there are many internal processes at work in the animal. We may simplify that idea, for example, by separating the idea of **growth** from the idea of **development** even though the two are usually taking place together — on parallel tracks, so to speak.

We speak of **growth** as referring to *increases in physical dimensions and body mass*, measured in various ways: height, weight, and so on. We speak of **development** as referring to *the sequence of passages from one stage to the next stage in the life cycle*. The two are not completely correlated; for example, an animal may stop growing early in life yet still develop to the point of sexual maturity. Similarly, an animal may put on weight and yet, if conditions are not right, cease development.

In natural environments, growth and development usually go hand in hand; otherwise natural reproduction would have ceased long ago. The fact that they can be separated artificially — experimentally — shows they are indeed separate and adds to our understanding of the true complexity underlying "everyday" events and sequences we observe in nature.

Fig. 6-16 serves to illustrate the points that (a) often, several of what we think are separate atmospheric variables[1] seem to determine maxima and limits, and (b) these atmospheric effects are at work in each stage of the life cycle. The results in the figure relate temperature and humidity — in the form of the TRe diagram from **Chapter 4** — with *biological responses* in the life cycle of a species of locust[6]. The biological responses here are matters both of *time* — number of days to pass through a **developmental sub-period**, or stage — and of *quantity* — number of reproductive units. Other data would show that biological responses in the life cycle are also matters of the *quality* — viability — of reproductive units.

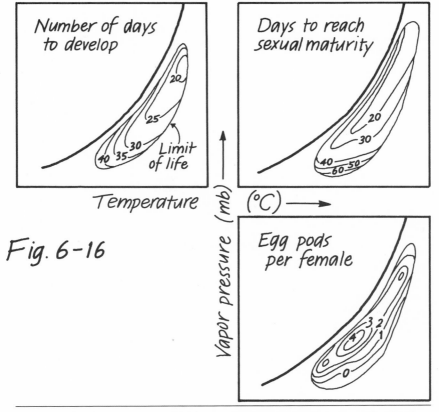

Fig. 6-16

(I) If fact, data from a variety of insect species suggest that the true optima in the egg and pupal stages are best expressed as values of a single variable: the wet bulb depression $(T - T_w)$, described in **Chapter 4**. That is, as discussed in Lowry (1969), the insect eggs and pupae "do best" when $(T - T_w)$ has a certain value, regardless (under naturally occuring conditions) of the combination of temperature and relative humidity producing it.

As suggested earlier, it is the smaller animals that are most sensitive to — most strongly coupled to — the vagaries of the atmospheric environment — the flows of heat and moisture in the short term. The kind of variability shown in these discussions of experimental results is most pronounced in animal life forms like insects and fishes — the smaller forms so strongly coupled to the thermal environments of air and water. This sensitivity is transferred to the lives of larger animals, of course, because the smaller are the food for the larger.

We have already, in very general terms, compared plants with animals in terms of their interactions with the atmospheric environment. We made note, among other things, of the fact that plants, like very small animals, are seldom at temperatures very far from the environmental temperature. It is that similarity of very small animals, insects, and plants that permits the application of an idea called the **Heat Units Concept** to all of these forms, both animal and plant.

The Heat Units (HU) Concept

We have considered, at some length, the Degree Day (**DD**) Concept in connection with **Fig. 5-9**. The concept says, in brief, that some biological response or other acts in proportion to the departure of the daily average temperature from some specified threshold temperature, (T_t). In the case of the heating or cooling of human habitations, the "biological response" is really the rate of consumption of energy for heating or cooling.

Extending that same idea to the biological responses of organisms — especially to the various aspects of development — is an extremely important conceptual tool for studying long-term relationships between smaller organisms and the atmospheric environment. In this more obviously biological context, Degree Days are more frequently known as **Heat Units** (**HU**), or **Growing Degree Days (GDD)** so that, generally speaking (**GDD** = **HU**) when we speak of the development of smaller organisms. Let's see how that works.

In connection with small animals and insects we spoke about stages in a life cycle and about the difference between growth and development. The following discussion applies as well to a great many plants. Suppose we did *a series of experiments connecting the temperature of the environment with the length of time a particular kind of organism takes to pass through a particular stage of its life cycle.* In the case of an insect, for example, this might be the time from egg laying to egg hatching. In the case of a flowering plant it might be the time from the first swelling of leaf buds in the spring until the appearance of the first flowers.

Common experience tells us the time is longer for cooler temperatures. In fact, in too-cool temperatures, the time would be "infinite" because the temperature would be below the threshold for development — the lower T_t in **Fig. 6-13a**. **Fig. 6-17a** shows typical results from such experiments.

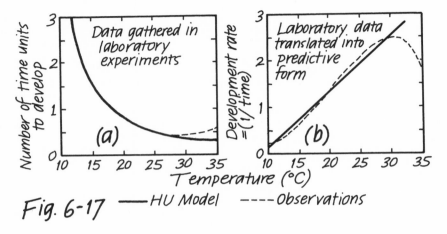

Fig. 6-17 ——HU Model ----Observations

The typical test data would show that, at low temperatures, the number of time units to develop would become very great, as just suggested. At high temperatures, however, these typical data would show *a minimum of development time at some intermediate temperature*, and that above that temperature the time would begin to increase again. The intermediate temperature, the optimum temperature (T^*), is near the value of 28 °C, mentioned so frequently in the last Chapter in connection with evolution in tropical conditions and human comfort. As we will see in a moment, the mathematical model that describes the HU Concept — the **HU Model** — makes the assumption — only a small departure from reality, as it turns out — that development time continues to decrease at higher temperatures.

In going from **Fig. 6-17a** to **Fig. 6-17b** we make a simple, but important, change in the way we express the biological response. The change consists of *dividing each development time (t) into one*, so that instead of plotting (t) against temperature, we plot (1/t) against temperature. In doing this we have converted "*the length of time an organism takes to pass through a particular stage of its life cycle*" into "*the **time rate** at which an organism passes through a particular stage of its life cycle*" — its **rate of development**. We cannot make direct experimental observations — in either the laboratory or in the field — of this rate. We must obtain it from observations of the time (t), as just described.

Now that we have plotted **Fig. 6-17b** we can see that it looks like — and means the same as — **Fig. 6-13a**. More particularly, the **HU Model** assumes that the truth lies in the ascending straight line and its departure from the actual plotted data, as in **Fig. 6-17a**. So what? What is all this manipulation about, you ask? Here's the answer.

The HU Model in **Fig. 6-17b** says that *the rate of development is proportional to the amount by which the environmental temperature rises above the threshold temperature.* Notice the statement implies that the rate of development is zero (it can never be negative) at any temperature below the threshold. In the particular case of **Fig. 6-17b** the threshold temperature is where the straight line crosses the temperature scale: approximately $8\,°C$ = $46.4\,°F$. This tells us, among other things, how we can convert observations of time (**t**) into values of threshold temperature (T_t).

A corollary, or implication, of the HU Model is much more important, for ecological purposes, than the statement we have just made. I will set it out, in bold type, by itself:

The number of heat units (HU) accumulated during any developmental stage, and therefore for an entire growth season, is constant. That constant applies from one year to another for one species or variety, or from place to place for that species or variety.

As a matter of simple definition, the constant value of accumulated heat units for a species or variety is called, naturally enough, its **species or varietal constant**. The implication of the Concept, you see, is that it is exclusively the *temperature* of the environment that governs development — as if there were some "master reaction" distilled from all the dozens of biochemical reactions going on inside the organism. We have already noted that, often, other atmospheric variables modify the temperature effect, but the Concept is admittedly only an approximation of reality.

I hope you can see, without further explanation, that **Fig. 6-18a** is the (**GDD**) equivalent of **Fig. 5-9** for (**HDD**) and (**CDD**). Here I have divided the time of one hypothetical season into stages in the life cycle of a particular organism, and have assumed for simplicity that the environmental temperature has only one, constant value during each stage. I hope you can see, also, that **Fig. 6-18b** is a plot of the accumulation Σ(**GDD**) with time through a growing season.

According to the HU Concept, as stated above, one species or variety of small organism — animal or plant — would develop the

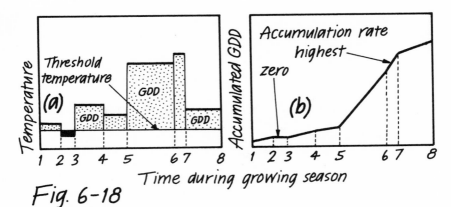

Fig. 6-18

Time during growing season

same amount between "Time 1" and "Time 8" in both Year#1 and Year#2, in **Fig. 6-19**, even though the weather sequences were quite different in the two years. In fact, one species or variety would develop the same amount between "Time 1" and "Time 5.8" (where the curves cross) in both Year#1 and Year#2. Careful examination will tell you, in passing, that Year#2 is the same as that in **Fig. 6-18**.

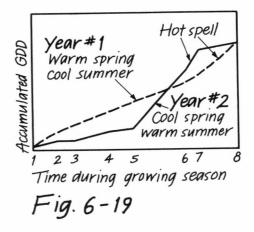

Time during growing season

Fig. 6-19

Having used the discussion of the HU Concept as a bridge between animals and plants, we can now change our attention to more detailed consideration of plants. Our later considerations of the application of the HU Concept in study of plant development and plant geography will apply equally well to very small animals and insects.

The varied roles of plants in atmospheric ecology

In terms of atmospheric ecology, plants may be thought of in several roles, all interconnected; for example:

(a) plants *change the albedo*, and other emissive characteristics of the surface, altering the surface's response to solar energy (see the *Gaia Hypothesis* in **Chapter 9**);

(b) plants represent *roughness elements*, altering the responses of the wind to a surface;

(c) plants represent *elements in a canopy*, contributing to the matrix of sources and sinks of heat energy and mass — water vapor, carbon dioxide, and other gases — within the transition layer between soil and air;

(d) plants act as *water pumps*, transferring water from the liquid state in deeper soil layers to the vapor state in air, even when the shallow soil is itself too dry to support surface evaporation;

(e) plants are the **primary producers** in the food chain, *capturing sunlight and converting it to organic materials* in concert with the atmospheric gases and minerals from the soil;

(f) plants provide the *food for certain animals* — herbivores — also known as **secondary producers**, and thus for the entire food chain, as will be discussed in more detail in **Chapter 9**; and

(g) plants provide means for the *architectural responses of most animals* to the atmosphere, as living canopy elements, and, when dead, as nesting materials.

Clearly, some of these roles are *active*, as when plants are acting as water pumps and as converters of solar energy to biological materials. The other roles are largely *passive*. In the rest of this Chapter we will give most attention to their active roles, concentrating on the physiology of *photosynthesis* (PSN), *respiration* (RESP), and *transpiration* (TRP). In addition, we will take a look, as we did with animals, at the effects of the atmosphere on their life cycles as expressed in their stages of *growth and development*.

Plant anatomy

Anatomically and bioclimatically, most plants — even most aquatic species — may be conceived of as having *sugar factories located in their leaves*, with the raw materials brought directly to the leaves from the atmosphere, and brought indirectly into the leaves from the soil by way of roots, trunks, and stems. Water is one of the raw materials, brought as a liquid to the leaves, mainly from the soil; but relatively small amounts of the water are used in the factories compared with the amounts brought in. The great majority is pumped on through the leaves, as a vapor, to the air.

The byproducts of the sugar manufacturing are either used within the factory or transferred to the plant's environment, mostly the atmosphere. The energy to drive the pumping and the manufacturing is supplied by the sun, as radiant energy, and by the surrounding physical environment, as heat energy.

Fig. 6-20 is a sketch of the principal structural elements in the leaves — the sugar factories and the water pumps. Within a waxy, waterproof **epidermis** a complex structure of **mesophyll cells** presents wet surfaces to air-filled chambers and passageways, *saturating the air*. Like an animal's lungs, the mesophyll cells are so structured that a huge amount of wet surface area, required for efficient gas exchange, is packed into a very small volume.

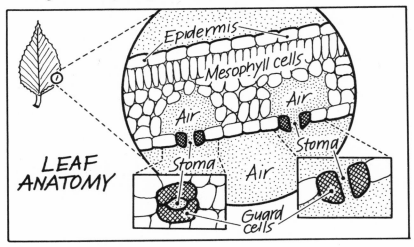

Fig. 6-20

The saturated air, in turn, is connected to the air outside the leaf through ports of variable cross-section, called **stomata**[J]. Each stoma's cross-sectional area is controlled by a pair of **guard cells**, which are like valves controlled, in turn, by processes linked to the atmospheric environment.

Water is brought from the soil to the leaf as a liquid, through roots, stems, and trunks. Some water molecules are used as raw materials for PSN within the leaves, but most evaporate at the wet surfaces of the mesophyll cells and move slowly outward, through the stomata, to the passing windstream in the process of transpiration. On balance, therefore, *water moves outward* from the leaf to the atmosphere through the stomata.

On balance, another gas — *carbon dioxide* — *moves inward* from the atmosphere to the leaf through the stomata. Still another gas — *oxygen* — *moves outward* from the leaf to the atmosphere through the stomata. Thus, the stomata provide the route by which most gaseous mass is exchanged between the leaf and the environment — some going in one direction and some going the other at the same time.

(J) The singular of *stomata* is *stoma*, though many writers insist on using the anglicized forms *stomate* and *stomates*.

As we will see presently, these three gases are involved in the sugar manufacturing processes of PSN; and the *feedback systems in the guard cells* operate so that the stomata are well open when conditions are right for rapid PSN, and closed when they are not.

As suggested in **Fig. 6-20**, a guard cell is mostly expandable, but has a non-stretch line of material along the side facing the other guard cell. When the cells fill with liquid water, they expand in such a way that the opening between the non-stretch lines *increases in cross-sectional area without changing its length*, so that the stoma is quasi-elliptical in shape. When the guard cells are empty of water, the stoma closes.

Stomata are typically 20 micrometers (μm — millionths of a meter) long and up to 10 μm wide at full opening. There are thousands of stomata per square centimeter of a leaf's surface, almost exclusively on the underside in most species.

Plants and the atmosphere: physiology

Photosynthesis (PSN) is the biochemical process in which light energy is absorbed by the chlorophyll (CHL) of green leaves and converted to sugar molecules, which are then transformed into a great variety of plant materials.

Respiration (RESP) is the biochemical process in which living plant materials are consumed within the leaf to obtain the energy for the plant's processes of maintenance, repair, growth, and development[K]. The same gases that are used and discarded in PSN are used and discarded in RESP — *the inputs to one process are the outputs from the other.* Both take place in the mesophyll cells, to which and from which gases flow through the thousands of air passages inside the stomata.

While PSN takes place only in the presence of light energy, RESP takes place all the time. Only when the mass of plant materials formed by PSN during a period of time (such as a day) exceeds the mass of plant materials consumed in RESP during the same time — only then will there be a net increase in materials appearing in growth and development. We will examine the details of these processes now.

Fig. 6-21 shows a flow diagram with which to follow the discussion of PSN and RESP. Photosynthesis, for purposes of this dis-

---

(K) This process of consumption is called "oxidation", and it is, chemically speaking, the same as slow burning, or combustion, such as would take place rapidly in a bonfire. Though our attention here is on leaves, RESP occurs everywhere in a plant, including the roots. It is *not to be confused with animal breathing*, for which the medical sciences use the same term: respiration.

cussion, is best separated into the **light reaction** and the **dark reaction**[P.16].

THE LIGHT REACTION :

$$H_2O + Carrier \xrightarrow[CHL]{Light} O_2 + Carrier{+}H^+$$

THE DARK REACTION :

$$Carrier{+}H^+ + CO_2 \xrightarrow{Heat} (CH_2O)_n + Carrier$$

These two equations are, of course, great simplifications, for purposes of the following explanations, of dozens of intermediate reactions that take place in the mesophyll cells. Among the simplifications is the fact that the dark reaction is *not balanced* as an equation should be. Inclusion of the subscript "n" — which puts the equation out of balance — is meant to show that sugar molecules are made up, generally speaking, of different numbers of chemical bits, each containing the group (CH_2O). For example, $(CH_2O)_6 = (C_6H_{12}O_6)$ is hexose sugar.

The *light reaction* uses light energy to split water molecules, attaching the hydrogen ions to carrier molecules, and releasing the oxygen in its molecular form. *The reaction takes place at a rate proportional to the intensity of light, and it is not sensitive to temperature.*

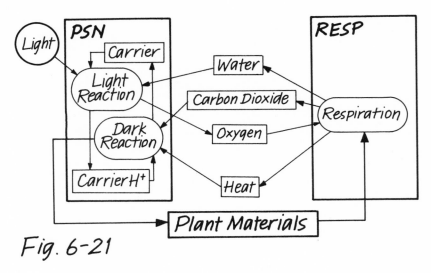

Fig. 6-21

The *dark reaction* uses heat energy to take **reduced carriers** from the light reaction, remove the hydrogen ions, attach them to carbon dioxide in such a way as to make sugar molecules — the building blocks of all plant materials — and send the **oxidized carriers** on their way back to the light reaction. *The reaction takes place at a rate proportional to the temperature (concentration of heat), and it is not sensitive to light.*

Combining the light and dark reactions of PSN as if they were one master reaction gives an equation which, paired with one for RESP, shows more directly the symmetry between them: the inputs to one are the outputs from the other.

PHOTOSYNTHESIS :

$$H_2O + CO_2 \xrightarrow[CHL]{Light + Heat} (CH_2O) + O_2$$

RESPIRATION :

$$(CH_2O) + O_2 \xrightarrow{Heat} H_2O + CO_2$$

This equation for RESP, too, is a great simplification for purposes of these discussions. For example, there are many kinds of carbohydrate materials other than the sugar group in the equation that serve as the fuel for RESP.

Now it's time to show how division of PSN into the light and the dark reactions can help explain the responses of PSN to the atmospheric environment. **Fig. 6-22** presents an important set of results from experiments that have been performed many times. They show that:

(a) the rate of PSN increases as the light intensity increases, up to a certain level called **light saturation**, and then it doesn't increase any more;

(b) at high light intensities, the response is different depending on the temperature and the concentration of carbon dioxide in the leaf's environment; and

(c) at low light intensities the temperature and the concentration of carbon dioxide make no difference.

To understand how these results are connected to the light and dark reactions, consider an experiment starting at the lower left corner of **Fig. 6-22**, where light intensity and PSN rate are both zero. The temperature and the CO_2 concentrations are held constant, let's say at the level called "low" in the figure.

As more light shines on the leaf, the rate of PSN goes up. In terms of the two reactions, this means that more light energy splits more water molecules, producing more reduced carrier molecules, so that the dark reaction has more hydrogen ions to attach to the plenty of CO_2 molecules around. There is plenty of water, heat, and CO_2 , so that PSN procedes at a rate as fast as the light energy permits. PSN is **light limited**.

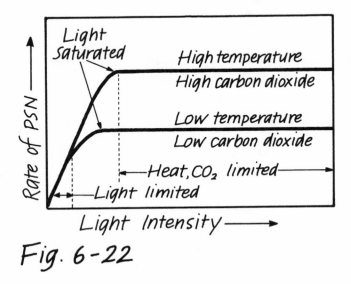

Fig. 6-22

As still more light shines on the leaf, the rate of PSN stops going up. In terms of the two reactions, this means that something besides the light is now limiting. It is now either the heat or the CO_2 that limits the rate of PSN. In strong light, a lot of reduced carriers are being sent from the light reaction, whose rate is *not* affected by temperature, to the dark reaction, whose rate *is* affected by temperature. *If the temperature is limiting*, the dark reaction cannot proceed at a rate faster than the temperature permits, no matter how fast the light reaction is sending reduced carriers.

If the CO_2 is limiting, the dark reaction cannot proceed at a rate faster than the number of available CO_2 molecules permits, no matter how fast the light reaction is sending reduced carriers. Thus, *there are three atmospheric factors capable of controlling, by their intensity or concentration, the rate of PSN: light, temperature, and CO_2*. The rate of PSN is determined ultimately by the rates at which the reduced and the oxidized carriers are shuttled back and forth between the light and dark reactions.

We've seen how PSN, in one sun-lit leaf, responds to the atmospheric environment. How about a whole plant? **Fig. 6-23a** suggests the answer. While a sun-lit leaf at the outer part of the plant canopy may become light saturated (that is, heat or CO_2 limited) toward midday, only then is the sun shining brightly enough to reach the leaves of the *inner canopy*. The result is that the whole canopy — the sum of all the leaves — continues to experience an increase in PSN as midday arrives.

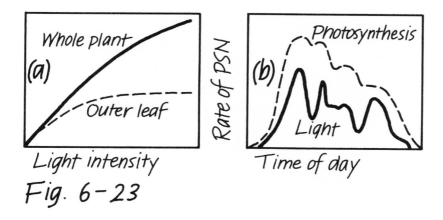

Fig. 6-23

Fig. 6-23b shows the usually observed response of a whole plant to changing sunlight: PSN rises and falls as the sunlight rises and falls, in this case because of partial cloudiness.

As we noted earlier, it is not the amount of plant materials formed by PSN that is important for plant growth and development. It is the *difference between PSN* — which forms the materials — *and RESP* —which uses up part of the materials. That relationship is given formally by

Net PSN = Total PSN - RESP

in which we have defined **Net PSN** and **Total PSN**. If (RESP) exceeds (Total PSN), then there is a net loss of plant materials. Since RESP goes on both day and night, while PSN goes on only in the light of day, a leaf could easily use up — in a hot spell — all of the new tissue formed the previous day, with a zero net gain. The plant will stay alive, but it will not grow or develop. If that condition continues for very long, the plant will die. *Only when Total PSN is substantial, added up over a considerable period of time* — such as a growing season — *will the plant reach maturity.*

We have seen the relationships of Total PSN to light, temperature, and CO_2 in **Fig. 6-22**. What about these relationships for Net PSN, the true measure of growth and development?

Knowing that RESP is a biochemical process that is sensitive to temperature but not to light, we should not be surprised to find out that RESP continues to increase as the temperature increases, as we would expect in any "combustion" process. With that in mind, we can see, in **Fig. 6-24a**, what happens.

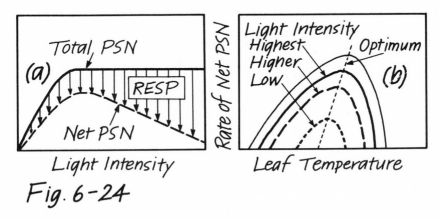

Fig. 6-24

As the light intensity increases on the leaf, the leaf warms up — usually above air temperature. As the leaf warms, RESP increases. Subtracting the ever-increasing value of RESP from Total PSN, as the light intensity increases, yields a curve for Net PSN that reaches a peak at some intermediate value of light intensity: the *optimum light level* for the air temperature at which the PSN is taking place.

There is an optimum light level for each temperature, and an optimum temperature for each light level. **Fig. 6-24b** shows a more frequently encountered presentation of these relationships among light, temperature, and Net PSN. It describes more nearly the actual responses of a plant *under field conditions*, while **Fig. 6-24a** describes the results of a set of laboratory experiments. The story that **Fig. 6-24b** tells is that too much light and heat over-balances PSN with RESP, yielding no net progress for the plant through its life cycle. In fact, this overbalancing appears under quite ordinary circumstances, as shown in **Fig. 6-25**.

In that figure, I have imposed on **Fig. 6-24b** a typical daily record of temperature and light level, the numbers for the hours being given in the sketch. At each hour there is a rate of PSN corresponding to the combination of temperature and light. The highest temperature and light levels shown appear at 1400 hours,

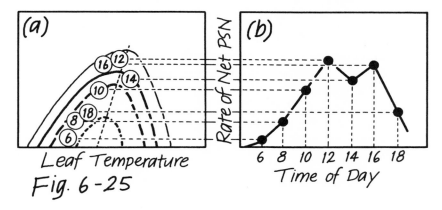

Fig. 6-25

but, as shown in **Fig. 6-25b**, there is a dip in the rate of PSN at that hour. This dip is a common occurrence in the lives of plants, and it is because at that hot, bright time of day RESP slightly reduces Net PSN.

Having considered PSN and RESP, we need now to consider the process of transpiration (TRP), that is so central to the plant's water economy, and that is so responsive to atmospheric variability. The key to understanding TRP is knowing how the feedback system, involving the variable stomata, works to control the rate of flow of water from the soil, roots, and stems, through the leaves, and out to the atmosphere.

As suggested in **Fig. 6-26**, light is the primary environmental control over the opening and closing of the stomata. In turn, this opening and closing is best expressed as a decreasing and increasing of **stomatal resistance** to vapor flow, as suggested by the familiar-looking relationship

$$\text{Rate of TRP} = \frac{(e_m - e_a)}{\text{Stomatal Resistance}}$$

where the difference in the numerator is no longer between two temperatures, but is now between *the vapor pressure in the mesophyll* (e_m) *and that in the airstream outside the leaf* (e_a).

As we have noted already, when liquid water flows into the guard cells, they swell in such a way that their stoma — the gap between them — gets wider. The leaf's biochemistry works this way: first thing on a sunny morning, increasing light causes PSN to begin and the carbon dioxide within the mesophyll to be used up. The stomata are closed from the previous night, so replacement CO_2 cannot flow in, and PSN slows down. The reduced concentration of CO_2 in the mesophyll causes water to flow into the guard cells, the stomata to open, and a fresh supply of CO_2 to flow in through the stomata. PSN proceeds as *positive feedback, governed by the CO_2 concentration, enhances the rate of PSN.*

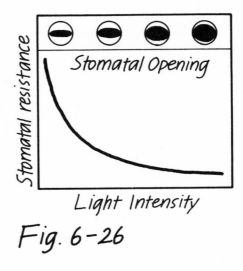

Fig. 6-26

At night, when PSN ceases, RESP again increases the CO_2 concentration in the mesophyll, and the guard cells empty of water, closing the stomata. The cycle is completed. The key to this positive feedback system, you see, is the concentration of CO_2 within the mesophyll, which governs the flow of liquid water into and out of the guard cells.

The stomatal resistance governs the flow rate of CO_2 into the leaf, and of water vapor out from the leaf as well. It governs the flow rate of oxygen outward to the atmosphere, though most of the O_2 produced by PSN is used rather soon by RESP, which goes on even when PSN does not, without leaving the mesophyll.

There is also *a negative feedback loop controlling the stomatal resistance.* If PSN proceeds rapidly under dry conditions — in air, soil or both — water vapor can flow out the wide-open stomata — rapid TRP — faster than the roots and stems can replace it in the mesophyll cells. Dessication of the mesophyll makes liquid water flow out of the guard cells into the mesophyll cells, closing the stomata, and reducing the vapor flow of TRP.

It seems that, although the complex feedback system governing stomatal resistance uses water and controls the rate of TRP, its main purpose — under ordinary circumstances — is to govern the flow of CO_2. So what is the purpose of TRP — the flow of water?

Why does so much more water flow through the leaves than is used there for PSN?

The principal role of water flow in a plant is to carry nutrients to, and metabolic products from, various locations. This flow of **chemical translocation** extends up from the roots and down from the leaves, through the plant's **vascular system**. In addition to translocation, this flow provides the **turgor pressure** within the vascular system that helps hold the plant parts erect to receive sunlight. Also, water must be constantly available to wet the surfaces of the mesophyll cells, so that carbon dioxide can be dissolved in it, thereby being taken from the air into the cells more efficiently. As you can see, water's flow to the mesophyll for splitting in PSN is only one of its jobs. Most of the water flowing in a plant is used for other things: to fill up the plumbing and circulation system, so to speak.

The anatomy inside the leaf, as we have noted, packs a lot of wet surface in a small volume, primarily to permit large masses of carbon dioxide to be absorbed for use in the dark reaction of PSN. As a side effect of having this structure, so efficient for gas exchange involving CO_2, water has a lot of surface from which to evaporate within the mesophyll.

While the evaporation of water cools leaves slightly, TRP must be viewed mainly as a "necessary evil" for having efficient gas exchange in the mesophyll. As such, it makes a supply of external water necessary for proper functioning of the plant. In many desert plants, as in many desert animals, physiological adaptations make it possible for water used in metabolic processes to be recycled to reduce the amount of external water necessary. Also in desert species, anatomical and physiological adaptations permanently increase the resistances to outward vapor flow, reducing the amount of external water necessary.

We have seen how TRP responds to light, as the stomata open and the stomatal resistance decreases. What about the response of TRP to other atmospheric variables? As with animals, the best way to answer such a question, in general terms, is by means of a mathematical model of the leaf's energy balance.

From **Chapter 4** recall that the radiation budget and the energy balance are shown in **Fig. 4-40**. It is clear from that figure that there are radiant exchanges and convective exchanges with the atmospheric environment, but no conductive exchange with the ground and no metabolic component to the heat load. In a manner similar to what we did for the discussion of the responses of animals to a changing atmospheric environment, we will modify the working equation of Footnote Q in **Chapter 4** without changing its meaning[P17]. Here it is:

(Radiant heat load) = **(Radiant Loss)** + **(Convective Loss)** + **(TRP)**

$$(Q_{in}) = (L_{out}) + (H) + (E)$$

The radiant loss is always outward and proportional to the (fourth power of the) leaf temperature. The convective loss is proportional to the temperature difference between the leaf and the air, $(T_L - T_a)$, so that it may, in fact, be *a heat gain if the leaf is cooler than the air*. The evaporative energy exchange, (TRP), is proportional to the vapor pressure difference between the leaf and the air, $(e_m - e_a)$, and is always a loss.

Now we can go on to discuss other responses of TRP to the changing atmospheric environment. **Fig. 6-27** shows two kinds of response to wind speed. In the calculations leading to the figure, the leaf itself — the system — is held constant, as we did for animals. In **Fig. 6-27a** we hold the air temperature and humidity constant and vary the radiant heat load; while in **Fig. 6-27b** we hold the radiation constant and vary the air temperature and humidity.

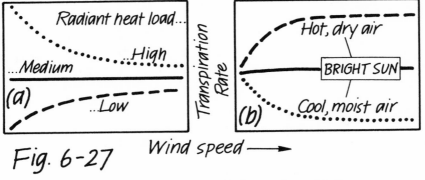

Fig. 6-27   Wind speed ⟶

One would think an increasing wind speed would always increase TRP by carrying the water vapor away more rapidly. The model tells us this is not always so, and experiments confirm it. Without going into detail, I'm sure you can see that the unexpected *decrease in TRP with an increasing wind speed* comes about under conditions when the leaf is much warmer than the passing air, so that both $(T_L - T_a)$ and $(e_m - e_a)$, and therefore (H) and (E), are very large[L]. Under these conditions, the increased wind speed cools the leaf by washing it with cooler air — increasing (H) — thereby decreasing (T_L) and (e_m), and thus $(e_m - e_a)$ and

(L) Remember from the TRe diagram, **Fig. 4-22**, that when T_L is large, e_m for saturated air is also large.

TRP. Said another way, the increased (H) removes the need for a large (E = TRP) to achieve equilibrium. As we see in the two parts of **Fig. 6-27**, (T_L - T_a), and therefore (H), is very large either with a bright sun or in cool, moist air, or both.

Finally, in a manner similar to that shown in **Fig. 6-9** for an animal, a larger leaf will heat up much more than a smaller one in the presence of the same increase in its radiant heat load. This is because the larger leaf has a thicker microscale boundary layer, and thus a need for a larger value of (T_L - T_a) to accomplish the same convective dissipation (H). This fact is certainly related to the obervation that large leaves are nearly always found in the under-story shade in sunny, hot climates; so that a large-leafed house plant tells you, just by its morphology, that it won't appreciate spending much time in direct sunlight. This fact is also related to the observation that it is the largest leaves — squash and the like — that wilt first when bright sun strikes your vegetable garden during a spell of high temperature.

In discussing the energy balance of a leaf, we have been concerned with the ways in which the temperature of a leaf responds to the changing atmospheric environment. And in discussing the three processes of PSN, RESP, and TRP, we have been concerned with very short-term physiological responses of plants to the changes in leaf temperature caused by the changing atmospheric environment. We need now to move on and examine the longer-term responses of plants to the changing atmospheric environment.

In consideration of longer-term responses, we need to recognize that very small numerical differences, in the short run, between the values of variables and process rates in two different plants— small differences in (T_e), (T_L), or PSN, for example — become very large when they are accumulated in the long run. *The small differences between two plants, caused by small differences between the environmental conditions to which they are exposed, may add up to very large differences in their development during the entire growing season.* It's like a savings account: at low interest rates, in the short run it seems the principal increases very slowly, but over a long time the change is very large.

Plants and the atmosphere: growth and development

Study of the HU Concept tells us about developmental responses of plants (and certain animals) to seasonal changes in the environmental temperature, but nothing about the responses to variations of temperature through diurnal periods.

Plant growth is frequently enhanced in an appropriately fluctuating temperature environment as compared with any single, constant temperature. To this fact, the term **thermoperiodism** has been applied; and when we consider the case in which temperatures fluctuate diurnally — as opposed to seasonally — we use the term **diurnal thermoperiodism**.

Fig. 6-28 tells the story of diurnal thermoperiodism. Plants have a different set of biochemical reactions operating during the night as compared with during the day, although there are overlaps between night and day, since some reactions operate continually. Photosynthesis and respiration are only a small fraction of the long list of processes that fit this statement. As a result of these different sets of reactions, most plants exhibit different sets of response rates between night and day, as shown in **Fig. 6-28a** where two separate "bell shaped curves", like the one in **Fig. 6-17b**, are shown.

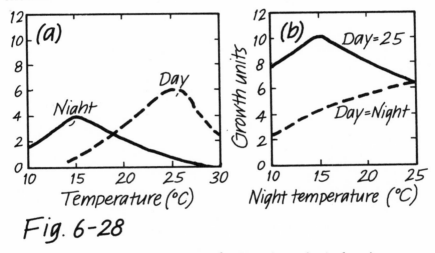

Fig. 6-28

The numerical example in **Fig. 6-28a** is hypothetical — it describes an imaginary plant — but with it you can see how the "data" in **Fig. 6-28a** yield the results in **Fig. 6-28b**. Here's how. Take, for example, the temperature combination (Night = 15; Day = 15) and get the number of growth units from **Fig. 6-28a**. You get (Night = 4 units; Day = 1 unit; Sum = 5 units). This result appears in **Fig. 6-28b** where (Day = Night = 15). The maximum rate appears for the combination (Night = 15; Day = 25) from which (Night = 4 units; Day = 6 units; Sum = 10 units). No constant temperature yields a rate as large as 10 units.

For real plants, biological (growth) responses of various kinds have been plotted as "contour lines" on a graph of night temperature and day temperature (Lowry, 1969). **Fig. 6-29** shows some of the results for tomato seedlings and for seedlings of two kinds of forest trees.

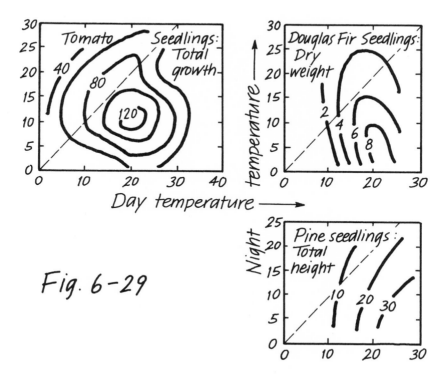

Fig. 6-29

There are several things to notice about these results:

(a) they are for very small plants — seedlings — that are particularly sensitive to the thermal environment — probably because they live near the surface where temperature contrasts are greatest;

(b) this sample of three species are all thermoperiodic, because we can see that the location of the maximum contours — the "topographic summits" — all lie away from the slanting lines that represent "night temperatures equal to day temperatures";

(c) the patterns of the contour lines are quite different from each other; and

(d) the growth responses used by the experimenters are, in some cases, height and, in others, weight.

As with animals, many plants too have "cold season requirements" (see **Fig. 6-14**) that affect their growth and development at a later stage. But here we have discussed daily variations during the growing season.

Before we go on to a consideration of the the timing of plant changes, and where different kinds of plants live, take a moment to consider in slightly more detail the fact noted earlier: *the HU Concept is only an approximation of reality.* Consider the ways in which the "errors" in the concept — the results of the mathematical approximations made in the HU model — cause it to depart slightly from reality, but consider also why the concept is so very useful in many practical applications having to do with *predicting plant (and animal) growth and development.*

First, the problems with the HU Concept.

(a) One of the points of **Fig. 6-13b** is that small animals and plants have different sets of optima and limits in different parts of their life cycles; so that a plant that "prefers a warm spring" will do better in Year #1 of **Fig. 6-19** than in Year #2, although the HU Concept says there would be no difference between years;

(b) if the plant species is particularly sensitive in part of its life cycle to some atmospheric variable other than temperature — rainfall, for example — the concept will be in error, because the concept assumes only the environmental temperature is operating;

(c) as noted in discussion of **Fig. 6-17b**, the straight line is only an approximation of the curved line, so that the HU Model — and thus the Concept — will not tell the truth in hot spells when the daily average temperature is above the optimum[M]; and

(d) since standard, published, temperature records used in applications of the HU Concept are seldom the same as the temperatures of the leaves and plant parts to which the biochemistry responds, this represents a source of error.

With all these sources of error, why do people have such success in using the HU Concept?

The answer is that, as in many things, *the various errors tend to cancel each other through the course of a growing season*, so that the HU Model and the HU Concept yield quite useful results in various kinds of predictions. As one example, there may be brief periods when air temperatures (standard measurements) are below soil temperatures, and the HU Model predicts the temporary cessation of development during these periods; *and that cancels* the occasion when, in a hot spell, the air temperature exceeds the optimum (T^*) and the HU Model predicts a continued increase in development beyond (T^*) — problem (c) above.

(M) Some researchers have suggested and used an HU Model with an "upper threshold temperature" — usually just above the optimum — which says all development stops when the environmental temperature exceeds its value. I think that the added precision of this variant is marginal, and certainly it does not add substantially to the explanation, in this book, of the HU Concept.

Let's see how some of these applications work. First, how do we use standard temperature observations to calculate heat units? Each day contributes a number of heat units to an accumulating sum, according to the following simple formula:

Heat Units for the Day $= (\mathbf{T_{average}} - \mathbf{T_t})$

as long as the average temperature is above the threshold. That is, as long as $(\mathbf{T_{average}} - \mathbf{T_t})$ is *positive*. We use the definition from **Chapter 5**:

$$(\mathbf{T_{ave}}) = (1/2)(\mathbf{T_M} + \mathbf{T_m})$$

These daily HU accumulate by simple addition, in a way suggested by **Fig. 6-19**. As a second example of how errors in the HU Model tend to cancel each other, the process of averaging the day's high temperatures and low temperatures, to get the value of $(\mathbf{T_{ave}})$, tends to "bury" most high temperatures — where error of type (c) lies — within a lower, average temperature used in heat unit sums. In addition to avoiding the error of type (c), this is realistic in the sense that shallow soil temperatures (see **Chapter 3**), to which many growth processes are responding, do not show the same day-night extremes as air temperatures.

Thus, it is the case that people have considerable success in using the HU Concept for various kinds of predictions. *The first of these applications* is the obvious one: predicting, in the current growing season, the timing of various stages in the developmental sequence, most especially the timing of 'harvest" in the case of commercial plants[6]. That application is one in which (i) species and (ii) weather information are known — predicted in this case — and (iii) a "biological response" is sought. As a basis for discussing the variety of applications of the HU Concept, consider the following table of possibilities.

Weather sequence known?	Organism known		Organism unknown	
	Response known?		Response known?	
	Yes	No	Yes	No
Yes	X	1	2	Y
No	3	Y	Y	X

X = Everything or nothing is known, so there is no problem.
Y = Two of the three factors are unknown, so there is no solution.

Application **#1** in the table includes the one we just described: predicting "response" when the particular organism and the weather are known. But there are other forms of this application, so we'll call this one **#1a**.

Application **#1b** would be the case in which we had climatic records, for a number of years, at a location where the success of the known organism was unknown, and we wish to estimate, based on knowledge of the climate, the likelihood it could be grown successfully there. The other side of that coin would be the case where the known organism is a *pest*, and we wish to estimate the likelihood it would *not succeed* there.

Application **#1c** would be the case in which we are able to monitor the weather in a location where the organism is developing, and we wish to estimate, day by day, the progress of its development, because we cannot observe the organism's development directly. A common example of this application is the tracking of the development of insect pests in remote terrain for which current weather data, but not development observations, are remotely available[N].

Application **#2** would include the case in which we have weather records for a place, and we already know an organism will grow there successfully. The problem is to select from among a list of similar organisms — varieties of grapes, for instance — which one would have the most consistently "good" response.

Application **#3** is one in which we wish to *infer past weather from known responses of known organisms*. This one is intriguing to me, mainly because it seems to hold such potential and has been used so seldom. It is the case, you see, of trying to reconstruct past weather events from known biological data. For example, which were the "good" months or years and which were the "bad", at a location where biological records, but not weather records, are available.

The applications in all cases rest on some sort of assumed connection between weather and a biological response, or outcome. For very small animals and plants, this connection is often assumed to be described in the HU Concept.

(N) To employ the HU Concept in this way, one needs the information in the format of **Fig. 6-17a**. For an example using the data in that figure, suppose a "time unit" were one week. Thus, a day for which the average temperature is 15 °C is one on which the insect developed by $(1/10.5) = 0.095$, or 9.5%, of its full development, because it would take $(1.5) \times (7) = 10.5$ days to develop if *every* day were 15 °C. Adding each day's percentage would predict full development on the day where the sum reached 100%.

Phenology and biogeography

To complete our examination of animals and plants, let's consider the ways in which weather affects the timing of a plant's life cycle, and the ways in which climate influences — perhaps even determines — where plants live. Though we consider plants, you'll be able to see the parallels for small animals. In particular, you'll be able to understand, for example, that the developmental progress of a chewing insect must be closely attuned — bioclimatically parallel, so to speak — to the developmental progress of its host plant.

*The study of the timing of events in a life cycle is called **phenology***. What we have called the "stages" in a life cycle are known in phenology as **phenological phases**, and one studies the timing of events — the transitions from phase to phase — by studying the lengths of the phases. We know when a phase is over by observation of a **phenological marker**, which is a distinctive characteristic of the plant; for example, the beginning of bloom of cherry trees is a marker for that plant.

Though traditional markers are outwardly visible, such as the first blooms, modern technology also permits the monitoring of *internal markers*, such as the ratio of concentrations of two biochemical substances. Oranges are considered ready for harvest, for instance, when the acid/sugar ratio reaches a certain value.

Though there is overlap, you can see that the set of markers — a sequence of events in development — must be different between *annual* and *perennial* species. For example, one knows exactly the planting date of a commercial annual — the date on which its annual cycle begins — but one can only infer the date on which annual development begins for a perennial.

The set of critical markers will depend also on the particular use of the crop, if it is a commercial plant being monitored. For example, *if grass is being grown for seed*, the date seed matures is important. *If it is being grown for hay*, the date may be much less important. *If flowers are grown for cutting*, the date of seed maturity is of no importance. *If the plant is the host for a pest* — for example pine trees hosting a shoot moth — very little about the

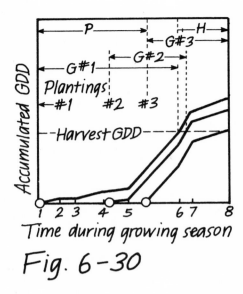

Fig. 6-30

flowering time will be of importance, and everything about the timing of early greening will be important.

With these comments in mind, consider several common phenological patterns in terms of the HU Concept. **Fig. 6-30** depicts *the progress of three plantings of an annual within one growing season at one location*. The plant is ready for harvest when it has experienced the accumulation of a certain number of growing degree days: "Harvest GDD" — its varietal constant.

In the figure, the three plantings experience different lengths of growing until harvest: three lengths marked "G." The spread of the three planting dates stretches over "P" days, while the spread of the three harvest dates stretches over a much shorter period: "H" days. This shortening comes, simply, from the fact that the season grows warmer as development proceeds. It is this fact that was of such value in managing the planting and harvesting of commercial green peas[7].

Notice, in **Fig. 6-30**, that *on any given date the lines of accumulating HU are parallel*, because all three plantings are experiencing the same number of HU on that date. **Fig. 6-31a** shows the case in which *three perennial species are experiencing the same weather in one season at one location*. There is only one curve of accumulating heat units.

The three species have different varietal constants: 'a', 'b', and 'c', and their "bloom dates" — the dates marked "T" — are also different. In **Fig. 6-31b** this same idea is extended to the case in which *two perennial species are each exposed to slightly different, neighboring* **microclimates** *within one growing season*. The species 'a' and 'c' and the cooler microclimate are the same as in **Fig. 6-31a.** Here both are also exposed to a warmer microclimate — for example the sunny slope of a ridge or a valley — as well as to the one on the cooler, shady slope.

Notice, in **Fig. 6-31b**, that *on any given date the lines of accumulating HU are **not** parallel*, because the warmer microclimate is accumulating heat units faster on that date. In the particular case shown, the difference in flowering dates, or delay, between slopes, for species 'a' — D_a — is much greater than that for 'c' — D_c. The difference might be the same or less, *depending on the particular sequence of weather events and the particular values of the two varietal constants.*

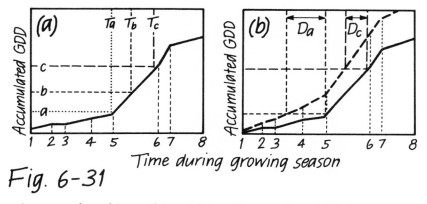

Fig. 6-31

Among other things, this explains the common perception that, in one season, the blooming of one species "explodes" almost simultaneously on all sites, while that of another seems very much retarded in cool places; and that the "exploding" may be exhibited by different species in different years. A closely related idea is that a delay may be an advantage for a species of chewing insect, because it extends the availability of food — and perhaps even the number of generations per year — for a mobile species whose individuals can change feeding and breeding sites.

"The bottom line" for a species is whether or not it can consistently — from one year to the next — pass through its complete life cycle, reproducing successfully. For a given place, that process of survival is connected not only to the bioclimatic requirements of that species, but also to those of predator and prey species — those that eat, and those that are eaten by, the species of concern. This web of life becomes pretty complex, but it is clearly connected with the atmospheric environment.

There have been numerous attempts to connect the climatic parts of a plant's life cycle with where it lives — its **biogeography**. We have already suggested how its **home range** may be connected with microclimate, and the maps in **Chapter 10** suggest the hemispheric scale of plant geography. **Fig. 6-32**, on the other hand, is a "map" on coordinates of annual temperature and precipitation, rather than altitude, latitude and longitude.

We will conclude this Chapter by commenting on this one example of the representation of plant geography. As with other mappings of this kind, we must recognize it is only a generalization of reality — an attempt to permit understanding through the process of simplification. Though you probably understand already — before you look at **Fig. 6-32** — the broad connections among temperature, rainfall, and vegetation types, still the figure "puts numbers on" the understanding. Even so, we know that, in reality, sharp boundaries, like the ones in the map, seldom exist. *Any scheme that divides Nature into regions and categories has this characteristic: boundaries are drawn where only "zones" exist in reality.* It is true for this kind of mapping, as it is for simple "classification."

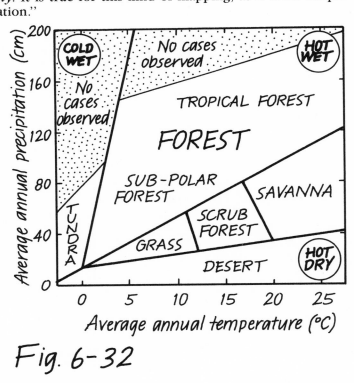

Fig. 6-32

Notes

(1) The information in **Fig. 6-4** comes directly from Fig.12.6 of Rosenberg *et al* (1983) and Fig. 6.3 of Oke (1978), which in turn come from the work of Max Kleiber, whose book on this

subject (1961) is a classic. The particular mathematical model in the figure is $M = (3.2)(W)^{3/4}$ where (W) is in kilograms and (M) is in Watts.

(2) This mathematical modification follows the same path used by David Gates (1980) in most of his writings, and the graphical treatments and insights are his as well.

(3) At this point we can note the relationship between **Fig. 6-11** and the following figure, for a human being, found in McGuinness *et al* (1980, page 89). This figure is related, in turn, to one found in Lowry (1969, page 249). Both describe a quiet, resting person in a room (no solar heat load: $\mathbf{R_{in}} = 0$). The intention of the figure is to show that, for a person without the ability to reduce his resting metabolic rate ($+\mathbf{M}$), increased evaporative cooling ($\mathbf{E}$) makes up for the fact that, as room temperature increases and ($\mathbf{T_s} - \mathbf{T_e}$) gets smaller, convective and radiant dissipation ($\mathbf{H} + \mathbf{L_{out}}$) gets smaller. The basic point is correct, although a zero value for ($\mathbf{H} + \mathbf{L_{out}}$) implies the slightly unrealistic condition that the walls and the air temperature both equal the body's core temperature. The more

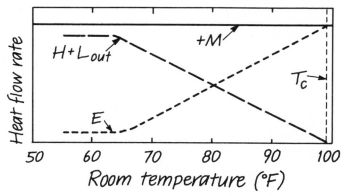

important point for this Chapter is that, during the evolution of animal thermoregulation, homeotherms that gave up the ability to vary ($+\mathbf{M}$) developed in its place an ability to vary ($\mathbf{E}$).

(4) A reasonable search for an appropriate introductory work on thermoregulation strategies in non-mammalian animals led to the conclusion there is no such thing. The best I could locate for the purposes of readers of this book are these two, neither of which is completely current: Mendelsohn (1964), and Bligh and Moore (1972). For information on current research in animal thermoregulation, a good person to ask is my former colleague, James. E. Heath of the University of Illinois.

(5) The swimming rate in **Fig. 6-14** is expressed as the more general term "rate of dispersal." This rate, it is clear, has a great deal to do with an animal's success in escaping a predator or in catching prey. In the sense of *migration*, it may have a great deal to do with whether or not an animal can reach a proper location for the next stage in its life cycle.

(6) Most of the examples of experimental results, such as **Figs. 6-14, 6-16,** and **6-24**, are taken from and discussed further in Lowry (1969).

(7) As described in Lowry (1969, page 197), there has been great commercial value in even very approximate predictions concerning the timing of the harvest of green peas. Another well-known example of this application is the annual prediction of the blooming date of the Japanese Cherry trees in Washington, D.C.

Peptalks

(P₁₄) The mathematical models will be presented, but they will be accompanied by graphical examples of results one may obtain using the models. The reader who is more comfortable with graphics than with mathematics may follow the discussion, and grasp the insights, by means of these graphical examples. Those readers ought to keep in mind, though, that (a) models permit quantification and diversification of relationships, but that (b) the relationships are only as real as the models.

(P₁₅) In the model, the factor (0.2) simply takes care of the particular *proportion* between (M) and $(W)^{3/4}$, which in turn depends on the particular physical units used for (M) and (W) — for example (kilocalories per hour) for (M), and (kilograms) for (W). As a simple example, say that (W = 150 kg). Then (M) = $(0.2)(150)^{3/4}$ = 8.6 kcal/hr, as shown in both halves of **Fig. 6-3**.

(P₁₆) Reading a chemical equation is pretty much like reading a mathematical equation. "Plus" means the same thing, and both sides of the equation must "balance." Subscripts tell you "how many" about the atoms or molecules just preceding; for example (H_2O) means two hydrogens attached to one oxygen — a water molecule.

(P₁₇) This modification will look very similar to the one we used to discuss **Figs. 6-8, 6-9,** and **6-10**. If you followed that one, this one will be easy.

7. Environmental design: the players and the game
Possibilities for design

Up to here, this book has been like a text on those technical subjects, from the field of atmospheric ecology, that my experience tells me designers and planners can make best use of. In this chapter and the next I want to give you my ideas about how designers and planners can proceed to use that information.

Up to here, I have been "the expert" on whatever was being discussed. In this chapter and the next, however, I recognize I am dealing with ideas on which you — designers and planners — are probably the experts. So, what follows in these chapters is just *my way of looking at how you and I can interact*: with you as the designer, and with me as the consultant. My way may not accord with yours, but you ought to know how I see things if you are going to make best use of me and what I know. At the least, you will know how I perceive these interactions as I explain my technical subject to you.

The Game of Environmental Design

I propose to discuss the planning and design process — and the building process — as if it were a serious game, with a game board, in which you and I are among the players. We are not the only players, and you might disagree with my description of the board and my list of players. Please be certain I am not making light of the profession of environmental design, nor of the serious concerns of developers and tenants. It is just that I believe viewing it as a game will help me present several of my ideas.

Let me hasten to say I recognize that **each project represents a unique game in itelf** — no two are exactly alike, as to the board, the players, or the rules. In fact, it is in this way that the process we are discussing is more like a *game* than it is like a *play*. One cannot, of course, generalize beyond a certain point. I try here to generalize as far as it seems to me to be fruitful.

A play has a script and a cast. For all practical purposes, the action is the same from one performance to another, and almost irrespective of who takes what role. The process of planning, designing, and building, on the other hand, is an infinitely dynamic and variable game. What happens at any point — and the ultimate result: the "final score" or who wins and who loses — depends on which players have made which moves at what times. In most games there *may* be little correspondence between the quality of the play during the game and the final score. If the quality of play is high in our game, however, the probability becomes very large that the final score will be "fair" and that there may be only winners and no losers.

The players

The players enter the game at different times, depending on the particular project. The order of entry I describe later is just one. Though the players are described as "one person", in reality some are just as often groups or teams. It is also true that sometimes individuals — or groups or teams — represent more than one player in the game.

The client. Someone decides that a project is to be built, and that the services of professionals will be required to build it. Those professionals are approached by the client, who is the ultimate source of the goals for the project. Inescapably, the client is and the project must be connected to the atmospheric environment.

The public. Perhaps the public good requires the building of the project, and the public declares that to be so. In this case, the public enters the game at the beginning. In particular, perhaps the public is to be the primary user of the project upon its completion. So, the public is also the client. Or perhaps the public, at some point, declares that the public good requires the project *not be built*, or that it be built in a certain way rather than in another way. In that case, the public enters the game at a later time. For any substantial project the public will enter, also, as the source of permissions and licenses.

This list of possible entries for the public is far from complete, but except in rare cases one of the players will be the public — the members of the society, taken as a whole, within which the project is to reside. Inescapably, the public will have to share part of the atmospheric environment with the client and the project.

The planner. The project must be built within a pre-existing context. The project may have to be injected into an already-developed environment, or it may be part of the primary development. In any case, the project will have to be placed on a physical landscape, on a social and political landscape, on a financial and economic landscape, and within a network of utilities and facilities.

I look upon THE PLANNER (a generic term in this discussion) as the player who (1) describes, to other players, the process of "connecting" the project to these different parts of the pre-existing context, (2) describes the alternatives available for making those connections, and (3) makes recommendations concerning the choices to be made from among the alternatives. Inescapably, the atmospheric environment is part of the physical landscape, set forth by the planner, to which the project will have to be connected.

THE PLANNER may be one person, but in reality he is almost always a group or a team. Members of the team are urban planners, architects, landscape architects, water resource planners, financial planners, civil engineers, environmental engineers, and so on.

The designer. Once the choices among alternatives have been made, the designer enters to create a detailed description of the reality that will be constructed within the constraints of the alternatives chosen: the physical, the socio-political, and the fiscal alternatives. Inescapably, the atmospheric environment will place constraints and opportunities before the designer as he creates the detailed description.

As with the planner, THE DESIGNER (a generic term in this discussion) may be one person, but in reality he is almost always a group or a team. Again, members of the design team are architects, landscape architects, civil engineers, environmental engineers, and so on.

The builder. When the detailed description of the reality that will be constructed has been created, and when all permissions and licenses are in hand, the builder enters to turn the description into a reality. Armed with different kinds of facts, experiences, and resources from those of the other players, the builder completes the play of the game. Inescapably, the atmospheric environment is part of the physical landscape with which the builder will have to contend while the project is being connected.

I spend little time here dealing with the concept of the builder, because, in my view of the process, most of the information concerning atmospheric ecology has been introduced, and used, by the time the builder enters[1].

The technical consultant. Since each project represents a unique game in itelf — no two are exactly alike — each project requires assembly of a different set of facts and experiences to draw upon in making choices and creating a design. The technical consultant brings to the process facts, experiences, or resources that the client, the planner, and the designer do not have.

Inescapably, facts and experiences concerning the atmospheric environment must be among those assembled to create a design. The planner and the designer decide which facts, experiences, and resources are relevant and important, and by implication which ones they themselves do not have. It follows that, if the technical consultant — in particular the atmospheric consultant — becomes a player, it is by invitation of the planner and the designer.

The game board

There are certain fixed components to any project — the components of the pre-existing context that are impossible or highly unlikely to be changed by the process. Two examples of such components are (1) the set of attitudes held by the public about ecological planning and (2) the physical landscape of terrain and shorelines. Perhaps you use the term **matrix** to describe this set of essentially fixed components. This is what I mean by *the game board*. It sets the context in which all the interactions for change — the moves — are constrained to take place. On the board, a multitude — perhaps an infinity — of moves is possible.

Move 1: The client approaches the planner

The game begins as the client selects for participation a player from the planning profession. Presumably, the client selects a player who meets his criteria for the game: professional competence and experience, financial compatibility, logistical appropriateness, and so on. Presumably also, the client places trust in the professional's ability to make sound choices from among alternatives.

I say "make sound choices from among alternatives" rather than "make sound decisions" because I want to emphasize, at each move, the notion of alternatives. Each decision — each choice — changes the list of alternatives to be addressed next. In my conception of the process, as in any dynamic game, the play of the game consists of recognizing and listing alternatives, choosing, and then moving on to the recognizing and listing of the next set of alternatives. In this view, the first move consists of the choice made by the client from among the list of possible planners.

Move 2: Formation of the planning/design team

After the first move, the client and the planner recognize and list another set of alternatives: the different ways that a team can be composed from among the list of possible players. The list contains architects, landscape architects, civil engineers, environmental engineers, and so on.

The list also contains various *technical consultants*, each of whom brings to the process facts, experiences, or resources that the client, the planner, and the designer *decide they need but do not have*. These decisions about what facts, experiences, or resources are needed from consultants are some of the choices among alternatives being made as part of Move 2.

In many cases the planner and the designer are the same "person" acting as two different players as these choices among alternatives are made in Move 2. That is why I combine them as "the planning/design team" at this point.

Although this book is primarily about the atmospheric consultant and his expertise, discussions with my colleagues from the planning and design professions make it clear to me that the best teams include all the basic kinds of planners and designers — at least architects and landscape architects — in Move 2.

To my way of thinking, *in games with play of high quality*, consultants become players in Move 2. That applies in particular for the atmospheric consultant, of course, on projects with special needs for his expertise. To say it another way, the best planners and designers know well the limitations of their own relevant technical expertise, and they seek to make up for that early in the game rather than later. You may recall I offered the thought earlier that, if the quality of play is high in our game, the probability becomes very large that the final score will be "fair" and that there may be only winners and no losers.

Move 3: The planner presents the project to the team

In Move 3 the players begin to take on their particular identities and to function according to their specialties. For example, the planner introduces the client, and then he presents, to the architect and the landscape architect, the regional and the local parameters of the problem, and a trial list of alternatives. Later, the team will improve upon the trial list of alternatives.

Fig. 7-1 gives me the basis for defining several terms having to do with the *spatial dimensions* of the project and its associated problems. We will need this vocabulary to discuss the next few moves of the game. The project is to be built on a **site** that lies within an **area**. The area, in turn, lies within a **region**.

The *region* lies on what I will call the **macroscale**, and is of a size, for example, an hour's drive across. It has a uniform overall climate, is probably all within the same major watershed, but it may have a variety of major landform types and vegetation types within it. It surely has a multitude of major political (i.e. governmental) and social units within it

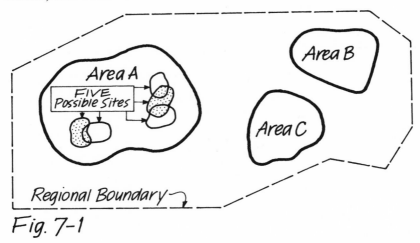

Fig. 7-1

Within the region there may be several areas potentially available to the client, so that a choice from among alternatives may be made on this size scale. An *area* lies on what I will call the **mesoscale**, and is of a size, for example, to consist of from several hundred to several thousand acres. Its physical landscape is likely to contain only minor variations of landform and vegetation, and it is likely to be within no more than one or two political jurisdictions and one or two social units.

Finally, within the area there may be several possible sites available for the project, so that a choice from among alternatives may be made on this size scale as well. The choice is most likely to be made on the basis of relationships to water bodies and drainage, position on terrain, access to existing transportation, and so on. The *site* lies on what I will call the **microscale**, and is of the intended size of the project.

As I noted earlier, in Move 3 the planner presents, to the architect and the landscape architect, the regional (macroscale) and the local (mesoscale and microscale) parameters of the problem, and a trial list of alternatives[A]. To my way of thinking, *in games with play of high quality*, consultants become players in Move 2. It may be, however, that in the process of refining a trial list of alternatives — especially if the type of project is unfamiliar to them — the architect and the landscape architect will not recognize their needs for various technical consultants until Move 3, so that the technical consultants will enter the game at that point.

Entry of the technical consultant

As I have just said, although entry of the technical consultant may take place at any one of several points in the game — as early as Move 2 and as late as Move 6 — the earlier it takes place the better. Unfortunately, some planners and designers do not know well the limitations of their own relevant technical expertise, and they seek to make up for that later in the game rather than earlier. In fact, even more unfortunately, at the very basic level of Move 2 many architects think of landscape architects as "technical consultants", and vice versa, to be invited into the game later.

Let me be specific. In my discussions of the game of environmental design I view architects and landscape architects — design professionals — as "design consultants" when they work in that capacity. A project with any considerable size will, almost automatically, have an array of technical consultants: for acoustics and noise, traffic, parking, biology, and so on. Thus, I recognize that the term "consultant" has several levels of meaning.

It appears abundantly clear to me that, though it may seem to a planner or a designer "unnecessary" and financially imprudent to use consultants from such "esoteric" fields as atmospheric ecology, a worthy consultant — "esoteric" or not — will increase immensely the chances of creating a more nearly optimal design, and will decrease immensely the chances of expensive errors later in the game. The choice to invite in an atmospheric ecologist, or not, and when the invitation is given, is one of the moves to be made by the planner or designer as part of the game.

(A) In rare planning problems, the client may not even have chosen the region; but I daresay in most cases the locational choice to be made is from several possible sites, the region and the area having already been determined. In many cases all locational choices have been made, and the site is already determined. Even without choices of region, area, or site, however, the region and the area have characteristics that must be considered in the plan and the design for the site. Therefore, information on all scales will be presented by the planner to the architect and the landscape architect. **Chapter 10** contains my catalog of essential concepts concerning differences among regions.

In **Chapter 8** we will discuss the matters of (a) recognizing when you need an atmospheric consultant, (b) some of the strategies and techniques available for using atmospheric principles in environmental design and management, (c) some of the techniques for analyzing atmospheric data and using them in presentations of alternatives, and (d) several examples of projects with atmospheric components appropriate for consultation with an atmospheric scientist. For now let us consider the identity of the atmospheric consultant — the atmospheric ecologist — and what he can add to the quality of play in our game.

Recalling my words in the **Preface** and in **Chapter 1**, I can tell you who he is. *Your atmospheric ecologist knows the physics of the atmosphere and the biology of how it interacts with the lives of living organisms, and he makes clear his interest in seeing to it that good scientific knowledge about the atmosphere shows up in the built environment.* He has other characteristics, of course, such as being able to explain his knowledge in terms you can make use of, and professional integrity of the kind any consultant must have. But my basic point is that your atmospheric ecologist is part physicist and part biologist (P_{18}).

Many atmospheric scientists — meteorologists and climatologists — know how the atmosphere behaves, but they know less well why it behaves that way. In terms most practical to you, the designer or planner, that kind of atmospheric scientist knows how the atmosphere behaves under a limited set of conditions he is familiar with — for example on size scales much larger than your project — but he may have difficulty anticipating its behavior under the conditions of your project, as yet unrealized. More often, atmospheric scientists know the atmosphere well but know little or nothing of its interactions with living organisms. You are looking for a very special sort of consultant. How do you find him? How do you recognize him? Can you work well with him?

How do you find him? Aside from word of mouth among your professional colleagues, you ought to be able to find your atmospheric ecologist by inquiring among several of the professions he is likely to be part of[2].

How do you recognize him? Well, he is familiar with the concepts and methods presented in this book. Naturally, I would say that, right? But it is true. Your atmospheric ecologist may not agree with some of what I say in this book, but he is familiar with the concepts and methods; and where he disagrees, he can explain his disagreements to you in language you can understand. That, you see, is a reason for studying this book. As I said in **Chapter 1**, you want to be able to tell if an atmospheric ecologist is "for real"$^{(P_{18})}$, and being familiar with this book will give you the confidence to make that judgment. Making that judgment is part of playing the game of environmental design.

Can you work well with him? I imagine you are in a better position than I to answer that question. For me the marks of your best atmospheric ecologist would be that he can readily comprehend the nature of your need for him; and, if that need is for something more than mere access to atmospheric data, he has several alternatives to offer you — in terms *you* can readily understand.

Let's get back to playing the game of environmental design.

Move 4: Addressing the regional scale problem

In my conception of this planning and design process, the regional scale problem consists of two *interactive* parts: (a) placing the project on the physical landscape by a choice from among alternative areas and sites, and (b) fitting the design to the chosen site.

(a) *The choice from among alternative areas and sites.* Assuming the region has been determined, as it usually is, this part of the game consists of making choices among possible areas and sites on the basis of local differences. The atmospheric knowledge required is concerned with mesoscale and microscale ecological differences, such as those in **Chapters 3** and **4** and in the following table.

(b) *Fitting the design to the chosen site.* Assuming the area has been determined, this part of the game consists of estimating, on the basis of local differences, which kinds of designs will and will not "work" — meteorologically speaking — in the various site alternatives available. The atmospheric knowledge required is concerned with microscale responses of various design types to the macroscale characteristics of the region, and with the ways in

---

Choices among possible areas and sites

- On flat land, sloping land, high ground, or low ground?
- Which side of a hill or ridge? For solar heating? Wind sheltering?
- How far up or down the slope?
- Near or far from a mature stand of trees that is to be saved?
- Near a major water body or not?
- Do characteristics of soil drainage matter or not?
- Does air quality ever matter? Seldom? Only at night?
- Near or far from a major highway? For transportation? Air quality?

---

which the mesoscale ecological differences modify those micro-scale responses. Choices are to be made on the basis of the infor-mation in **Chapters 5** and **6** and the following kinds of questions about differences.

Fitting the design to the chosen site

- Would the topographic position allow net benefits from passive solar heating?
- Would the regional climate permit the use of open water for summer heat management on site?
- Does wind force loading — wind speed — limit the form of a building in consideration of heat management? Does wind direc-tion limit the form?
- Would the site provide existing major vegetation for manage-ment of the solar heat load in summer?
- Would snow loads, expected as part of the regional winter cli-mate, permit the use of a low pitched roof and cantilevered con-struction? Covered walkways to out buildings?
- Do local trends in air quality suggest limitations in the kind of vegetation — existing or planted — appropriate for the site?

Move 4 ends with (a) the choice of one from among alternative areas, (b) listing several sites judged appropriate within that area, coupled with (c) listing the design types not excluded for the sites chosen. To the extent these questions, and others like them, are relevant to the making of choices, to that extent there is a need for an atmospheric ecologist to participate as a player in Move 4.

Move 5: Addressing the microscale problem

In my conception of this game of environmental design, the microscale problem consists, first, of choosing one from the list of appropriate sites, and then examining, in detail, the remaining feasible design types for that site. Presumably, all feasible design types will be responsive to the mesoscale and macroscale charac-teristics of the site. The solution of the microscale problem then becomes one of merging four *interactive* parts: (a) choice of one design type from among the feasible alternatives, (b) choice of materials, (c) choice of placements of components, and (d) articu-lation of the fine tuning components. As before, *in games with play of high quality*, the game is kept open and dynamic by mak-ing informed choices from carefully drawn lists of alternatives. Premature or (it is the same thing) blind choices, in particular those related to the integration of the principles of atmospheric ecology, may foreclose potentially valuable and important options in design.

With the site having been selected, this part of the game consists of choosing one design type from the list of alternatives appropriate for the site, and then of making choices concerning materials, sizes, placements, and so on. The atmospheric knowledge required is concerned with processes on all size and time scales — from macroscale to microscale, and from minutes to months. The information in **Chapters 2 through 6** represents the knowledge involved. Choices involve application of the atmospheric information through answers to questions such as those following.

Atmospheric questions to be answered

- In a typical summer, which hours on which dates experience the highest air temperatures of the year? What are those temperatures?
- What are the values of the solar zenith angle and the solar azimuth angle during those hours?
- Given the most desirable sight lines to take mandatory advantage of a distant view, should protection of the relevant large windows from summer sun and winter wind rely more on building design or on landscape design?
- Given that an enclosed patio is to provide a means for capturing and retaining midday solar heat in spring and autumn, what is the best shape and orientation for the floor of the patio? Would open water in the patio, for reflecting extra sunlight to an absorbing surface, be counterproductive because of its evaporative cooling?
- Given the probable directions of wind and sun during those hours with the largest values of daytime wind chill, what is the best placement of thermostats to integrate passive solar heating with supplementary heat?
- Given the macroscale climate of the site, should control of snow drifting be a consideration in the landscape design?
- Given the site's position with respect to the surrounding terrain, is local summer cloudiness more likely than one would expect from the records available for the nearby airport station? If so, is this worth including in the design for summer heat control?

Move 6: Changing the landscape

Move 5 sets the design within the plan, and in Move 6 — the last move of the game of environmental design — the builder changes the landscape as he turns the description of the project into a reality. Presumably, the planner and the designers will have taken into account not only how the landscape will influence the project but also all the principal ways that the completed project will influence the changing of the landscape.

In games with play of high quality, the planner and the designers will have taken into account the principal ways that the project will influence the landscape *during construction*. During the construction period major existing vegetational components may be sensitive to surface scarification and temporary changes in drainage patterns. The influence of soil moisture content — very wet or very dry — on certain engineering characteristics of the soil during construction, or the probability that excessive muddiness will influence the impact of construction vehicle traffic on the quality of the finished landscape design are matters in which the atmospheric ecologist will be able to provide assistance.

During Move 6 the builder is, to a greater or lesser extent, at the mercy of the weather. His operations will surely be affected by a weather sequence including out-of-the-ordinary events such as strong winds or hot, dry air. He needs the services only of a weather forecaster — not an atmospheric ecologist — to pick his way through this weather thicket. On this level, there is no particular need for the builder to have entered the game much before Move 5.

On another, more subtle level, however, the builder should have entered the game much earlier. If the design involves special materials or construction techniques, it may be so that, while these materials and techniques will serve well *given normal (average) weather conditions during construction*, they may result in a badly flawed project after its completion, if abnormal weather events and sequences occur during construction. With the help of the atmospheric ecologist, the builder may well enter the game at Move 3 or 4 to avoid costly errors and litigation by relating the calendar of construction to the varying seasonal probabilities of various "unlikely", but potentially damaging, weather events and conditions.

While an experienced builder may believe his experience — and even statistical knowledge about the climate — makes it unnecessary to consult an atmospheric ecologist, the atmospheric ecologist will be more aware than the builder of ways in which the *actual* probabilities might differ from those that might have been developed, statistically, for a time period much earlier. Discussion of the frequency of tornadoes in **Chapter 10** makes this point abundantly clear. *Since Nature and her weather are known to be fickle, not only from day to day and year to year, but also from one group of years to another, she must be considered as a player in the game of environmental design. In games with play of high* *quality,* the atmospheric ecologist provides a hedge against the unexpected effects of fickle Nature.

Summary: Who won the game?

So, after all of the players have made all of the choices from all of the lists of alternative possibilities — after all the moves have been made on the game board — how can we tell who won the game, or whether there were even any losers?

In this chapter I have presented twin propositions concerning this question. First, I claim that a mark of *games with play of high quality* is that there are no losers. The completed project "works", not only functionally for the client and for the users, but also for the public, whose landscape is shared with the client and the users.

Second, I claim that a mark of *games with play of high quality* is that the planner and the designers know well — and admit — the limitations of their own relevant technical expertise, and they seek to make up for that earlier in the game rather than later. *Planners and designers have plenty of professional expertise of their own.* The best ones do not try to operate as if they were also experts on all relevant technical matters. The best ones recognize it is false economy to try to be expert on all relevant technical matters, most especially matters of the physical environment, and in particular the atmospheric environment.

In **Chapter 8** I shall try to outline my own thinking about how you — designers and planners — can tell when you need the assistance of an atmospheric ecologist, how to present your needs to him, and how to listen to his responses.

In writing this book I take it upon myself to present you — among the best planners and designers — with several ideas:

(a) there are nearly always, in any major project, important ways in which the project and the atmospheric environment will interact;

(b) firm knowledge of the principal concepts of atmospheric ecology is well within your ability to grasp;

(c) armed with your firm knowledge of the principal concepts of atmospheric ecology you can recognize these potential interactions; and

(d) you can make efficient and effective use of the services of a technical consultant in atmospheric ecology — get him to be an important player in the game of environmental design.

Notes

(1) In fact, when an Environmental Impact Statement is required for the project, matters pertaining to atmospheric ecology and matters pertaining to the activities of the builder are injected very early into the game.

(2) Your atmospheric ecologist is definitely *not* what most people refer to as a "Weather Man" — the kind on TV. Rather, he is most likely to be an atmospheric scientist, an ecologist, or a geographer; though he may be a geologist, a soil scientist, an environmental engineer, or an applied biologist, such as a forester or an agronomist. A good search plan is to ask at the headquarters of several professional societies, and also to scan professional journals for articles on relevant subjects written in a style you can comprehend. Ask your science librarian about journals to scan. In any journal you will find addresses of the professional organizations associated with the subjects of the journal. The American Meteorological Society (45 Beacon Street, Boston, Massachusetts 02108) maintains a list of Certified Consulting Meteorologists, and publishes the business cards of advertising members in each monthly issue of the Society's *Bulletin*.

Peptalks

(P₁₈) As I suggested in the **Preface**, designers and planners don't often seek out a consulting atmospheric ecologist because, among several reasons, they think they will never be able to understand the *physics* involved. To many people, these days, "physics" has to do with atomic structure, cyclotrons, nuclear weapons, and other forbidding things. In the context of this book, however, physics refers to matters such as the flow and storage of heat in soil, evaporation and condensation of water, and the force of wind against a wall. The only reason I bring up the term "physics" at all is that knowledge of it separates that potential consultant who can deal with these ideas in several forms — because he knows the underlying principles, and can apply them in unfamiliar situations — from the one who can deal with these ideas only as a phenomonologist — in the form of particular phenomena he has heard, observed, or read about. These remarks that I have made about physics apply as well, in my view, to "biology."

8. What could it be like here, and how?
Strategies for design and management

In **Chapter** 7 we considered the people involved in the design and planning process — the players — and their roles in the game of environmental design. We considered those things in "the ideal case" and without much detail about exactly what each player would do in each of his moves. In particular for the atmospheric consultant, we got only as specific as considering several groups of questions of the kinds the designer and planner would be likely to ask of him.

In this chapter we will be more specific about the integration of the atmospheric ecologist into the game of environmental design: when you need him, how to approach him, and how to recognize a well formulated atmospheric component ready for presentation as part of the plan and the design. Although we will be more specific than we were in **Chapter** 7, we still must deal somewhat in generalities.

Understanding the problem

Before we proceed, let us be certain we understand the nature of the problem to be presented by the planner/designer to the atmospheric ecologist. I imagine my description of the problem will hold true for other kinds of environmental consultants as well, but I will formulate my description mainly in terms of the field I know best. To those readers who think only synthetically — integratively — my habit of breaking the problems, strategies, and solutions into small subunits may seem tedious and unnecessary. My experience tells me, however, it is a way to be more certain that all aspects of the problem and its solution get appropriate attention, and that alternatives are not omitted from the lists of alteratives repeatedly mentioned in **Chapter** 7.

As described in **Chapter 7**, the problem consists of fitting the design to the chosen site, in the sense of Moves 4 and 5 in the game of environmental design. But that is too general a statement of the problem. In slightly more operational terms, when the site is chosen, along with that choice comes a set of macroscale and mesoscale constraints — the *geographical and climatological parameters of the problem*, set by Nature. In addition to Nature's contribution of the geographical and climatological parameters come the *laws of physical and biological science* that must be understood and made use of for the solution.

In terms of atmospheric ecology, then, the problem is one of energy management. **The problem is to design a physical system, usually with feedback, that will control *efficiently* both the energy inflows and the energy outflows so that the effective temperatures (T_e) of the system are within prescribed limits at all those times that are important for the use of the system**.

Thus, the atmospheric ecological problem is to design a physical system that permits the user to control the energy balance of that system so as to meet certain physical and biological criteria, particularly those of human comfort. As discussed in **Chapter 5**, Nature — the environment — brings certain requirements to the problem, and Man brings certain requirements to the problem. The solution consists of the construction of a system with the correct variable resistance.

We may note in passing that there are three kinds of time variability — all mentioned in **Chapter 10** — in the atmospheric environment that the system must respond to in its task of producing acceptable system temperatures:

1) diurnal — day to night — changes within one weather type[A];
2) changes between weather types within one season; and
3) changes between seasons within a year[A].

The tools used by the designer in solving the problem consist of a list of strategies and techniques. The professional contribution of the designer is his experience with using these strategies and techniques *and his understanding of the physical and biological prin-*

(A) See **Chapter 10** concerning weather types. If you watch the TV weather news, you know that ''the weather map'' changes from hour to hour and from day to day as weather systems come and go. Generally speaking we can say that, for a given location, the weather type changes with each major map change. Seasonal changes are clear and obvious, but longer term changes — ''climatic change'' — are much more subtle. This aspect of climate is mentioned in connection with **Fig. 5-13**, with the role of the builder in **Chapter 7**, and in connection with probability later in this chapter.

ciples that make these strategies and techniques appropriate in different interacting combinations. That understanding is what I mean to impart in this book.

In summary, the problem is a biophysical energy management problem. We know that the problem has been solved when the prescribed effective temperatures (T_e) are achieved, in harmony with the other constraints of layout, access, aesthetics, economics, and so on.

Understanding the alternatives

As in **Chapter 7**, I suggest to you that in dealing with the atmospheric component of your design problem — and with your atmospheric consultant — you should continually *keep in mind the alternatives* you have in dealing with atmospheric ecology. My intention is that the following materials be cast in the form of alternatives.

What do you need to know ? This book contains what is probably, to you, an overwhelming multitude of facts and concepts, all waiting to be selected in the process of planning and design. I imagine you would see a similar book on geology or plant nutrition in the same way. You need some sort of road map connecting the various facts and concepts as an aid in formulating the atmospheric component of your plans and designs.

I cannot tell you how you ought to think about these things — each one of us thinks about complex matters in different ways — but I will suggest there are three kinds of charts — each one a "matrix" — that would be useful for just about any project with a recognizable atmospheric component. I cannot make a universally useful set of charts for you, but understanding the charts I present here will be a good way for you to see how to make a set of charts for your own use, according to your own ways of thinking, making use of the materials in this book.

The first kind of chart connects weather and climate variables to each other. Weather and climate variables can be divided into several basic kinds, in the context of this discussion, according to the following scheme.

Observed or calculated from observations:	Available in published records ?	
	Yes	No
Observed	A	B
Calculated	C	D

Later in this chapter we will discuss sources of data and calculations with data. For now, here are *some examples* of each of the four categories. These examples, of course, are *not a complete list*, but are only a means for explaining the four categories.

A) Observed and published. The National Climatic Data Center (NCDC) publishes the maximum and the minimum temperature for each day from a good network of stations.

B) Observed but not published. The largest daily value of momentary wind speed — the so-called "peak gust" — is recorded at major airport stations. These values are available in original records from the station, but in the sense that they must be obtained by special request, they are not "published."

C) Calculated and published. The NCDC publishes the daily values of heating and cooling degree days — HDD and CDD (see **Chapter 5**) — but these values have been calculated for you by the Center — they are not themselves observed. In addition, these values are calculated by the NCDC for only a few stations — mainly major airports. Thus, for example, if you had daily temperature records from a small airport near your project, whether the records were or were not published, you would have to calculate your own values of HDD and CDD from them.

D) Calculated but not published. Values of effective temperature (T_e) are obtained in the manner discussed in **Chapter 5**, but you will not find them published regularly for a network of stations.

Here is one example — an incomplete, suggestive example — of a chart of the first kind. This chart connects observed and calculated variables, but it does not include information on whether or not the variables are published.

As you see, there is a variety of circumstances surrounding the availability of the weather and climate data you might need. In some cases, you might be able to use the already-published data directly in your presentation. In other cases you might have to use other published data as a basis for your own calculations, because the final values you want are not themselves published. In still other cases you might have to discover other sources of data — some published and some not — or even make your own specialized observations.

I want to give you a rule of thumb: *never use published data directly unless it is exactly what you need*. Always take whatever steps may be necessary to get exactly what you need. I have become accustomed to viewing *published* data as "answers to questions they hope someone will ask." If the published data

CHART OF THE FIRST KIND	Observed variables included in the calculated variable				
Calculated variable	Hourly Temperature	Daily Maximum Temperature	Daily Minimum Temperature	Hourly Wind Speed	Hourly Solar Radiation
Monthly HDD	-	X	X		-
Monthly CDD	-	X	X		-
Daily average temperature[B]	X	X	X		-

Daily total of solar radiation on a south facing wall	-	-	-		X

Hourly Effective temperature (T_e) (wind chill)	X	-	-	X	-

answer questions you are not asking, get the answers you need. Some of the methods for doing that are discussed later in this chapter.

The second kind of chart connects weather and climate variables to atmospheric concepts. Here is an example — an incomplete, suggestive example — of such a chart. You can see at once from the three listed what I mean by **concepts**: they are ideas, or techniques, for comprehending and describing various kinds of environmental interaction among several individual processes.

You can also see at once that I have listed only a very few of the weather variables I might have listed. For two examples, I have not listed either the variables needed to make calculations on various aspects of the concepts of solar geometry and insolation (see **Chapter 4**) or even some of the variables —such as those for soil conditions — needed for the concept of the microenvironmental energy balance.

(B) As you might imagine, one can calculate (estimate) the daily average temperature from 24 hourly values, or from one value of the maximum and one of the minimum. The former is slightly more accurate but for many purposes the simplicity of the latter is preferable despite the potential for less accuracy.

**CHART OF THE
SECOND KIND**

Atmospheric Concept	Weather variables included in the concepts				
	Solar Radiation	Air Temperature	Humidity	Wind Speed	Species Varietal Constant
Microenviron- mental energy balance	X	X	X	X	-

Weather stress index	-	X	X	X	-

Average spring bloom date for a particular species	-	X	-	-	X

The third kind of chart connects a set of outcomes you wish to produce with a set of means for producing those outcomes. For example, you ought to be able to think of several ways to make use of physical principles to produce, *in a design*, warmer spring soil in a particular place, or less winter wind loading on a sunny wall. *Examples of this kind of chart are found in **Table 4-5** (microenvironmental energy balance) and **Table 5-1** (human thermoregulation).* As noted, I will leave the design of particular charts to you, to reflect your personal approach to these problems.

With these three kinds of charts, I suggest, you would begin to formulate the atmospheric component of a particular project by assigning importance rankings — priorities for that project — to each of the atmospheric concepts. With this list of priorities for concepts, you could then work backward and determine which weather and climate data you will need to acquire for your presentation. In addition, the process of assigning priorities will help you to formulate questions in your mind with which to approach your atmospheric ecologist, or to decide NOT to approach him.

How to know when you know enough. Up to now in this chapter, I have offered some ideas about what might be called "checklists" for assembling and using weather and climate data for your design and presentation. But checklists can go on forever. For some projects you might need only a short checklist, and for others you might need a long and complex one. *How can you tell when you have checked enough?*

The answer, of course, is that I cannot give you a general answer, and that knowing when you have checked enough is an art that comes with experience — your personal experience. I can, however, offer a couple of guidelines.

First, it seems to me important to recognize that "knowing enough" differs from one player to another in the game of environmental design. As a general rule, I suggest the following table to indicate where the major (**M**) and the minor (**m**) atmospheric expertise should lie on the planning/design team.

Player	Scales of expertise Macro	Meso	Micro	Chapters in this book
Regional planner	**M**	**M**	**m**	Macroscale: **4,5,9,10,11**
Project planner	**M**	**M**	**m**	Mesoscale: **3,4,5,6,11**
Project architect	**m**	**M**	**M**	Microscale: **2,3,4,5,6,8,11**
Landscape architect	**M**	**m**	**M**	All-scale
				processes: **2,3,4,9,11**

As usual, you will very likely differ with me on this partitioning of expertise, probably because, in the world outside of textbooks, one person plays several roles and must have the expertise to go with all his roles. Again, the point is not whether I am right or wrong. The point is that no one can "know everything" on a project. This table is simply a way to make that idea explicit in the context of this chapter. As we discussed in **Chapter 7**, of course, knowledge often must be available on a project by way of the technical consultant — in our case the atmospheric consultant. We will get back to that in the next section.

But how can an individual player know when he knows enough? I think there are 5 steps involved in reaching the point where you know enough:

1) develop a list of potential atmospheric impacts[1] on the project, prioritized according to their importance;

2) develop a description of the nature of each impact — a statement (in words, diagrams, or equations) of which atmospheric elements are involved and the processes by which each impact would take place;

3) develop a list of major alternative design responses to each potential impact;

4) develop a description of the likely design-environment *interactions* associated with each major alternative; and

5) develop a basis for choosing among the alternatives.

When you have completed the tasks in this list, you are at the point where you can make the decisions you need to make. It is easier said than done, as the saying goes, but when you have completed the tasks in this list — AND NOT BEFORE — you know enough. Making those decisions is beyond the "getting ready" stage that we are discussing in this chapter.

When do you need the consultant? Where, in the context of this section, does the atmospheric consultant enter the game? After the lengthy discussion of the entry and role of the atmospheric consultant in **Chapter 7**, it is relatively easy to answer that question now.

If you recognize several items on the list of major impacts in Step 1 of the 5-step list we have just developed, you need the atmospheric consultant early in the game. If your list for Step 1 contains few or no major impacts, you may not need the atmospheric consultant until later in the game. In this case of few or no items in Step 1, however, you ought to be prepared for the possibility that there are potential impacts you yourself do not recognize[2]. In any case, having the 5-step list in mind as you progress will help to clarify how you approach the atmospheric consultant, and it will enhance your ability to listen to his response.

Perhaps more important than either asking or listening to the atmospheric consultant, *the process of completing the set of tasks in the 5-step list* — probably with the assistance of the atmospheric consultant — *will permit you to know, with confidence, when you have a well formulated atmospheric component ready for presentation as part of the plan and the design*.

Overview: A paradigm for the management of a system

This chapter has consisted mostly of my ideas about checklists and guidelines for formulating alternatives, for making choices from among alternatives, and for reaching various objectives in the game of environmental design. I have been able to be more specific than I was in **Chapter 7**, but most of what I have had to say is still pretty general. Later in this chapter I will provide you with much more specific information about ecological alternatives, but before I do I would like to discuss two very general ideas connected — in the larger sense — with the game of environmental design: (i) management of a system, and (ii) the balance of nature.

Although our discussion of the game of environmental design has pictured it as a dynamic process, nothing much of what we have said about the *product of planning and design* — the project — suggests that it, too, is dynamic.

The essence of a dynamic system — feedback — was defined and illustrated in **Chapter 2**, and in **Chapter 5** the notion was presented that the feedback loops in Man's thermoregulatory system appear in the conceptual extension of the body and its clothing to architecture: the thermostatted components of buildings. In this sense, thermostatted bodies and buildings are automated **managers** of the heat balances of their respective systems.

Other features of bodies, clothing, and architecture are not fully automated, and require the conscious responses and participation of a human manager. For example, our bodies respond to hyperthermia by means of the feedback loop leading to perspiration and evaporative cooling. That is automatic. But if the evaporative cooling is not enough, our bodies send signals leading to conscious changes of posture, location, or some other response. We become participants in the management of our own heat loads. That management may lead not to changes of posture or location, but rather to opening or removing clothing. And so on. The body-clothing system is dynamic in its management of heat energy. The analogs of these examples in architecture are lowering sunscreens, removing storm windows, opening fixed windows, and so on.

Few landscape projects, I imagine, are automated in the sense of being thermostatted, but many are managed. The idea is easily comprehended if one extends the notion of "managed landscape" to include an agricultural cropping system. If you think about it, plowing, irrigation, and so on are forms of **intermittent management** of the heat load of a soil-atmosphere-plant system — a landscape unit. Intensive agriculture makes use of various forms of **permanent management** of the heat load: walls, windbreaks, and so on.

We have now introduced the term **management** into our discussion as being related to the dynamic nature of both the game of environmental design and of its products. As I have done with other concepts, I want to break the concept of management down into its components, in order to provide a generalized view — an overview — of it to aid discussions that follow. The breakdown of the concept is in the form of a **flow chart** — a paradigm — that links three other ideas we have considered: (i) the manager and his role, (ii) the model of the system, and (iii) the process of listing and then choosing from among alternatives.

Phase A: formulation of the problem. In my paradigm we begin with the manager and a description — in text, graphs, mathematical equations, or some combination — of the present state of the system. With the aid of the model of the system[C] the

(C) The essentials of any model consist of (a) a list of the components, elements, or players, and (b) a list of the functional relationships connecting the components. If you distill the description of any model — for example, the rules of a game — you will find it comes down to these two.

manager lists "all possible" future states of the system. The model may exist in many forms, but for the sake of discussion let us say it exists in two mutually supporting forms: (i) a perception in the mind of the manager and (ii) as a computer program. While the computer model may generate all the possible future states[D] more quickly and effectively, the manager may use his perceptual model more effectively — by means of intuition — to sort the list. Thus, the two forms of the model are mutually supporting.

PHASE A : FORMULATION OF THE PROBLEM

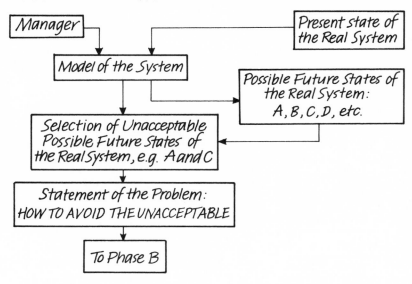

From the list of all possible future states of the system, the manager selects those that are *unacceptable*. This sorting consti- tutes one example of the choosing from among alternatives so often mentioned in **Chapter 7** as part of the dynamic game of environmental design. In the context of this paradigm, the state- ment of the problem consists of this list of unacceptable future states together with the implied statement that they are to be avoided.

(D) Since we are carrying on this discussion in the abstract and theoretical mode, I use the theoretical term "all possible" simply to suggest that the computer will doubtless generate a longer list than will the manager's perceptual model, though the computer cannot list any future states the manager does not make it capable, in its program, of recognizing.

Phase B: making the decision. In my paradigm, the manager enters Phase B with the list of unacceptable future states, from Phase A. Again with the aid of the model of the system, he lists all possible solutions — methods for avoiding the unacceptable future states. But some of these possible methods may be discarded when the manager applies his ethical system and finds some unacceptable. A solution involving the application of poisons, for example, may be ethically unacceptable.

PHASE B : MAKING THE DECISION

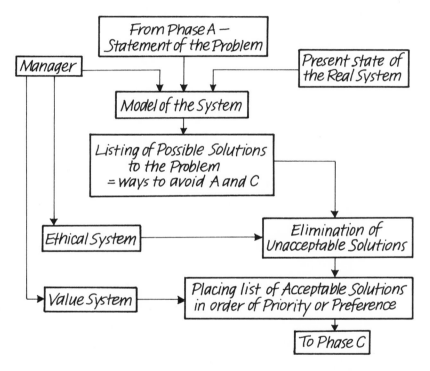

When the manager makes appropriate specialized cost/benefit analyses, his value system may find other possible solutions unacceptable. Use of a rare metal with highly beneficial thermal characteristics, for example, may involve costs and procedures that make it unacceptable according to the manager's value system[3].

Having sorted the list of possible solutions to find the ones that are acceptable, the manager then places them in an order of priority or preference. I cannot describe the general criteria for making this ordering — it is a matter peculiar to the manager and beyond the scope of this paradigm. *Expertise in making such an ordering, however, is the primary reason the manager has been assigned that role.* Making this ordering, of course, constitutes the act of "making the decision" about the solution to the problem. It is one more example of the choosing from among alternatives so often mentioned in **Chapter 7** as part of the dynamic game of environmental design.

Phase C: carrying out the decision to produce change. You might be tempted to end the process at this point, and to ignore the acceptable solutions that the manager has given lower priorities. I would instead prefer, in the names of reality, completeness, and the desire to emphasize the dynamic nature of management, to enter Phase C.

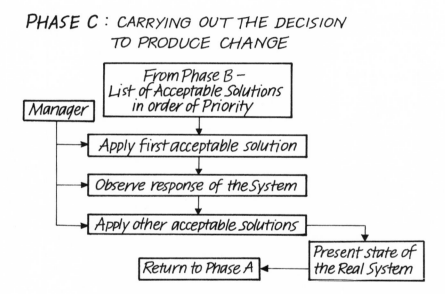

PHASE C : CARRYING OUT THE DECISION TO PRODUCE CHANGE

Manager

From Phase B —
List of Acceptable Solutions
in order of Priority

Apply first acceptable solution

Observe response of the System

Apply other acceptable solutions

Return to Phase A ← Present state of the Real System

In Phase C the manager implements the solution that, in his *decision*, he gave highest priority. If he is an effective manager, he will then observe the response of the system, since his decision may have been in error. If he was in error, the system may show less of a response — or, worse yet, an unexpectedly different kind of a response — than his calculations told him would result.

If an alternative solution must be applied — if his previous decision is *not irreversible* (see the next section) — he has alternatives available. This is my conception of the manager and what he does. As I have suggested, it can be applied to both the planning and the design process and to the project resulting from that process.

Overview: What's all this about The Balance of Nature?

The moment the planning and design team makes the results of its work public (or even before!) it is open to the charge that it will *upset the balance of nature*. Unfortunately, too often the levelling of this charge is the most important and visible way in which the public — labelled "environmentalist" — enters the game as an active player.

Since the levelling of this charge occurs so frequently, I want planners and designers to have a clear notion of what the phrase — upset the balance of nature — means to me. I want you to be able to see the phrase in an ecological context and to anticipate the situation in which you will need to respond to the charge as one of your moves in the game of environmental design. I certainly hope you understand that I do NOT believe the environmentalist is always an unwelcome player in the game of environmental design. In fact, I consider myself to be an environmentalist.

The term "balance", in the sense of **Chapter 2**, refers to the fact that all energy (or mass) flowing in a system can be accounted for. It does NOT refer to the idea that "nothing is changing." More particularly, it does NOT mean that the storage in each of the reservoirs of a complex system remains unchanged. That circumstance is described (see **Chapter 2**) by the term "equilibrium", in which INFLOW = OUTFLOW for each reservoir.

Using one's own financial assets as an analog, it is clear that "balancing" refers to an accounting in which you and the bank and whoever else is involved agree that the flows and storages, over a certain time period, are such and so. It does NOT mean that the amounts in the reservoirs called checking, savings, etc. have not changed during that time.

Thus, in the sense of **Chapter** 2, *Nature's accounts are always in balance but seldom in equilibrium*.

I am virtually certain that most environmentalists, in using the term "balance of nature", have an equilibrium, not a balance, in mind. They have in mind an ecosystem where the flow rates in the pathways and the storage in the reservoirs are not changing, even allowing for the idea that energy and mass may be flowing through the system. They have in mind a "steady state" in which the web of interactions is in "delicate balance."

A corollary of this view of the balance of nature, then, is that an ecosystem cannot be "natural" unless it is in equilibrium[(E)], and that, therefore, *change* (read "disequilibrium") *is bad*.

In my view, change is not inherently bad nor equilibrium good. *What is bad is change that is too rapid, and, in the face of imperfect knowledge about the system, irreversible.*

"Too rapid" implies several kinds of things. *First*, it implies that knowledge of the system is too imperfect to permit making even short term predictions with confidence; requiring, therefore, (see Phase C in the previous section) that constant monitoring and appropriate "midcourse corrections" — feedback in the form of alternative solutions — are needed to keep things within bounds.

Second, "too rapid" implies that the time lag to incorporate the corrections — alternative solutions —is so short that any corrections are "too late" and no longer appropriate. *Third*, these two imply that one has a basis for saying what is "appropriate" in the sense that knowledge of the system permits making a list of alternative responses that are rationally related to the list of clearly understood goals and criteria.

I trust you can see from this commentary the advantages for the planner/designer to have "knowledge of the system" and that that knowledge is what this book is all about. Anticipation of the consequences of manipulation of a microenvironment can come, for example, from an understanding of the concept of the microenvironmental energy balance. Said another way, understanding of the concept of the microenvironmental energy balance can reduce the chances that a system's response to manipulation will be *unexpected* (see Phase C).

Change in ecosystems is always taking place *naturally* — that is, without the hand of man. For examples, species are becoming extinct and species diversity in the standing crop (the "biota") is changing in the face of disease, physical disruption from fire, etc. What the hand of man introduces is different rates of change and a different list — sometimes shorter and sometimes longer — of the possible ways in which change can take place.

It seems to me that what environmentalists — including designers and planners, in the context of this book — should be working for, rather than no change, is a set of agreements to assure the following:

(E) **Chapter 9** contains a more extensive view of the concept of "ecosystem." In particular for the present discussion, while "natural" doubtless implies "equilibrium" in the minds of most environmentalists, "unnatural" — man-made —environmental systems may also be in equilibrium in the technical sense of the term.

(a) a list of mutually and clearly understood short term and long term goals and criteria for acceptability,

(b) a schedule of criteria — among them reversibility — for choosing among alternatives (Phase B),

(c) a list of rational alternative responses to a problem or a need (Phase C), and

(d) *Enhanced knowledge of how a system works before it is changed*. This last, of course, is the purpose of this book.

SUMMARY: if you don't really know how a system you are changing or managing works, be a gradualist and avoid anything irreversible!

Strategies for energy management in human microenvironments

We have indulged ourselves in some idealizing about the game of environmental design and several philosophical overviews of that game. It is time to get down to discussion of a few practical methods available to planners and designers. First, I want to make clear the difference between the terms **strategy** and **tactics**. "Strategy" refers to one's overall, long-range plan for reaching a major objective or goal. The goal of a game is to win, and one's strategy, for example, might be "defense."

The goal of the game of environmental design is to produce a project that functions well for all players: the client, the users, and the public. One's strategy for a commercial project, for example, might be "least cost construction and return rate no less than a certain value." This strategy is stated in fiscal terms, but it also implies many things about the strategy for energy management. "Tactics" refers to one's short-term responses to dynamic changes in the state of a system or play of the game. "Tactics" usually implies "methods." I intend to use these terms in what follows, making their meaning clearer to you as I go along.

As an example of a strategy for energy management in a human microenvironment — say a residence — passive solar heating of the building is to be used to the maximum extent possible. Another example is to use evaporative cooling of the building by means of close-in landscape elements. Strategies need not be exclusive, so that one may have a **mixed strategy**.

But strategies for energy management can be classified more broadly, and I suggest the following as a useful classification to keep in mind at the planning stage. In considering the list, have **Fig. 2-1** in mind. Clearly, one may use a combination of these in a mixed strategy, within one season and between seasons:

1) *control* — either increasing or decreasing — *of the inflow rates*, which includes
   a) passive routing of environmental energy impinging on the system, and
   b) active use of supplementary energy for either increasing or decreasing the inflow rates;
2) *control* — either increasing or decreasing — *of the outflow rates*, which includes
   a) passive routing of energy stored within the system, and
   b) active use of supplementary energy, usually for increasing outflow rates.

To make certain of your understanding of this classification, I offer a few specific examples. Both the passive solar heating of the residence (increasing +R in **Fig. 4-42**) and the evaporative cooling of the building by means of external landscape elements (decreasing +R in **Fig. 4-42**) come under the classification of (1a) — passive routing of inflowing environmental energy.

The use of fuels for heating (increasing +M in **Fig. 4-42**) comes under the classification of (1b) — active use of supplementary energy for increasing the inflow rate. The use of insulation in walls and ceilings comes under the classification of (2a) — passive routing of outflowing energy stored within the system (decreasing "-H" in **Fig. 4-42**); and electrically driven air conditioning comes under the classification of (2b) — active use of supplementary energy for increasing outflow rates (increasing "-H" in **Fig. 4-42**).

Strategies and methods for energy management in plant microenvironments

Generally speaking, the classification of strategies for energy management in plant microenvironments is the same as for human microenvironments: use of passive/active modes in the control of inflow and outflow rates. The scheme is implied in the contents of **Table 4-5**, where it is clear at once that active — energy intensive — strategies are much less often used for plants than for animals and people.

Table 8-1 connects *strategies* for energy management with intermittent, *tactical methods* in plant microenvironments. In doing so, it puts **Table 4-5** in context with the specific methods to be described presently. This minicatalog of methods derives mostly from agricultural management, rather than from the practice of landscape architecture; still, I trust the insights it provides will be useful, if only as a checklist.

As with any such classification, some of the entries in the table may be questioned. For one example, a white wall next to a high-valued crop will trap solar energy by reflection, but it will also store heat for nighttime release, as a wall in an urban canyon will do (see **Fig. 5-21**). For another example, nurse cropping and shelterbelting operate both through shading (mainly inflow of shortwave radiation) and through reduction of the ventilation rate (mainly outflow of sensible and latent heat).

Table 8-1 Methods and strategies for the management of heat energy in plant microenvironments

Method	Component of the Surface Energy Balance primarily affected	Strategy
Changing insolation		
Heat trapping by reflection	Incoming shortwave	Passive control of inflow
Shading	Incoming shortwave	Passive control of inflow
Changing surface characteristics		
Changing the albedo	Outgoing shortwave	Passive control of outflow
Plowing	Evaporation rate	Passive control of outflow
Mulching	Evaporation rate	Passive control of outflow
Changing the shape of the soil surface		
Soil packing	Soil heat flow	Passive control of inflow
Furrowing	Incoming shortwave	Passive control of inflow
Use of supplementary energy		
Irrigation	Evaporation rate	Active control of outflow
Flooding	Soil heat flow	Active control of inflow
Spraying cold water	Sensible heat	Active control of inflow
Burning fuels	Sensible heat	Active control of inflow
Wind machines	Sensible heat	Active control of inflow
Soil heating cables	Soil heat flow	Active control of inflow
Changing ventilation of the surface		
Hot capping	Sensible heat	Active control of outflow
Nurse cropping	Sensible heat	Passive control of outflow
Shelterbelting	Sensible heat	Passive control of outflow

Table 8-1 NOTES: (Also see Lowry, 1969, page 216; and Rosenberg *et al*, 1983)

Heat trapping by reflection — usually accomplished with light colored walls located on the side away from the sun. Usually permanent, but can be an intermittent tactic.

Shading — permanent (passive, strategic) or removable (active, tactical) vertical or overhead elements.

Changing the albedo — See the discussion later. Usually accomplished with dark or light, sprinkled or sprayed powders, or opaque plastic sheeting. Usually permanent, but can be an intermittent tactic.

Plowing — darkens the surface, but the more important effect is the introduction of air pockets into the soil. These pockets form barriers to both heat flow and moisture flow. An intermittent tactic.

Mulching — See the discussion later. Mulching is accomplished with natural materials or plastic sheeting. Natural materials — until they deteriorate — act as both heat and moisture barriers (as with plowing) while the effects of plastic sheeting depend on whether it is opaque or translucent. Opaque sheeting acts mainly as a moisture barrier, whereas translucent sheeting acts like a greenhouse: a barrier to both heat flow and moisture flow. Usually permanent, but can be an intermittent tactic.

Soil packing — usually accomplished with rolling machines in orchards. It does the opposite of plowing. It decreases barriers to both heat flow and moisture flow.

Furrowing — See the discussion later.

Irrigation — See the discussion later.

Flooding — used occasionally in bog-type croppings, such as cranberry, to introduce a water mass too huge to experience freezing during one or two nights. The water may be drained away after the freeze period is over.

Spraying cold water — delicate, near-freezing fruit blossoms are sometimes misted so that the release of latent heat (see **Chapter 4**) will retard (not prevent) cooling when the mist freezes on the blossoms. The method is very risky, not to say expensive. Too small an application rate results in evaporative cooling — the opposite of what is intended. Too large an application rate results in damage as branches break under the weight of ice. Definitely an intermittent tactic.

Burning fuels — See the discussion later.

Wind machines — See the discussion later.

Soil heating cables — this expensive method is successful with small stands of high-valued crops.

Hot capping — covering individual, or small groups, of small plants with plastic or paper covers protects them from wind and radiative cooling on cold spring nights. Definitely an expensive, intermittent tactic.

Nurse cropping — interplanting of tall and short species. The nurse crop is taller during the season when protection of the valued crop — mainly from wind — is important. Later, the valued crop may outgrow the nurse crop.

Shelterbelting — See the discussion later.

Methods for plant microenvironmental management: quantitative examples

The following methods are from **Tables 4-5** and **8-1**. The list is by no means complete, but the following presentations will supply you — unlike **Tables 4-5** and **8-1** — with several kinds of *quantitative* information on commonly used methods.

Changing albedo. When the shortwave reflectivity — the albedo — of a system's surface is changed, the fraction of the incoming solar radiation that is absorbed also changes. Lightening decreases the heat load on the system and cools it, while darkening increases the load and warms it.

As noted earlier, this change of color can be accomplished by application of either powders or sheets. What is more, one may apply powders and sheets to only parts of the system's surface. Plastic sheets usually, though not necessarily, go only between plant rows. **Fig. 8-1a** shows what happens to the solar heat load on a row crop when either the soil between the rows, or the plants themselves, but not both, are covered with a white powder[F]. In **Fig. 8-1a**, the upper curve represents the incoming solar radiation — the insolation — for each daylight hour, while each of the lower curves represents the amount of solar energy *absorbed* (the heat load) by the plant canopy. Whitening only the soil between rows seems to have a slightly greater effect — but in the opposite, warming direction — as compared with whitening the canopy, which achieves the desired cooling effect. Clearly, whitening only the soil increases the side-reflected sunlight on the plants.

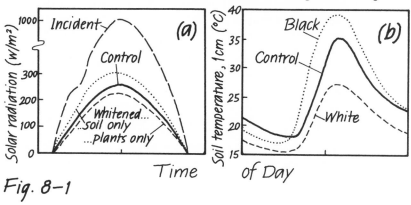

Fig. 8-1

(F) This example is from western Nebraska, where the growing season climate can be extremely droughty. This method has been tested mainly as an emergency response to hot, dry weather. Otherwise, as in a rainier climate, dusting leaves would be prohibitive.

In **Fig. 8-1b**, from a different study, the curves represent the soil temperature under each of three treatments: no change (control), lightening (white), and darkening (black). Darkening warms while lightening cools. The midday changes from an untreated soil, on this plot at this time of year, are between 5 and 8 degrees C.

Fig. 8-2 shows that, as expected from the discussions in **Chapter 3**, the temperature effects of changed albedo become smaller at greater depths. Only the cooling of white powder is shown, but darkening would show the same depth effect.

Mulching. Covering the soil surface with one of a list of various materials is called "mulching." As noted earlier, there are several variables involved: materials, thickness, and timing being the principle ones. "Natural" materials deteriorate, though at different rates, and in that sense they are temporary. Man-made materials — usually plastics — tend to persist and be more permanent.

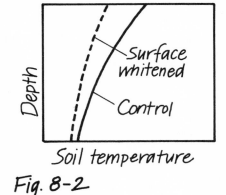

Fig. 8-2

Mulching is a technique very familiar to landscape architects, I am sure; but perhaps the following examples will emphasize not only the variety of materials but also — I suspect less familiar — the variety of results. With your knowledge of the concept of the energy balance (for example, **Fig. 4-45**) you will understand the discussion easily.

Fig. 8-3 shows the results of various mulches in terms of the midday temperature profiles just above and just below the soil surface. **Fig. 8-3a** shows the results for summer for three kinds of mulching material, while **Fig. 8-3b** shows the results for autumn at the same location, but for a different set of materials. The incomplete sampling of times and materials is from a larger list of tests, but it certainly suggests the variety of results. In both panels, the vertical line marked "C" is the temperature of the air farthest above the surface *at that time over the unmulched plot* — a kind of "reference" temperature.

Comparing the shapes and locations of the shaded areas with the "control" plot will permit a quick evaluation of the results for a particular material. For example, here are my evaluations. The comments about the changes in the energy balance are based on data I have not presented.

a) In summer (**Fig. 8-3a**) the *black plastic sheeting* warms the air near the surface, but cools the soil just beneath the surface, without much effect higher and deeper. The mulch has increased the amount of solar radiation absorbed at the surface (**R**), but has decreased the evaporation rate (**E**) effectively to zero. The extra absorbed heat has been diverted into heating the air (**H**) and away from evaporation (**E**) and heating of the soil (**B**). The same comments apply for the black plastic in autumn (**Fig. 8-3b**), but the sunlight is not so strong and the energy flows, and thus the temperature contrasts, are less

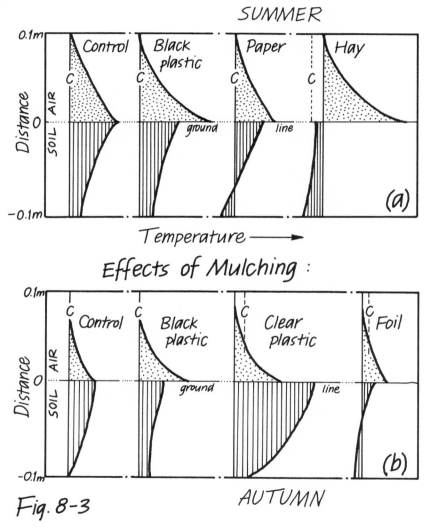

Fig. 8-3

b) In summer (**Fig. 8-3a**) the *paper* and the *hay* both cool the soil markedly, even though they both reduce the evaporative cooling. Less soil cooling, but cooler soils? Isn't that against intuition? The energy balance shows that the two mulches are more reflective than the bare soil (smaller **R**), and that they act more effectively as barriers to soil heat flow (**B**) — insulation — than they do as moisture barriers (**E**). So, less energy is absorbed at the surface, and less is used in evaporation (**E**), but even so, not much heat can flow into the soil through the insulator. These effects are exaggerated under the hay, as compared with the paper, because the hay is a thicker layer of insulation. In summer, the plastic and the paper scarcely affect the midday air temperature only 10cm above the surface, but the hay warms the air at all levels (the air temperature at 10cm is warmer than the reference temperature of the line marked "C").

c) In autumn (**Fig. 8-3b**) the differences of the results *with clear plastic sheeting* from those with the black, opaque sheeting are very marked. The effects of the black sheeting have already been discussed. The clear plastic does not result in the absorption of any more sunlight (same **R**), and it also reduces the evaporation to zero (**E** = 0). The large differences arise because the clear plastic acts like a greenhouse: it reduces both the warming of air (**H**) and the evaporative cooling (**E**), so that most of the absorbed sunshine goes into heating the soil (**B**). It is both a heat barrier and a vapor barrier between the soil surface and the air above. Even the air above the surface is cooled at the expense of warmer soil.

d) In autumn (**Fig. 8-3b**) the *aluminum foil* brings a general cooling to air and soil, simply because it is so reflective that little sunlight enters the system to do any heating of anything.

Different mulching materials change different parts of the energy balance, and so produce a wide variety of temperature effects. Since they are all "labor intensive" methods, and so are to be used only with high-valued plantings, *one should be very certain as to the results to be expected before selecting a mulch.*

Fig. 8-4 will serve to illustrate the fact that, not only is there a difference in effects between mulching materials (**Fig. 8-3**), and between depth in the soil (**Fig. 8-2**), but also between day and night. Referring to the effects of various "natural" mulches — corncobs, sawdust, and straw — the study showed that, though the mulch produced cooling at both 25mm and 75mm depths at midday, the mulched soil was actually slightly warmer than the bare soil (control) at 25mm just before dawn.

Finally, **Fig. 8-5** shows something you probably already know: the effects of "natural" mulching on evaporation are greatest with

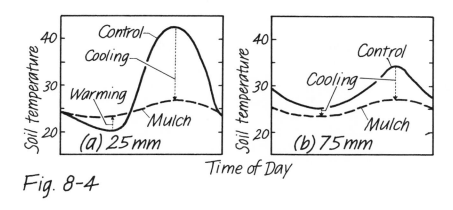

Fig. 8-4

small thickness, the addition of more mulch having little effect after the mulch is already rather thick. In the sense of cost/benefit for this costly method, this information can be valuable.

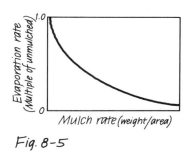

Fig. 8-5

Changing the shape of the surface. Furrowing the soil surface changes the patterns of insolation, and thus the patterns of energy absorption and heating. Obviously, the effects depend on the orientation of the furrows — north-south or east-west. Several separate radiative and thermal processes interact. The north-south furrows in **Fig. 8-6** show several kinds of effect on soil temperature. (a) Near sunrise the coldest soil is in the bottom of the furrows — exposed to the cold night sky and sheltered from the warming wind. (b) In mid- and late afternoon, the warmest temperatures are near the tops of the furrows, but (c) in the furrow bottoms the soil is warmer than it would be without the furrow, as seen by the soil cooling as you move horizontally from the furrow bottom.

You might suspect a furrow to act as a small valley — in particular, as a frost pocket (**Fig. 3-26**). It does in some sense, but the system is too small for substantial flows of air. That is, circulation cells (**Fig. 4-29**) don't form, and so heat is not transferred by moving air. Unlike the effects in larger valleys, temperature differences in furrows result not from density-driven flows of air, but mainly from differences in the angles at which the sunlight has recently

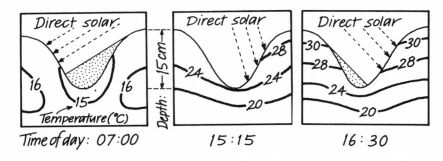

Fig. 8-6

been striking the soil surface nearest the point where the temperature is measured.

Supplementary energy: irrigation. While watering dry soil as an aid in the water economy of plants is well known, the addition of water for its thermal effects is less often practiced. In recent years warming of soils by addition of heat carried in irrigation water has been thoroughly investigated. The thought has been that water carrying excess heat away from nuclear power plants, rather than simply being allowed to cool in open ponds, could increase crop yields, or lengthen the growing season, in nearby fields if it were to be used for irrigation.

In other situations, irrigation with cold water has been used to lower the temperature of excessively hot soils, again for the benefit of crop yields. As with mulching, several interacting effects may produce unexpected results. The effects are that even when warm water is added to soil, watered soil cools by evaporation, relative to dry soil, and heat is conducted more efficiently — from warmer soil to cooler — in moist soil as compared with dry. For one thing, then, extra warmth of warm water, if it is not dissipated at the surface by evaporative cooling, may be quickly conducted to deeper layers, producing essentially no warming of the soil. When cold water is added, of course, only cooling takes place, since evaporation works in the same direction.

Another factor is that "warm" and "cool" are terms that must be used in comparison with some reference temperature. Even if the temperature of irrigation water is *warm as compared with that of deeper soil* — which is constantly near the daily average temperature of near-surface soil (see **Fig. 3-4**) — but *cool as compared with the midday temperature of near-surface soil*, the result will be a general cooling of all the soil. The opposite of this combination is irrigation with ground water for freeze protection in the cold season. Compared with the temperatures that would otherwise be experienced in near-surface soil, water from deeper layers is warm, and produces a short-term hedge against freeze. Soon, however, the evaporative cooling would undo the slight warming, so the method is a last resort and must be used only for short periods of cold.

If the evaporative cooling that counteracts the beneficial effects of warm irrigation can be prevented, warm irrigation can indeed be a benefit. Two ways to prevent the evaporative cooling are (i) the containment of warm water in underground pipes — not true "irrigation"[(G)] — and (ii) covering the surface with clear plastic sheeting.

Typical results of non-evaporative warm irrigation on crop yields look like this. The numbers in the table are expressed in terms of the percentage of the harvest weight of the heated crop, which is 100 for both plants. Although the differences are dramatic, the method is expensive and the value of the plantings — in this case crops — must be substantial.

Days after growth begins	Tomato			Broccoli		
	Unheated	Heated	Difference	Unheated	Heated	Difference
20	5.6	9.7	4.1	-	-	-
40	29.0	41.1	12.1	1.0	11.2	10.2
60	76.6	100.0	23.4	8.2	34.7	26.5
80	-	-	-	28.6	61.2	32.6
100	-	-	-	38.8	85.7	46.9
120	-	-	-	45.9	100.0	54.1

Supplementary energy: burning fuels. Protection against damage due to frost, in orchards with high-valued crops, is often achieved by the burning of fuels at selected points within the canopy. The fuels range from propane through oil to wood. Before society's concerns for air pollution, old tires were sometimes burned. In any case, **Fig. 8-7** suggests the way the supplementary heat is distributed through the lower air layers. The

---

(G) This form of warming is indistinguishable in its effects from the use of buried electrical soil heating cables. In terms of expense, however, the two methods may be quite different, depending on economic factors.

left panel shows two temperature profiles: one before heating and one after. The right panel shows the temperature difference — difference in heat content due to the supplementary heating — between the two profiles.

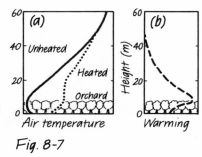

Fig. 8-7

What makes the method feasible is the fact that there is a temperature inversion in the lower layers of air — at least up to 60m in **Fig. 8-7**. An inversion (see **Chapters 4 and 11**) is the condition of cold air beneath warmer air, and it usually accompanys the clear, calm nights that bring the danger of freeze to an orchard.

As will be explained in **Chapter 11**, an air layer resists mixing and overturning — it acts as a lid — when it is in a condition of inversion. This being the case, a manager can release the supplementary heat in the lowest, coldest part of the air layer — within the orchard — and expect the heat not to be mixed (and diluted) through a very thick part of the layer. The heat is retained within the canopy where it is needed.

The major variables over which the manager has control in this situation are (a) the time burning begins, (b) the spacings between burning locations and (c) the rate at which fuel is burned at each location. If the air were not calm, or nearly so, there would be no inversion in the first place, and this method of protection would be of no use to the manager.

Supplementary energy: wind machines. Under the same starting conditions as were just described for an orchard in danger of freeze damage, a manager could make use of the fact that there is already a supply of warmer air above the treetops. Rather than create warmer air by burning fuel, he can mix the colder and warmer air with a wind machine — often an old airplane engine with propeller. While the propeller turns, the whole machine also turns slowly on a vertical axis. **Fig. 8-8** shows how this takes place under conditions with a very slight wind drift from right to left.

In **Fig. 8-8a** the same kind of temperature inversion is shown as in **Fig. 8-7**. The value "D" is the difference between the lowest temperature in the orchard — at treetop level — and that at the height of the wind machine. The larger is D, the "stronger" is the inversion, and the warmer is the supply of warmer air relative to the colder.

In **Fig. 8-8b** the warming response — the temperature rise after the mixing begins — is shown related to distance from the wind machine and to the strength of the inversion. As in the case of burning fuels, the major variables over which the manager has control in this situation are (a) the time mixing begins, (b) the spacings between machines and (c) the rate at which the machines rotate around their vertical axes at each location. If the air were not calm, or nearly so, there would be no inversion in the first place, and this method of protection would be of no use to the manager.

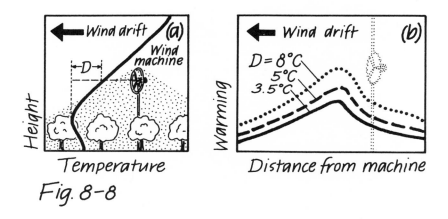

Fig. 8-8

Shelterbelting. As shown in **Fig. 3-23**, barriers across wind-flows reduce the speed for a distance — 10 to 15 barrier heights — downwind. As noted in the discussion of **Table 8-1**, shelterbelts — barriers formed of living trees — produce changes in convection around downwind plants, which in turn reduces heat and moisture losses from the plants to the passing airstream. Clearly, this is an advantage if the plants are stressed by cold and/or drought. Under hot conditions, however, the plants can be further stressed in the lee of a shelterbelt. **Fig. 8-9** shows, in proportional terms, how various measures of environmental moisture respond to a shelterbelt.

Sources and limitations of published atmospheric data

As a planner and designer who is sensitive to atmospheric ecology, you will need to know where to find weather and climate data for preparation of various presentations. I have assembled most of the basic information you will need about published data sources in **Appendix B**.

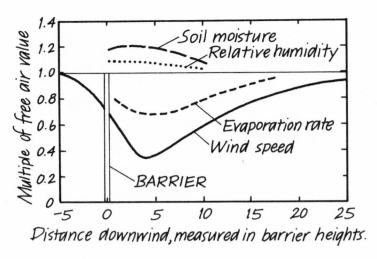

Fig. 8-9

After reading **Chapters 3** and **4** you should be well aware that observations from one place may have very limited applicability in describing conditions at another place, even though they may seem "very close together." Most weather observations in the world are taken for the benefit of the air transport industry, and the variables chosen, together with the locations of the observatories, reflect that.

Limited applicability or not, the published data are usually what we must work with, unless we decide to invest in specialized observations of our own. In either case — using what is available or making specialized observations of our own — an atmospheric consultant has a contribution to make to a proper game of environmental design.

Chapters 3 through **6** will tell you that published observations, described in **Appendix B**, seldom include much information about *microclimates* and *local bioclimates*. However, **Chapters 9, 10,** and **11** will show you that there are many concepts, valuable for planning and design, that pertain to *mesoscale* and *macroscale climates and bioclimates* (see the discussion of **Fig. 7-1**), and that the data in **Appendix B** are suitable for applications of these concepts.

An atmospheric consultant can often use the published data to make good estimates of microclimatic conditions not observed directly, but the methods he would use are beyond the usual capabilities of designers and planners. Because well informed designers and planners oftentimes need to prepare their own descriptions of mesoscale and macroscale climates, however, in the remaining sections of this chapter I have presented a modest catalog of methods (and cautions) you will be able to use for that purpose.

Calculations and manipulations with data

I realize there are several very fine references that will show planners and designers how to assemble weather and climate data into meaningful components for analysis of a project site. Givoni (1969) is complex and theoretical, while Olgyay (1963) and Loftness (1982) are more concise and practical. You might know of still other references, but this section provides (I believe) some additional useful insights into the general nature of weather and climate data — as opposed, for example, to generalized data that might be treated in a textbook on statistical methods — and some additional thoughts on statistical methods useful in analysis of a project site.

Kinds of variables. Most atmospheric variables change their values continuously, as do dry bulb temperature and dewpoint temperature, or wind speed and direction. These variables not only change their values continuously, but, in going from one value to another, they take on all values in between. Other variables — such as the total rainfall from one storm at one place — may not change continuously in time, but by their nature they may potentially assume any value over a range of values.

These two kinds are both called **continuous variables**. We sometimes forget this when we record values only at specific times, or to certain degrees of precision (for example, "to the nearest degree").

Other variables are **naturally discrete variables**, such as the number of hourly observations in a month for which the dry bulb temperature was below freezing. Discrete variables can have only certain values, and no values in between, but the possible values need not be whole numbers (as in this case of hours). The simplest kind of discrete variable is the **binary variable** that can have only two values: on/off, yes/no, or above/below.

We often convert continuous variables into discrete variables by the way we observe and record them. For example, I have already mentioned recording a temperature to the nearest degree. For another example, in North America we record sky cover in "tenths" — from zero to ten, the fraction of the sky hemisphere occupied by cloud — but in Europe sky cover is recorded in "octas" — from zero to eight, the fraction of the sky hemisphere occupied by cloud. European clouds behave no different from American clouds, but our discrete variables do.

Variables, in addition to being continuous or discrete, can be **bounded** or **unbounded**. Temperature is doubly unbounded, because it can (theoretically) have high (positive) values of any magnitude and low (negative) values of any magnitude. Wind speed, however, is semi-bounded since it cannot have a value less than zero. Sky cover is doubly bounded — above and below.

While both wind speed and wind direction are continuous variables (though both are usually recorded as discrete) wind direction is also a **circular variable**. Wind speed is semi-bounded, while wind direction is doubly bounded. In graphical presentations, this poses no difficulty; but in various techniques of statistical analysis, intended for continuous and/or unbounded variables, wind direction's circular nature can pose problems. For example, if wind directions are recorded in degrees (usually they are published in tens of degrees) then 0 and 359 are numerically at extremes of the scale; but actually they are side-by-side. Computers making calculations don't readily understand things like that.

I mention these various characteristics of data simply as an extra dimension for your thought about using data: an additional vocabulary with which to use a consultant, and an extra way of thinking about visual presentations of data.

Frequency distributions. When a mass of data is assembled into categories, with the number of times each category is observed, the result is called a frequency distribution. Here is a frequency distribution, expressed both as to the number of observations (**frequency**) and as to the percentage of all observations (**relative frequency**).

Value	0	1	2	3	4	Total
Frequency	28	153	92	247	42	562
Frequency (%)	5.0	27.2	16.4	44.0	7.4	100.0

The value of the variable itself (first line — called the *measure*) and the frequency variable (second line) are both discrete and doubly bounded, since the total number of observations has a maximum value. The percentage (relative frequency) variable (third line) is bounded but continuous. Because it is continuous, someone has had to make a decision about the *precision* to be used — to the nearest tenth of one percent — one part in thousand.

There are examples of decisions about precision that are ridiculous. A classic one is to express, to one part in ten thousand (two decimal places on a percentage scale) the average snow depth for a month over an entire state. Though calculating it is quite a straightforward exercise, of what earthly value could that number be? Be sure, when you present, to have a reason for whatever precision you choose. Similarly, be sure you understand — even question — why a consultant chooses whatever precision he chooses for a variable.

Another form of frequency distribution is the **relative cumulative frequency** (RCF). The RCF is shown as follows. It is simply the sum of the relative frequencies up to and including the value under which it is tabulated. It tells you the fraction of all observations that are at or less than the value tabulated.

Value	0	1	2	3	4	Total
Frequency (%)	5.0	27.2	16.4	44.0	7.4	100.0
Cumulative Frequency (%)	5.0	32.2	48.6	92.6	100.0	

The obvious reason for using percentage (relative frequency) is that it reduces all sets of observations — no matter how numerous — to a common scale. The reason for using the RCF (relative cumulative frequency) will be clear in the next section.

Central tendency: mean (average), mode, and median. If you had to express the set of observations above in a *single number*, what number would you choose? You would almost certainly choose a number somehow "typical" — representative — of all the numbers in the data set. Though not the only choice, the average value — the **arithmetic mean** — is the one we almost always use. It is one of the measures of *central tendency* that we use to say "what is typical."

You know already that, to calculate the average, you "add up all the values and divide by the number of values." That's the way you calculate, for example, the average size of an electric bill, or the average temperature for a month. When there are a lot of observations of a discrete variable, as there are in the table of 562 observations above, the fast way to get the arithmetic mean — it always gives the same answer as "adding up all the values and dividing by the number of values" — is shown in this table.

(Frequency) x (Value)	28 x 0 = 0	153 x 1 = 153	92 x 2 = 184
	247 x 3 = 741	42 x 4 = 168	
	Sum = 1246	Mean = 1246/562 = 2.217	

You calculate all the products (Frequency) x (Value) — the pairs of numbers in the first two lines in the first table above — and then add them all up and divide by the total number of observations.

Another measure of central tendency is the **mode**. It is the value of the variable — the measure — that occurs most often. In the set above, that value is 3. In the table of relative frequency, it is the value with the largest percentage. In this little data set, with only a few categories (also called "classes" — 0,1,2,3, and 4) the mode isn't a very useful number. If there are many more classes, the mode is often quite useful at expressing what is typical.

The third common measure of central tendency is the **median** — the value that has half of the observations less, and half of the observations more than itself. It is the value that corresponds to RCF = 50.0. In our data set that value occurs somewhere slightly above 2, since the RCF for 2 is 48.6. That looks sensible, since the mean (average) is 2.217, but for these discrete data, such a number cannot itself be an observation. In the set we are discussing, it is not possible to observe a value of 2.217 — only values of 0,1,2,3, and 4.

Have you ever known a family with the average number of children — 2.2? I suggest no single resolution to the dilemma. All I am doing is exhibiting characteristics and cautions for the weather and climate data you will encounter. The best resolution to the dilemma in this case seems to be the mode — 3 — since that is one of the values it is possible to observe[P,9].

Central tendency: running, or moving, average. A variation on the theme of the arithmetic mean involves knowing time — the sequence in which the observations were taken. Here is part of the observations tabulated above, listed in groups of five, in the order they were observed. I have listed only 40 of the 562 observations, so this is a **subset** of the larger set. We will have a further discussion of *the importance of sequence* presently.

(i) 0,0,1,1,3	(ii) 4,2,4,1,3	(iii) 2,0,0,1,0	(iv) 3,4,4,2,1
(v) 1,1,2,1,4	(vi) 3,4,2,1,1	(vii) 0,1,2,4,3	(viii) 2,1,2,2,2

As I am sure you know, I have made up these observations just to provide a set of data to discuss. The first point to be made is that this subset has some characteristics similar to the whole set, and some characteristics quite different. For one thing, the modal value is 1 for the subset. Without dwelling on the differences, you can see them most clearly by comparing values of the RCF for both.

Value	0	1	2	3	4	Total
Frequency	6	12	9	6	7	40
Frequency (%)	15.0	30.0	22.5	15.0	17.5	100.0
Cumulative Frequency (%)	15.0	45.0	67.5	82.5	100.0	

But we need to discuss the running average. Let us say that each observation represents one day, so that each group — (i), (ii), etc. — represents a sequence of five days. A *5-day running average* is obtained as follows:

a) take the average of the first group of five (observations 1 through 5) and tabulate it under the middle (third) day;
b) take the average of the second group of five (observations 2 through 6) and tabulate it under the middle (fourth) day;
c) take the average of the third group of five (observations 3 through 7) and tabulate it under the middle (fifth) day;
d) and so on.

Here are the results for the first five 5-day groups: 1.0, 1.8, 2.2, 2.8, and 2.8. I trust it is clear to you how you would calculate 3-day, 4-day, or 6-day running averages, and so on. This technique "smooths" the data over time. It is often helpful in finding any "cyclic" behavior in a sequence, if there is any present.

Central tendency: weighted average. The arithmetic mean is only one kind of weighted mean: a mean in which each observation gets equal *weight*, or importance. It is often desirable to weight different parts of an average differently. In **Chapter 4** we examined the weighted average for soil heat capacity (see *(P_{11})* in Chapter 4). In the case of the running average, let us say we want to accentuate the middle value(s) of each group as we take running averages. For examples, in a 3-day running average we assign weights of [w = 1, 2, 1] for the three values; in a 4-day running average we assign weights of [w = 1, 2, 2, 1]; in a 5-day running average we assign weights of [w = 1, 2, 3, 2, 1]; and so on.

Dispersion. Central tendency — what is "typical" — tells only part of the story about a set of observations. Two sets can have the same mean but quite different variability around the mean — dispersion. The simplest way to express dispersion is with the **range** — the difference between the largest and the smallest values observed. Thus, we have climatic data, for one station, giving the *diurnal* (daily) temperature range, the *monthly* temperature range, the *annual* temperature range, and so on[4].

Sometimes the range is exaggerated by the occurrence of one or two extreme values we consider very "atypical." The conservative technique in a case like this is simply to ignore the 1 or 2 largest, and the 1 or 2 smallest — don't ignore too many in a small data set — and then with what is left, take the **quasi-range**.

Another standard measure of dispersion — like an exaggerated quasi-range — is the **inter-quartile range**: the difference between the value where RCF = 75% and the value where RCF ′= 25%. It is the range that includes the middle 50% of the observations within it[H].

Sequence. In dealing with running averages, we have raised the matter of the sequence in which observed values of a variable occur. In my estimation, any proper description of the climate must include information on the three major aspects of (i) central tendency, (ii) dispersion, and (iii) sequence.

The importance of sequence in a variable such as temperature is that, if any patterns of repeating sequence can be detected, they offer the basis for some degree of *predictability*. **Fig. 8-10** suggests two kinds of sequence pattern: (a) periodic variation, or periodicity, and (b) "stepwise" variation.

In Fig. **8-10**, the small dots connected by lines represent the sequential plotting of observed data, and the underlying heavier lines suggest the kinds of pattern often disclosed by calculation of

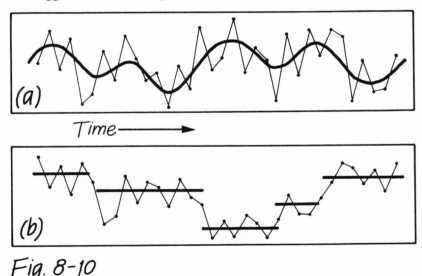

Fig. 8-10

(H) Statisticians concern themselves with variances, standard deviations, sample variances, standard errors of the mean, and other complicated measures of dispersion. If you know about them, so much the better; but I believe I would only make more confusion if I discussed them here.

running averages. If the pattern is uncovered, and the analyst has reason to believe the particular behavior is real, or will repeat, there is a measure of predictability in the data that might not be readily apparent without the running average.

The subject of the analysis and interpretation of sequences — called **time series** — is extremely complex, and the additional aspect of predictability is far beyond the scope of this book. I wish principally to introduce these terms to your professional vocabulary, and to show a linkage between the analysis of time series and the search for predictability. Presently, in discussions of wind behavior and air pollution, I will demonstrate an additional aspect of sequences that is important for *explanation*, as opposed to prediction.

Return period, residence time, and turnover time. There is an interesting concept, used often in ecology, for describing a bio-geo-physical system, in particular one aspect of the climate of a place. It contains information about *sequences* expressed in terms of an *average*. Depending on the particular form and interpretation placed on it, it is variously called the return period, the residence time, or the turnover time.

"Return period" is a term usually associated with relatively rare, individual events, such as floods, strong winds, and so on. Suppose, for a simple numerical example, you had a lengthy record of floods[1] in a certain watershed. Each year was designated as a year *with*, or a year *without* a flood — a binary variable. Your record showed 5 floods in 80 years. You would calculate the return period as (80/5 = 16 years), because, *on the average*, assuming no particular *sequence* of repeating flood years, there would be 16 years between floods — a flood event would *return* every 16 years, on the average.

Now, look at a more complex version of the return period by considering the following table of 30 intensities, arranged in descending order of intensity. These events could be floods, wind speeds, or something else. They may not be all of the observations in an entire record, but they are all of the "extreme values" in the record — rare events in some sense — and we no longer are dealing with a binary variable. Another fact about the data in the table: these observations are not just one per year. This record has been observed during 22 years.

---

(1) For each of these rare events, you would need to specify a **threshold value** which the event must exceed in order to qualify. For floods: a certain water depth, flow rate, or whatever. For winds, a certain 2-minute average, peak gust speed, or whatever. And so on.

Intensity	Serial	Intensity	Serial	Intensity	Serial
111	1	76	11	67	21
106	2	74	12	67	22
105	3	74	13	66	23
97	4	72	14	65	24
91	5	71	15	65	25
89	6	71	16	64	26
87	7	70	17	62	27
84	8	68	18	60	28
84	9	68	19	60	29
79	10	68	20	60	30

Fig. 8-11 shows a frequency distribution — in the form known as a **histogram** — of these data.

It is possible, of course, to use the simple binary version of the problem we just discussed by saying something such as "any intensity equal to or greater than ($\geqslant$) 80 qualifies to be counted as a rare event." There are 9 of them in the table, so the return period of the events, defined this way, is (22/9) = 2.44 years = 2 years 3 months and 20 days.

But let us look at the problem in its expanded version — taking account of the information about the range of intensities. *Suppose we ask a design question*: "What maximum intensity must the structure I am designing withstand if the planned life of the structure is 2.5 years?" It is the same as the question: "What intensity

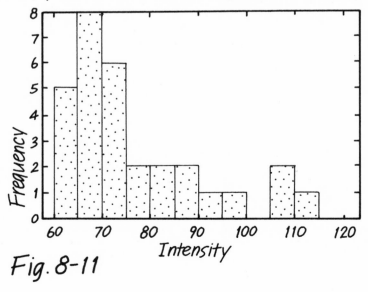

Fig. 8-11

has a return period of 2.5 years?'' To answer, do the inverse of what we did before: (22/2.5) = 8.8. In the table at serial number 8.8 we find 84, which is the intensity to be designed for, or used as the basis for multiplying by a ''safety factor.''

Residence time and **turnover time** are essentially the same thing, and both are relatives of the return period. Look at **Fig. 2-1**, where you see a reservoir with an inflow rate and an outflow rate. Assuming the reservoir's storage is in equilibrium — (inflow = outflow) — we say the residence time of whatever is flowing through the reservoir is (storage/flow rate). This fraction answers the question ''On the average, how long will a molecule (or a calorie) entering the reservoir remain in the reservoir before it departs?'' It is, you see, another example of *information on sequence expressed as an average*.

Fig. 2-5 gives a numerical example, though the storage is not quite in equilibrium. In the lefthand reservoir, the residence time is (10/13 = 0.77 time units); in the middle reservoir, the residence time is (10/984 = 0.01 time units), while in the righthand reservoir, it is (1000/876 = 1.14 time units).

Probability. Most of us know about probability in a sort of intuitive way. ''Probability'' is usually expressed as a percentage or as something like ''3 in 10 it will happen.'' Incidentally, this last statement is equivalent to the statement ''the **odds** are 3 to 7 it will happen'' or ''the odds are 7 to 3 it will not happen, or against it.''

Go back to the problem with 5 floods in 80 years, and assume *a flood year had no more than 1 flood*. The probability a year will have a flood is then *estimated* to be (P = 5/80 = 0.0625), or 6.25%, and the probability a year will be flood-free is (1 - P) = 0.9375, or 93.75%. I say *estimated* to be sure that you and your client both know that, with only a sample of years, we cannot know the true probability for certain[J]. Before going on, notice that *the probability is one divided by the return period*: (0.0625 = 1/16).

The probability we just discussed is called **simple probability**. Suppose we wish to inquire about the **joint probability** that a year, at that place, will have both a flood and a tornado. For now, we will assume that having a flood and having a tornado are **independent events** — having one doesn't make having the other

---

(J) Now that we are beginning to assemble a lot of long records as scientific observation progresses, we realize one of the aspects of ''climatic change'' is that geophysical systems, like climate, are not **stationary** — their probabilities change slowly through the years. It is hard to account for this in estimating probabilities, but that problem is far beyond this book.

any more or any less likely. We will look at this independence again in a moment. If the simple probability of a flood is (P_F = 0.0625), and the simple probability of a tornado (P_T = 0.125), then the joint probability of a year experiencing both is [(P_F)x(P_T) = (0.0625)x(0.125) = 0.0078], or about 4/5 of one percent. Notice, you cannot multiply percentages and get the answer. That is, (6.25%)x(12.5%) = 78.125%, and 78% is most certainly not the answer!

Contingency. Dealing with contingency in the context of probability is accomplished through the idea of the **conditional probability**(P_{20}). Suppose you have reason to think that floods and tornados are *not independent* — a tornado is more likely in a year (or a month) with a flood, and vice versa. Suppose, to extend our example, that there have been 10 tornados in 80 years, and 8 of them occurred in years with floods. To keep things simple, we will tabulate the five flood years according to whether they have had at least one tornado (Yes - N tornados), or none (No) this way:

Flood Year 1: Yes - 2 tornados	So that 2 tornados occurred in other years.
Flood Year 2: Yes - 3 tornados	So that 3 out of 5 flood years
Flood Year 3: No	have had at least one tornado.
Flood Year 4: Yes - 3 tornados	So that 8 out of 10 tornado
Flood Year 5: No	years have had a flood.

Now we have two conditional probabilities: (i) the probability of at least one tornado *given that* a year has a flood; and (ii) the probability of a flood *given that* a year has a tornado. The symbols and the calculations for these two conditional probabilities are:

(i): ($P_T \mid P_F$) = (3/5) = 0.60   (ii): ($P_F \mid P_T$) = (8/10) = 0.80.

Since these conditional probabilities are much larger than the simple probabilities (P_F = 0.0625) and (P_T = 0.125), we know at once that floods and tornados tend to occur together, at least within the same year. When the conditional probabilities are much smaller than the simple probabilities, we know that floods and tornados tend NOT to occur together.

The Markov chain. We have ignored the fact, in the analysis just completed, that tornados seem to cluster — more than one in a year when any occur — while floods do not. We can analyze this aspect of "sequence" by use of a Markov Chain. Since the flood/tornado data are limited, I will use the sequence data from an earlier discussion, shown below.

(i) 0,0,1,1,3	(ii) 4,2,4,1,3	(iii) 2,0,0,1,0	(iv) 3,4,4,2,1
(v) 1,1,2,1,4	(vi) 3,4,2,1,1	(vii) 0,1,2,4,3	(viii) 2,1,2,2,2

Suppose, for example, that the values — 0, 1, 2, 3, and 4 — are *daily* ratings of air pollution severity, with 4 being the most severe. We want to see if our suspicion is correct: that severe air pollution days tend to come in clusters. We say that we suspect severe air pollution days exhibit **persistence**. The truth of the suspicion may be important — sequence may be important — in problems of *public health planning*, if foul air, once it arrives, tends to remain longer than a single day. The same idea holds also for the question of whether or not very hot and humid weather — it is related to heat stroke — exhibits persistence in an area. How do we find out if our suspicion is correct?

The 40 days tabulated in sequence represent, let us say, a typical month and a half in the middle of the air pollution season for a certain city. For illustration, we tabulate the **transitions** from one day to the next, *for the first eight days (seven transitions)*, as follows. We say (0>0) means "a 0 rating day is followed by a 0 rating day"; (4>3) means "a 4 rating day is followed by a 3 rating day"; and so on.

(0>0) ; (0>1) ; (1>1) ; (1>3) ; (3>4) ; (4>2) ; (2>4).

There are five classes of air pollution severity here, but only data from 40 days — 39 transitions. If we had enough data, we would try to analyze all five classes, but with so few I am going to make the severity measure into a binary variable: (0, 1, or 2 = not severe) and (3 or 4 = severe). After making that **variable transformation**, I will construct a **transition matrix** and enter the 39 transitions, as follows.

Today	Tomorrow		
	Not severe	**Severe**	**Total**
Not severe	21	5	26
Severe	7	6	13
Total	28	11	39

Note the totals in the bottom line are not exactly the same as the totals in the righthand column, but they are off by only two. This is because the first day and the last day are each part of only one transition; all others are parts of two transitions, and the grand total is 39 both ways. From this table we form another table containing simple and conditional probabilities:

Today	Tomorrow		
	Not severe	Severe	Total
Not severe (N)	$P_1 = (P_N|P_N)$ $= (21/26) = 0.81$	$(1-P_1) = (P_S|P_N)$ $= (5/26) = 0.19$	$(P_N) = (26/39)$ $= 0.67$
Severe (S)	$P_0 = (P_N|P_S)$ $= (7/13) = 0.54$	$(1-P_0) = (P_S|P_S)$ $= (6/13) = 0.46$	$(P_S) = (13/39)$ $= 0.33$

We can make a wide variety of inferences and calculations using this Markov transition matrix, but I will suggest only a few[5]. My intention here is only to introduce you to the subject, so if you are interested in using Markov chains, look in your library under the mathematical topic of "finite numerical methods."[5]

We have the suspicion that severe air pollution days tend to come in clusters. Can we confirm that by looking at the transition matrix? Look first at the last line, for which today was severe. *Given that today is severe*, what is the probability tomorrow will be too? In other words, what is the probability the cluster — the "spell" — of severe days we are in today will continue at least until tomorrow? We see at once that that probability is $(P_S|P_S) = 0.46$, while the probability the spell will end tomorrow is $(P_N|P_S) = 0.54$ — somewhat greater. Does this show there is no clustering? Not necessarily, because, although the probability the spell we are in today will continue at least until tomorrow is less than the probability the spell we are in today will end tomorrow, there are only a third to a quarter as many severe days overall — severe days are relatively rarer.

Looking at the data another way, we can say that *if there were no clustering* the 13 transitions in the second line would be distributed in the proportion (26:13) — the overall proportion of days. That is, *if there were no clustering*, the numbers in the second line would be more like 8 and 5 rather than 7 and 6, so there is a weak tendency to clustering. The answer is just not really clear from the matrix without some further testing, beyond the scope of this book to describe. What is very clear from the matrix, however, is that days that are NOT severe tend very strongly to come in clusters — to exhibit persistence.

The principal utility of this discussion of Markov chains is the way of thinking about sequences, transitions, and persistence, and the way in which the probabilities of different "chains of events" can be estimated[5]. As with most things in statistics, the estimates can be made with ever greater confidence the longer is the record — the larger is the sample — of observations available. One six-week period is scarcely enough for really studying air pollution, but it has been enough for illustration. A second way to have greater confidence in these estimates is to be sure that the data are all from one season — in this example, they are all from "the air pollution season." Adding similar data *from the same season but in other years* would be the best way to increase the sample size.

Regression and Correlation. These two terms are frequently encountered in statistical analysis of data, but are often misunderstood — often being considered to be the same thing. A complete discussion of the details of these methods is beyond the scope of this book, but I do want to make a few general remarks indicating what they are, what they are not, and how they might be used in solving design and planning problems. A complete discussion of these two topics can be found in any modern introductory statistics textbook.

"Regression" refers to the family of methods one uses to examine the way in which the value of one variable in a problem — rainfall rate, for example — *is associated with* the value of one or several other variables — for example, atmospheric pressure and wind direction. I have been careful to say "is associated with" rather than "depends upon" because sometimes, when systems are not fully understood, which depends on which is not known.

The most common *visual* presentation of the ideas considered in regression is the **scattergram**, in which two related values — one from each variable — are plotted as a single point, or dot, on a graph with two coordinate axes. If there is a sizeable number of data pairs — a large sample — and the scatter of points forms a recognizable visual pattern, we usually assert that we have uncovered the form of the association, or relationship, between the variables scaled on the axes. The "pattern" may be linear, curved, two intersecting lines, or a complete absence of pattern.

If a recognizable pattern is detected, an investigator usually converts the data, mathematically, into a regression equation, or **regression model**, of the relationships detected. With these models, one makes predictions, or estimates, of unobserved values of one variable from known values of the others. Often, investigators are disappointed when two variables do not show a nice, neat, recognizable pattern in a scattergram. They seem not to understand that sometimes the absence of a pattern may be very important information.

"Correlation" (co-relation) refers to the family of methods one uses to examine the degree to which the association between variables is strong or weak. Most commonly, a "**correlation coefficient**" is calculated for a set of data pairs, and the value of that coefficient is taken as a measure of the strength of association[K]. Again most commonly, the "least squares method of linear regression" is used to calculate both the regression model and the value of the coefficient. Many investigators fail to recognize that, while the "least squares" part of the method gives the measure they seek, the "linear" part of it assumes beforehand that the scattergram will be a straight line. If the true relationship is in the form of a curved line, for example, the association might be excellent; but the "linear correlation coefficient" will be small, and the investigator, who incorrectly assumed beforehand that the scattergram will be a straight line, may conclude the association is weak. In such a case, reality may be recaptured by use of the more complex, but more appropriate, "least squares method of *curvilinear* regression."

When projects are planned and designed by well known structural engineering methods, or methods based on well known physical theory — "by the book" so to speak — the behavior of the project system is usually well correlated with the variables used in the design. Such projects are usually static or commonplace, or both.

Novel and dynamic projects for which there is no "book" are the ones in which, it seems to me, the methods of regression and correlation become useful. They become useful in planning and designing for variable traffic flows, for intensity-limited energy demand (when peak demands temporarily exceed capacity), for wind-rain relationships, and so on, in novel or non-standard projects for which only approximately representative projects are already in existence and generating data on performance. Regression and correlation become useful also in evaluating the performance of recently activated projects — Phase C of the paradigm for management of a system.

Wind roses and trajectories. As with regression and correlation, I wish here only to mention, describe, and generally evaluate two aspects of the methods commonly employed in analysis of wind data. For a more complete treatment of these methods, see Stern *et al* (1973, Appendix E).

A "wind rose" is a graphical form of frequency distribution for the circular variable of wind direction — like a circular version of

---

(K) Complete and perfect association yields a value of 1.0 for the coefficient — positive if direct and negative if inverse. If the correlation coefficient is 0.80, for example, one concludes that the variable cast in the mathematically independent role in the calculation explains *the square of* (0.80) = 0.64, or 64%, of the variability of the variable cast in the mathematically dependent role in the calculation. Notice that the same conclusion is reached from a value of -0.80 for the correlation coefficient, since a negative number squared is positive.

the histogram in **Fig. 8-11**. It shows (at least) the relative frequencies of *wind direction*[L] in the record from a certain observatory. It sometimes shows, also, the relative frequencies of *wind speed*. The precise forms of wind roses are many — sometimes they are works of graphic art — but they nearly always appear as bars of various lengths radiating outward from a center, roughly in the form of a "rose." In the sense of the discussions earlier, a wind rose depicts *dispersion of the data* on a circular, discrete, doubly bounded variable.

My point in bringing this up is that, like any frequency distribution — like **Fig. 8-11** — a wind rose is static and gives no information on sequence. So what? If you are laying out the directions of runways at an airport, or the air intakes for a passive cooling system, what you want to know about is frequency. If you are doing public health planning for the impacts of persistently stressful weather conditions, as suggested earlier, you want to know about the sequences of ventilating wind speeds, rather than about their average or frequency, or about their direction. But if, as in problems to be discussed in **Chapter 11** on air pollution, you are interested in *where materials move in the atmosphere* (and in flowing water as well), you must have information on speed and direction sequence[L]. In *none* of these problems, please notice, are you interested in the *average wind direction*.

A "trajectory" is the path followed, through space and time, by an identifiable parcel of mass moving in a fluid. If one releases a balloon, so constructed that it will rise and fall only because of wind rather than because of buoyancy, the ballon will trace a trajectory across the landscape. If a swarm of these balloons is released, their *average trajectory* can be *estimated* by the analysis of the *sequence* of wind speed and direction (velocity) observations taken near the point of release. Rather than using words, let me demonstrate, with a set of imaginary wind velocity observations, the importance of sequence in many important problems involving the analysis of wind data. Just picture in your mind's eye the differences, between Case A and Case B, in travel of pollutants — their trajectories — being continuously emitted from a stationary point source. The entries in the table are wind velocity observations, in sequence (speed/direction), taken at the source of the pollutants.

Obviously, these data were invented just for illustration. For one thing, they have only two wind directions represented: W (from

---

(L) As was mentioned in **Chapter 2**, air currents are named for the direction from which they come. Furthermore, when information on both speed and direction are combined *in one observation*, the observation is of **velocity**.

the west) and E (from the east). For another thing the total travel eastward equals the total travel westward in both cases. The difference between the cases is the difference in their sequence. Otherwise the data are the same in both cases.

Sequence		Total flow	
		E	W
Case A	5W, 6W, 6E, 3E, 3E, 4E, 5W, 8E, 7W, 9E, 10W	33	33
Case B	5W, 6W, 5W, 7W, 10W, 6E, 3E, 3E, 4E, 8E, 9E	33	33

They illustrate several things. You can perhaps see these things best by tracing the trajectories for yourself with pencil and paper. First, if a balloon were to travel with the wind in these two cases, being released at the start of either sequence, it would return to the starting point in both cases. That means that *the relationship between starting point and ending point is the same regardless of the sequence.*

However (the *second* point), where the balloon travelled during its flight — its trajectory — would be quite different with the different sequences. In Case A, the balloon would have travelled as far as 11 units to the east and ten units to the west, passing the starting point once in between and spending about half its time within 2 units of the starting point. In Case B, however, the balloon would have travelled 33 units directly to the east, and then 33 units right back to the starting point, never having been near the starting point in between. If the balloon were a gob of polluted air, you can easily see the importance of sequence.

Sample design problems with atmospheric components

The list and kind of projects, on which an atmospheric consultant would be a contributor to the work of the planning/design team, is endless. I would like to suggest, by a short list, something of the variety of projects on which I see an atmospheric consultant could make a contribution. I expect, among the projects on my list, to suggest two kinds of atmospheric component: (i) a rather straightforward kind that involves details an experienced planner/designer could address, and (ii) a less obvious aspect that an atmospheric consultant would be in a much better position to anticipate. My plan is that this short list will include regional planning, architectural, and landscape architectural components, increasing in scale from the very local, to the scale of the urban region, and to the scale of the geographic region.

1) *The large rehab commercial block*. Details of this project are in **Appendix C**. The building is a large commercial building in the shape of a "U", with interior skylit court facing east. To the immediate west lies a major topographical ridge — it rises 600 feet within a quarter of a mile of the block. The building was formerly a regional warehouse for a catalog sales concern. It is to be remodelled into a trade center, with offices, display areas, and attractive space for receptions and meetings.

The design question requiring an answer is whether or not to enclose the east-facing court with glass, making it a huge atrium space. On the one hand, the enclosure might increase the annual heating bill by requiring the heating of more space, especially since heat would tend to rise above the level of the atrium occupants. On the other hand the glass might decrease the annual bill by permitting passive solar heating of the space, and by reducing the total surface area exposed to exterior winds. The principal source of weather data for the area is for the regional airport 5 miles to the east.

2) *Trees to cool urban canyons*. The design question requiring an answer is whether or not street trees planted in urban blocks will cool the canyon spaces by evaporative cooling as they transpire. The question is an old one (see **Chapter 1**). My own answer is in **Appendix C**.

3) *The sick trees on the parkway*. Many of the lovely trees on the parkway — all of one species — are apparently very sick. They have been maturing in place for many decades and are a much-loved aspect of the approaches to the city, especially when they show their fall colors. The parkway is laid out from northwest to southeast. A transect perpendicular to the roadway shows the following sequence: (a) a quadruple row of trees (parallel to the roadway), (b) 2 lanes of traffic, (c) a double row of trees, (d) 4 lanes of traffic, (e) a double row of trees, (f) 2 lanes of traffic, and (g) a quadruple row of trees. There are distinctly more problem trees in the north plantings — (e) and (g) in the transect — than in the south.

The trees remaining — not yet cut down — show signs of both disease and insect infestations, which are probably related to disease. Naturally, as the landscape architects have worried about what to do, the list of possible environmental factors explaining the problem has quickly grown sizeable: (a) greater water stress on the trees exposed to the direct sun, (b) nutrient deficiency, (c) compaction of soil, and interference with water economy, (d) air pollution from regional sources and from the parkway traffic, and (e) mechanical damage from lawnmowers.

At first glance, only the first environmental factor would seem to be related to the fact that there are distinctly more problem trees in the north plantings. A case could also be made, however, that the parkway traffic could be such a factor, since days when pollutant concentrations build up are days when there is little ventilation, but on these days what little air movement there is is from the southwest toward the northeast.

One immediate reaction is that a monoculture — all one species — is a well known invitation to the kind of problems the parkway trees face. The "solution" would be to cut the remaining trees down and begin again with several species. For obvious reasons, other solutions are being sought.

Assume for our purposes here that the immediate question requiring an answer is whether it is the water stress on south-exposed trees or the localized air pollution causing the pest problems. Here is an example of a different kind of task for consultants. Clearly, the question cannot be answered with certainty *without measurements being made* to see if one hypothesis or the other is favored. A "theoretical" answer would just not suffice. An atmospheric consultant of the kind I described in **Chapter 7** would be able to guide the program of biological and environmental measurements and tend to the interpretations of the results.

That may sound like a platitude, but when the program of measurements is being prescribed, all concerned will want to be certain that (i) only necessary measurements be taken, (ii) they are designed and taken properly, (iii) they are taken only when appropriate and necessary, and (iv) they lead to as clear an answer as possible about which hypothesis is favored. The contents of this book will go a long way toward providing the insights to set up and carry out such a plan of measurements.

4) *Climate, humidity, and air conditioning.* Using "air conditioning" in the common way — "cooling air with coincident moistening and/or drying" for comfort of inhabited space — we may discern three major options: (i) passive cooling, (ii) active evaporative cooling, and (iii) active refrigerant cooling. In one sense, this is a local design problem. In a more important sense, however, it seems to me to be a planning problem on a regional or national scale.

Passive cooling involves reallocating energy already present rather than using supplementary energy. It modifies existing atmospheric conditions by means such as reflection, shading, heat storage in structural mass, and natural ventilation. *Active evaporative cooling* involves exposing outside environmental air to wetted materials, and then using energy to circulate the resulting moistened and cooled air through the inhabited space. *Active refrigerant cooling* involves both dehumidifying and cooling air from the inhabited space by means of refrigeration. Energy is used to circulate air and to circulate and compress a refrigerant[6]. In the process, both heat and water mass are moved from the inhabited space to the exterior environment.

Over large, geographical, climatic regions the choice from among these three is probably obvious, at least on the basis of economics. *Along the boundaries* of these large regions, however, the choice may not be clear. I propose that the three options lie on spectra from (as already suggested) passive to active, and from technologically undemanding (simple) to technologically demanding (complex). On what atmospheric — climatic — spectrum do they lie?

I have discussed the rudiments of defining this atmospheric spectrum in **Appendix C**, where guidelines to thinking, rather than actual calculations, are featured. I suggest, in extension of my remarks about **Fig. 5-12**, that calculations concerning this atmospheric spectrum might some day become important in light of a national urban housing policy as it relates to a national energy conservation policy.

5) *The prevailing westerlies.* The following quotation from Landsberg (1981, page 255) could as well have been included in **Chapter 1**, **Chapter 11**, or **Chapter 12**. It echoes comments made just above (including the problem of the parkway trees) about analysis of wind information. I have presented it here as an example of a design problem with a clear atmospheric component — a problem with a considerable historic dimension, as well as being a simple, crisp example of what happens when interdisciplinary communication fails to take place. Landsberg was a principal and influential proponent, especially in international circles, of the themes of this book concerning interdisciplinary communication.

"Each profession can only add a narrow facet to the total design spectrum. The climatologist has in the past, more often than not, been left out of the planning process. In fact, sometimes the most primitive, if not ludicrous, reasoning about atmospheric conditions has been incorporated into urban plans. In many eastern United States cities the planners have arranged industrial zones on the eastern fringes. They had heard about the "prevailing westerly winds" and thought this location would carry the industrial air pollutants downwind from the urban area. But the prevailing (or most frequent) wind direction is not the relevant factor. Rather, the wind directions for the so-called stagnation weather patterns are pertinent. These are often easterly [*from* the east] in that area, rather than westerly and would carry, with low mixing heights, the polluting plumes back over the city."

6) *Human well being and the texture of the urban fabric*. My final example of a design problem with a clear atmospheric component also centers on wind and refers to the scale of the cityscape, rather than to a localized project. The example is based on an analysis by Oke[7]. The design question requiring an answer is about the *optimal* height, ground plan, and separation dimensions of blocks of urban buildings.

The physical dimensions of city blocks determine four *competing* aspects of the well being of people using the canyon spaces among the blocks: (i) shelter, (ii) thermal comfort, (iii) dispersion of foul air, and (iv) solar access. Clearly, shelter implies more dense clustering, while dispersion and access imply a more open geometry. The implications for thermal comfort, of course, depend in part on the nature of the regional climate. One can take the attitude that the physical dimensions of city blocks — the urban fabric —are determined building-by-building rather than by plan, and so the design question of optimizing is moot. Or one can take the attitude that not all cities need be shaped building-by-building in the future. Oke's answer to his own question is interesting.

Using mathematical and engineering information of the kinds suggested in figures of this book, such as **Figs. 3-20**, **3-22**, **4-41b**, **5-4**, **5-16**, **5-17**, and **5-22**, Oke reached the conclusions that (a) the (H/W) ratio (see **Fig. 5-22**) should be in the range of 0.4 to 0.6; while the (roof area/lot area) ratio should be in the range of 0.2 to 0.4; and (b) traditional European urban forms are climatically more favorable [for human well being] than more modern forms, especially North American ones. In particular, sky scrapers are a "poor climatic form" on the basis of dispersion and solar access near the ground. Of course, the reasons that existing sky scrapers are numerous, though bioclimatically far from optimal, are clear to all.

Presentation of designs and plans

We have come to the end of the main part of this book — the part in which I have exhibited and explained the natural processes and local consequences of atmospheric ecology, and in which I have connected that knowledge of process to the work of designers and planners. Before going on to the "Supplements" — the processes and consequences at regional and global scales — I want to close with a reprise of some things I have already said.

Probably the best context for what I want to say here is that of the preparation for a principal presentation, by the designer/planner, to clients and team members. As professionals, you know better than I do what goes into this preparation; but, as we have considered several times, the *atmospheric* components of presentations are often given short shrift because of the absence of well-developed channels of communication between the design professional and the atmospheric ecologist.

One example of the results of poor communication is the one, just mentioned, concerning placement of industrial areas downwind of the "prevailing westerlies." Here, apparently, the design professionals proceeded to make major design decisions, without consultation, on the basis of only a very shallow understanding of a major concept about the atmospheric environment. Published data were available for analysis that would have avoided the planning error. It is very likely the planner, in his shallow understanding of the concept, made use of the wrong data — probably seasonal averages — when the correct data — daily and hourly observations — were available.

Another example of a less-than-successful decision with a central atmospheric component is Candlestick Park, in San Francisco. Built in the lee of a major hill, in a climatic region with a strong afternoon sea breeze nearly every afternoon and evening, especially in the summer (see **Fig. 3-24**), the ball park has suffered ever since opening day from strange wind behavior. While it has become (some will grudgingly admit) a major home-field advantage to the local ball clubs, it is often very uncomfortable for the fans. The point in the present context, however, is that designers went to considerable pains to make use of technical consultants well versed in wind problems. They used wind tunnel models to simulate wind effects, and the team had every reason to believe the design would be successful. It wasn't. What happened?

One could say that the design team didn't ask the right questions, but, in all honesty, they went to considerable expense, having been given a pre-determined site and knowing the problem was a clear example of very complex interactions among structure, terrain, and the atmosphere, to ask and answer the right questions, and in the right way. For one thing, they recognized that physical modeling was necessary because (i) published weather data were simply not up to the task and (ii) mathematical modeling would have been too crude because of the inherent complexity of the problem.

So here we have two examples — the prevailing westerlies and the winds at Candlestick Park — that can represent for us two ends of a spectrum of faulty design response to the atmospheric environment. In the latter case (as it seems to me from a distance) no clear, avoidable error was committed by the design team. Some solutions to complex problems simply don't work out as well as the team predicted, because of the inherent complexity. As is sometimes said "whether we like it or not, there are questions to which there are no answers." If there was avoidable error, it seems to me, it was at the front end, in insistence that the site was suitable, or at the after end, in not making plans for flexible response measures — Phase C management — after the structure was built.

Not so in the former case — industrial siting. The error was eminently avoidable, since both consultants and appropriate data were available. The design team didn't start with the right atmospheric questions, let alone reach out to a proper consultant. More specifically, they confused atmospheric goals — having the atmosphere carry unwanted effluent harmlessly away from population centers — with atmospheric impacts, alternatives, and questions — "Under what conditions will the atmosphere achieve the goal set for it, and *under what conditions will it not?*"

How does that relate to the preparation for a presentation, by the designer/planner, to clients and team members? Now to my point — my reprise. My answer to my own question is that, when planning and design decisions are made, especially when the decisions will have major effects — large in space and long in time — going through the five steps[M] to get to the point where you know enough is well worth the extra effort.

I want to make that point in another, particular way, based on my own experience with presentations. Very seldom is the atmospheric component of a presentation omitted completely — it is often included merely as a "hedge against tradition" — but quite often it is given short shrift. To my way of thinking, if it is unimportant, it should be omitted with a brief, reasoned explanation of why it is deemed unimportant. If it is not unimportant, it should never be given short shrift. In either case, the judgment about importance should, especially in the case of major, fundamental plans and projects (such as the industrial siting), involve someone knowledgeable in the fickleness of the atmosphere.

Assuming the designer/planner follows the five points to the letter, there remains another point I want to repeat about the proper use of appropriate data: never use published data directly unless it

---

(M) Professionals sometimes think of "cookbooking" as beneath their professional position — as reducing to a pat list an art form that is what they offer as professionals. I prefer to think of my five points as a checklist, rather than as a cookbook, to be used for the same reasons that an airline pilot uses a takeoff checklist before his aircraft leaves the gate. The stakes are high, and the insurance of a checklist, properly and consistently used, is very cheap. Anyway, to recall, the five steps are: 1) list potential atmospheric impacts, 2) list which atmospheric elements are involved in each impact, 3) list major design alternatives for each potential impact, 4) describe the likely design-environment interactions for each major alternative; and 5) develop a basis for choosing among the alternatives.

is exactly what you need — always take whatever steps are necessary to get exactly what you need. In the case of Candlestick Park there were no published data appropriate to the task. Even special observations at the site would fail because the structure had to be in place before the interactions would appear. The designers had to do what they did: study a windtunnel model.

In the case of the downwind siting of industrial areas, the planners (almost certainly) accepted monthly — or even annual — averages of wind directions as their basis for decision. The hourly observations (see the discussion above about wind *sequences*) were in the published record, but the planners did not pose, and then try to answer, the question *"Under what conditions will it not?"*

As final examples of inappropriate use of published data, I remind you of the cases of (a) the average snow depth for a month over an entire state, and (b) use of relative humidity as a measure of the moisture component of a site's regional climate. Dewpoint describes *moisture content* of the atmosphere better than relative humidity, and the effective temperature or the Weather Stress Index (both incorporating information on moisture) best describes a a climate's *human discomfort*, and the implied need for energy to achieve comfort.

Finally, I have presented these examples in the context of the presentation, because, in my experience, *the time of preparation for the principal presentation is the time the definitive thinking is done*. Like a candidate, or a debater, trying to anticipate the questions he will be asked, so that he can fashion the perfect, convincing response, the designer/planner preparing for a presentation considers all sides — all alternatives — and critically examines, in detail, the implications of each of the possible responses. From the list of possible responses he chooses, professionally and with proper consultation, which ones he will accept and which ones he will reject. And on his ability to formulate, anticipate, and choose successfully he lets his professional reputation rest. Unlike the candidate, or the debater, he will not be so readily able to escape the consequences of his choices: they will be there for all to see for a long time.

Notes

(1) The term "impact" used here, and in following considerations, implies that the system encounters something, and that the encounter somehow represents a potential disruption of the system. The encounter causes a potential departure from the "equilibrium," or the "optimum," state of the system. In addition to the following discussions of the concept of management

and "The Balance of Nature," I want to comment here that the equilibrium, optimum, state of the system is the planned, managed-for, and desired state. In the case of a human habitation, that would be the state of minimal use of supplementary energy to achieve stable internal conditions near those best for human comfort.

(2) As discussed in **Chapter 7**, the entry of a technical consultant takes place when a designer/planner recognizes he has reached the limits of his own understanding of a technical problem and its alternative solutions. The understanding I am referring to applies to the tasks in each of the 5 steps of this discussion. The entry of a technical consultant takes place only at the invitation of the designer/planner, and I fully recognize that your willingness to issue that invitation — at whichever Move in the game — constitutes an act of considerable professional integrity. The answer to the question "When do you need the consultant?", therefore, is that you need him when you no longer feel honestly comfortable and confident that you are still within the limits of your own technical expertise on the subject of the design component. I suspect you have known that answer all along. What I suspect you have not known (see the **Preface**) is where to find an appropriate atmospheric consultant, and how to understand what he would tell you. It is these last two needs that this book is intended to fill.

(3) I realize that unacceptable states, unacceptable solutions, and value systems can get hopelessly snarled in such a discussion. But keep in mind my purpose in setting out this paradigm in the form of a flow chart: breaking the concept — the process — of management down into subunits so that you will see the three elements in action in a dynamic game: (i) the manager and his role, (ii) the model of the system, and (iii) the process of listing and then choosing from among alternatives. A likely snarl occurs to me: we have not yet suggested that a future state may be judged "unacceptable" simply because it is somehow *suboptimal* — less acceptable than the present state or a known possible state. It may be acceptable in the sense that it could be lived with, but unacceptable if the process has improvement as its primary goal. A smaller snarl is undone when we recognize that "no change" is a decision, whether or not it is acceptable.

(4) These things can get complicated when you use published data and/or talk to a consultant. For example, unless you are careful you might miss the differences among (a) the diurnal temperature range for a single day, (b) the average diurnal range for a date, (c), the average diurnal range for a month, and (d) the

difference between monthly average maximum and monthly average minimum temperatures. While I won't dwell on such complications, I will suggest that, for some applications, the differences may be important. First, I will define each one:

(a) the diurnal temperature range for a single day: the difference between the maximum temperature and the minimum temperature for a particular day, let us say it is a September day.

(b) the average diurnal range for a date: for example, if you have data for 10 years, you will have (in the whole set) ten values of (a) for September 16. (b) is the average of these ten ranges.

(c) the average diurnal range for a month: you can think of this average in any one of three *equivalent* ways.

(i) In your 10-year sample you have 10x30 = 300 September days, each with its own value of (a), and an average value of these 300.

(ii) In your 10-year sample you have 10 Septembers, each with its own average value of 30 values — and the average of these 10 September averages.

(iii) In your 10-year sample you have 30 September dates, each with its own value of (b), and the average of these 30 September averages.

(d) the difference between monthly average maximum ($T_{Max,ave}$) and monthly average minimum ($T_{min,ave}$) temperatures: ($T_{Max,ave}$) is an average of 300 daily values, and ($T_{min,ave}$) is an average of 300 daily values. (d) is the difference between them, and it is actually a *fourth equivalent* to (c).

So what? Suppose you need to plan for the design or the operation of a facility that closes for the season just after Labor Day each year, and you need to know the effect of the day-night temperature swing near Labor Day. You may find a published value for **(c: i, ii,** or **iii)** — the average diurnal range for a month — or for **(d)** — the difference between monthly average maximum and monthly average minimum — but they include information from the mid- and late-September dates when cooler, stormier weather may be common. What you need — and they probably will not be published, so you'll have to calculate them — is values for **(b)** — the average diurnal range for each of the first few days of September.

(5) Though doubtless not the latest book on the subject, I have found Kemeny *et al* (1966) to be very readable on this subject. For the more mathematically adept reader, here are a few of the calculations you can make with a Markov transition matrix (named for a Russian mathematician). Noting each formula is a

string of products, you see where the name "Markov chain" comes from. All the calculations are preceded by "The probability that a spell of —."

— non-severe days will last *exactly*
  n days $= (P_1)^{n-1} (1-P_1)$
— non-severe days will last *exactly*
  4 days $= (0.81)^3 (0.19)$ $= 0.100$
— severe days will last *exactly*
  m days $= (1-P_0)^{m-1} (P_0)$
— severe days will last *exactly*
  3 days $= (0.46)^2 (0.54)$ $= 0.114$
— non-severe days will last *at least*
  7 days $= (P_1)^{7-1} = (0.81)^6$ $= 0.282$
— severe days will last *at least*
  5 days $= (1-P_0)^{5-1} = (0.46)^4$ $= 0.04$

(6) See McGuinness, Stein, and Reynolds. (1980, Chapter 5)

(7) The example is based on an article by Professor Oke (Oke, 1987a), who is perhaps the world's best known urban climatologist since the passing of Professor H. E. Landsberg.

Peptalks

(P_19) Perhaps you think these comments are getting a bit esoteric. As noted, I mention these various characteristics of data simply as an extra dimension for your thought about using data. If you already know these things, so much the better. They represent what my experience tells me will be most useful to you as a designer/planner.

(P_20) In discussion of the next two subjects — contingency and the Markov Chain — the math may look a bit forbidding to you. It is just a matter of counting and keeping clear in your mind what you are counting. The mathematical symbols are not essential. It is the ideas that count, and the effort to understand will be worth it.

9. Mother Earth
Earth as an ecosystem

In this chapter we begin the final section of the book, in which the contents are very condensed versions of three major topics in atmospheric ecology: Earth as a planetary ecosystem (**Chapter 9**), the large scale climate and life zones of Earth (**Chapter 10**), and Earth's atmosphere as a garbage dump — air pollution (**Chapter 11**).

I have labelled this last section "Supplements" because its contents are not likely to be of direct, or even frequent, use in the everyday practice of planners and designers in the ways I hope the contents of the first eight chapters will become. This **Chapter 9** takes a stand-off look at Earth in relation to our sun and to our sister planets, as a basis for you to understand and appreciate the uniqueness, the fragility, and the robustness of our spaceship home.

Chapter 10 is an extension of **Chapter 9** in the sense that it also takes a stand-off, *macroscale* look at Earth's atmosphere and at the various sorts of habitats for life found within its folds. My principal intention in presenting the contents of **Chapter 10**, please note now, is to make the point that the climate and life zones of Earth are where they are, in the forms they are, for quite understandable reasons. *"Habitats" are not sprinkled at random across Earth*, but form a comprehensible pattern that, once understood, can supply insights to the designer/planner who may find himself involved in a project in a part of Earth quite unfamiliar to him. Making this point could be a book in itself. My version is quite concise, though it may be demanding if you choose to see it through.

Chapter 11 contains the concepts my experience tells me will be of most recognizable use to you as regards our atmosphere's potentials and limitations for assimilation of our aerial effluents. These effluents of ours are added to those of "Mother Earth" as they flowed into and out of the atmosphere before Man began to multiply. That is, Earth's atmosphere has always contained gases, liquids, and particles other than the combination we call "air", but it seems our additions to those "natural" materials have caused problems for us and for Mother Earth.

In these remaining chapters, I intend to follow the scheme I set out in the beginning: first we will discuss "what" about topics as they arise, and then we will examine "why" about them.

Our solar system

Somewhere in the universe, way over at the edge of a minor galaxy we call the "Milky Way", is a very minor star we call "our Sun." There are millions of galaxies and billions of stars, and who knows how many stars have planets grouped in what we call a "solar system?" Our solar system has nine planets, of which Earth is the third closest — 93 million miles, or 150 million kilometers — from the sun. **Table 9-1** shows how they line up, all dimensions given as multiples of the dimension for Earth. There is no simple relationship among distance, mass, size, and period.

Table 9-1   A Comparison of the Planets in the Solar System

Planet	Distance from the Sun	Mass	Radius	Period (time to circle the Sun)
Sun	0.0	-.--	109.42	-.--
Mercury	0.39	0.06	0.38	0.24
Venus	0.72	0.81	0.95	0.62
Earth	1.00	1.00	1.00	1.00
Mars	1.52	0.11	0.53	1.88
Jupiter	5.20	319.40	10.91	11.86
Saturn	9.54	95.15	9.12	29.46
Uranus	19.18	14.67	3.85	84.01
Neptune	30.06	17.22	3.86	164.80
Pluto	39.44	0.18	?.??	247.70

Note: These data are based on information from Goody and Walker (1972, page 14) and Tarbuck and Lutgens (1976, page 362).

Sun and Earth seen from space

Closer to home, look at the way Earth moves around the sun, in **Fig. 9-1**, as it would be seen by a space traveler approaching our solar system. Careful observation would disclose seven key facts about this movement, in addition to Earth's average distance from the sun, which is the average radius, mentioned earlier, of Earth's slightly elliptical orbit:

(i) Earth circles its orbit in 365.25 days;

(ii) Earth spins on its own axis once every 1.00 days;

(iii) Earth's axis is tilted from the plane of its orbit by 23.5 degrees;

(iv) Earth's axis is always pointed at the same point in space;

(v) Earth's northern hemisphere points most directly at the sun in June, and Earth's southern hemisphere points most directly at the sun in December;

(vi) Earth is closest to the sun in January and farthest in July; and

(vii) Earth's farthest distance is about 7% greater than its closest distance.

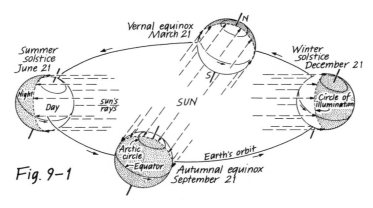

Fig. 9-1

Fig. 9-2 shows a different view of Earth from space. This view is presented in many books, and it provides important information. Unfortunately, it suggests that Earth wobbles on its axis between summer and winter, and back again to summer. Earth doesn't really do that, so I want to explain the basis for the presentation, since it does provide important information. My way of explaining this involves an imaginary experiment, so get your imagination into gear.

Imagine you are seated in a space chair, travelling around the orbit with Earth always just in front of you. As you travel around the orbit in the course of one year, *with the sun always to your left*, Earth and the sun's incident beams would appear to you as they do in **Fig. 9-2**. In this presentation, the axis appears to change orientation with season, even though we have said in Fact (iv) above that it does not[(A)].

Why go to all this trouble? For one thing, it emphasizes Fact (v) about the two hemispheres pointing most directly at the sun in opposite seasons. In showing three other features, **Fig. 9-2** demonstrates several other important facts. These *features* are (i) the line — called the **circle of illumination** (nearly the same as what NASA calls the **terminator**) — separating the sunlit hemisphere of Earth from the hemisphere in shadow, and (ii) the small, local hemispheres located at each point on Earth's surface. Each of these small hemispheres is the **vault of the sky** as seen from that place. The third feature is (iii) the set of latitude circles known as the **Equator**, the **Tropics**, and the **Polar Circles**, defined implicitly in the following.

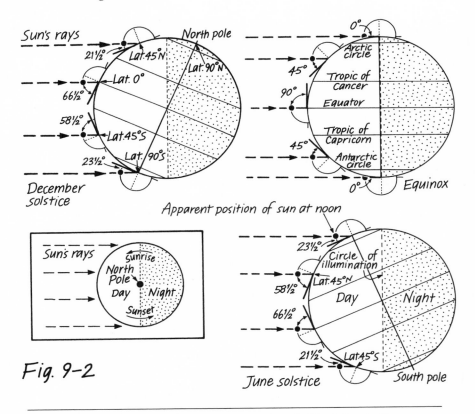

Fig. 9-2

(A) In actual fact, you would see the views exactly as in **Fig. 9-2** only if your space chair were slightly below the plane of the orbit from December through June, and slightly above the plane of the orbit from June through December. This is such a nitpick I have put it in this footnote.

The *facts* that these features demonstrate are:

(1) The circle of illumination passes through the poles on only two days each year, called the **equinoxes** (March 21 and September 21);

(2) On any day of the year, the circle of illumination divides each latitude circle into sunlit and shadowed parts in the same proportion as the lengths of day and night at that latitude on that day; so that

(3) On every day of the year the Equator has days and nights of 12 hours;

(4) On the *equinoxes* every point on Earth has days and nights of 12 hours;

(5) On two days each year — both of the *equinoxes* — the noon sun is directly overhead on the Equator;

(6) At the poles, the sun is above the horizon either zero or 24 hours each day, the change taking place on the *equinoxes*;

(7) On the longest days of the year — the summer **solstices** (June 21 and December 21) — the noon sun is directly overhead on the Tropics;

(8) Only on and between the Tropics is the noon sun ever directly overhead;

(9) On the shortest and longest days of the year — the winter and the summer *solstices* — the Polar Circles, and all of the Earth poleward of the Polar Circles, are in continual darkness or continual sunlight, respectively.

Where, when, and how much the sun shines on Earth

In **Chapters 3** and **4** we examined, in considerable detail, the variations one finds in *solar geometry* — how the sun looks from Earth's surface. These variations are based on latitude, date, hour, and the orientation of the surface on which sunlight falls. Also in **Chapters 3** and **4** we examined *insolation* — the time rate at which solar energy arrives on a given area of Earth's surface — expressed in units of (energy/(area-time)]. That is, in **Chapters 3** and **4** we examined statements about "what" concerning solar geometry and insolation.

In **Fig. 9-3** we see insolation, on a global scale, for two sets of circumstances: (i) at the top of the atmosphere, which is the same as "at Earth's surface without any atmosphere", and (ii) under clear skies at Earth's sea level surface[1]. The units attached to the sets of lines are:

(Percent of the maximum extraterrestrial insolation arriving during 24 hours), the area of that maximum being marked by the area enclosed by the line labelled "100." With the coordinates being "Date" *versus* "Latitude", several things are at once apparent (compare with the lists of facts in the previous section):

(i) The combination of date and latitude at which the noon sun is directly overhead — marked "Midday Sun" — reaches 23.5 degrees, North and South latitudes, on the *solstices*, and crosses the Equator twice, on the *equinoxes* (Facts v, vii, and viii in the previous section);

(ii) The large dotted areas marked (zero) represent the times when the sun is below the horizon for 24 hours each day (Facts vi and ix in the previous section);

(iii) The presentation for "the top of the atmosphere" is nearly symmetrical for the two halves of the year, except for the fact that Earth is closer to the sun in January than in July, so that the line marked "100" appears in the Southern Hemisphere and not in the Northern Hemisphere (Facts v, vi and vii in the previous section).

Those are statements about "what", and now we consider statements about "why" concerning insolation. For any combination of date and latitude — any point on **Fig. 9-3** — the variations of insolation under conditions of "no atmosphere" are due (i) to variations in orientation to the solar beam of the horizontal receiving

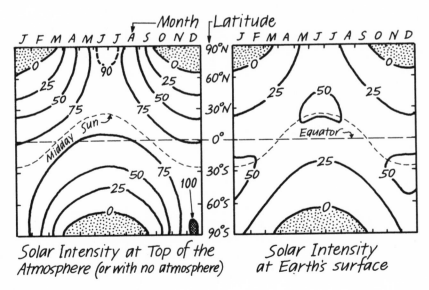

Solar Intensity at Top of the Atmosphere (or with no atmosphere)   Solar Intensity at Earth's surface

Fig. 9-3

surface (the solar zenith angle, see **Table 4-2** and **Fig. 4-7**) and (ii) to daylength. Otherwise, for any combination of date and latitude — the same point on the two halves of **Fig. 9-3** — the differences in insolation produced by the presence of the atmosphere are due to changes, during the course of a day, of (iii) the solar path length (see **Fig. 4-7**) — in addition to (i) and (ii) just mentioned.

Several comments will complete discussion of **Fig. 9-3**:

(a) The presentation for "with an atmosphere" is nearly symmetrical for the two halves of the year, except for the fact that Earth is closer to the sun in January than in July;

(b) The maximum value on the presentation for "with an atmosphere" is less than 70 (see Note (1) at the end of the Chapter) because only at noon on these dates is the sun directly overhead; and

(c) The maximum value on the presentation for "with an atmosphere" is located in midlatitudes rather than in the polar regions of the Southern Hemisphere.

Earth's atmosphere: composition and pressure

In the atmosphere of this particular planet, Earth, the mixture of gases is quite different from the mixture in any other planet we know about. As all other nearby planetary atmospheres still do, earth's atmosphere used to contain a large fraction of carbon dioxide in eons past — when spewing volcanoes and the weathering of rocks were the most active processes determining the mixture. Since carbon dioxide-using and oxygen-releasing green plants have populated Earth, the mixture has changed slowly — over a billion years or more — into a mixture with very little carbon dioxide and a large fraction of oxygen.

Today's atmosphere is usually divided, in discussions of its composition, into *"dry air"* and *"water vapor"*, which together is a moist, or wet, atmosphere. The division would doubtless seem senseless except for the fact that the proportions of gases in the mixture called dry air, because of the constant stirring and overturning of the atmosphere, are constant everywhere in the inner 9/10 or so of the atmosphere. In contrast, the proportion of water vapor is highly variable from place to place and time to time.

The proportions of the several gases in the mixture called dry air are:

Gas	Symbol	Percentage	Cumulative percentage
Nitrogen	N_2	78.084	78.084
Oxygen	O_2	20.946	99.030
Argon	A	0.934	99.964
Carbon dioxide	CO_2	0.033	99.997
Other	**	0.003	100.000

** Trace constituents are: neon, helium, krypton, xenon, hydrogen, methane, nitrous oxide, and radon. These data are from Anthes *et al* (1975)

Water vapor is present at about a 2.5% proportion in the humid tropics, and less than 1% in midlatitudes, depending on the weather for a particular day.

Earth's atmosphere (with a cloudless sky) typically depletes the energy in the solar beam by about 30% when the sun is directly overhead[1], and proportionally more when the beam travels slantwise through the atmosphere. It is appropriate to examine a few ideas about the physical distribution in the atmosphere of the materials doing that depleting.

We can conceive of this distribution by imagining that the atmosphere — any atmosphere, not just Earth's — is made up of a set of layers, *each containing the same amount of mass* (molecules of air), and stacked outward from the surface, one atop the next. Because these layers are of a **compressible fluid**, they each are compressed in proportion to the number of layers above them. Thus, the innermost layers are the most dense — they contain the most mass for a given thickness — while those at the outer edge of the atmosphere are vanishingly less dense. The results of this stacking are shown in **Fig. 9-4**.

In **Fig. 9-4a** the pressure exerted by the atmosphere — a measure proportional to the mass of air above the point of measurement — is related to height above sea level, both on linear scales. In **Fig. 9-4b** the pressure is on a logarithmic scale[P_{zi}]. Since, as we noted, pressure measures "mass above", it is clear from the figure that most of the mass is very near sea level, and that, in fact, the atmosphere has no top. The air just continues to get thinner as one moves away from Earth. Numerically, the mass distribution can be understood by relating it to familiar heights above sea level:

Cruise altitude of a typical transcontinental jet aircraft	80% of mass lies below
The top of Mount Everest, the highest on Earth	70% of mass lies below

The top of Mount Whitney, California, highest in the 48 states	45% of mass lies below
The top of Mount Katahdin, Maine, highest in eastern states	15% of mass lies below
The Mile-high City of Denver, Colorado	15% of mass lies below
A typical peak in the Shenandoah Mountains, Virginia	10% of mass lies below

Though it seems a great height to one standing at sea level, the jet's cruise altitude is only about 0.17% of earth's radius, and the top of Mount Whitney half of that. Seen in this way, Earth's atmosphere seems very thin. In fact, a typical layer into which atmospheric pollutants are mixed during a day contains only 10% of the atmosphere's total mass — certainly a finite volume.

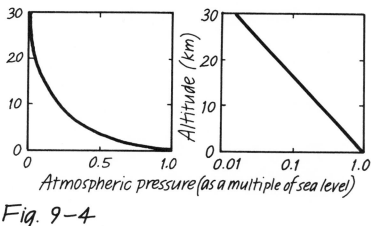

Fig. 9-4

Earth's thermal climate — the foundation of life

Earlier we considered how Earth is the third closest planet in our solar system. Our nearest neighbors are Venus (closer) and Mars (farther). **Fig. 9-5** shows how the sun's radiant energy "thins out" as it travels farther and farther outward, from the sun into space. Each one of the expanding shells in the figure contains the energy radiated *during one time unit* from the sun, so that each one of the shells contains the same amount of energy[2].

Each shell passes Venus before it reaches Earth, and then it expands on to Mars and beyond. Clearly, the density of radiant energy (the inverse of thinness) decreases at greater distances, and the amount of energy intercepted by a planet is this density multiplied by the cross-sectional area of the planet. This density is called the **radiant flux density**. Already, we see that there are at least four factors determining the insolation on Earth: (i) the sun's size, (ii) the sun's temperature, (iii) Earth's distance, and (iv) Earth's size.

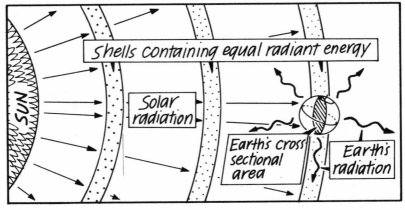

Fig. 9-5

Just because a certain amount of solar radiation is intercepted by Earth during each time unit — let us say, in a minute — doesn't mean it is absorbed by Earth. In connection with **Table 4-3** we saw that various surfaces reflect different fractions of incident solar radiation — measured by their albedo — and that these albedos vary typically from 0.7 for the tops of clouds to 0.2 for oceans, fields, and forests. As you would expect, the average albedo of the earth-atmosphere system — the reflected fraction of solar radiation intercepted by Earth — is somewhere between the largest and the smallest values of the albedos in **Table 4-3**. The average value for the earth-atmosphere system is near 0.3 — about one third. It has this value partly because of (v) the composition of Earth's atmosphere and partly because of (vi) the distribution of land and water on Earth, so that these two factors are added to the list of factors determining the insolation on Earth.

Finally, the fact that Earth — or any planet — has an atmosphere at all is determined mainly by (vii) its mass — the combination of its size and its density. Less massive planets, such as Mars, cannot retain atmospheres by the *gravitational attraction* dependent on mass. A minimal mass is required to keep all the gas molecules from drifting slowly out into space.

The seven factors we have listed only suggest the complexity of the list that governs the *radiant inflow* to the earth-atmosphere system. The temperature of the system, as you know because of our discussions of **Figs. 2-5** and **4-38** and the second law of radiation (**Chapter 2**), depends on the thermal equilibrium between this inflow and the *radiant outflow* from the earth-atmosphere system. This outflow, in turn, depends on the composition of Earth's atmosphere, already factor (v) on our list, and in particular the strength of its **greenhouse effect**[3]. The strength of the greenhouse effect, in turn, depends on (viii) the particular, unique history of Earth's atmosphere.

The way I have spun the tale, there is a list of at least eight factors determining the nature of the thermal equilibrium between the radiant inflow and the radiant outflow of the earth-atmosphere system. Each of the dozens of geophysical variables acting in these eight factors has a particular numerical value, and all of them interacting together determine the particular temperature of this particular planet.

The point of all this about geophysical variables and factors is that the particular temperature of this particular planet permits the simultaneous presence at Earth's surface of water in all of its three states — vapor, liquid, and solid — especially liquid. This is a fact that we take for granted, but it is surely so rare as to be almost unique in the universe. This fact, you see, is the very basis of life as it exists on Earth.

Earth's annual biophysical energy balance

We have just seen that Earth's thermal equilibrium is such that the rate at which radiant energy enters the earth-atmosphere system from space just equals the rate at which radiant energy leaves the earth-atmosphere system to space. Furthermore, the average temperature of the system is the one necessary (see the second law of radiation in **Chapter 4**) to emit that correct, equal amount of energy, and have it pass through the atmospheric greenhouse, out to space. That average temperature happens to be about 15 °C (59 °F, or 288 °K), which is between the freezing and boiling temperatures of water at sea level pressure.

Obviously, that average temperature for the whole system isn't observed everywhere on Earth all the time — it's just an average. There are different average temperatures at different latitudes and longitudes, different elevations and altitudes, and in different seasons. The reasons for those differences will be considered in **Chapter 10**.

Fig. 9-6 shows — on the average for the whole earth for a whole year — how the sun's energy enters and then follows many pathways within the system before it departs again to space, at a rate equal to that entering. In the figure, the system is divided into four layers — space, the stratosphere, the atmosphere, and the earth itself. Furthermore, the energy is divided into three kinds with three kinds of arrows — shortwave radiant, longwave radiant, and convective (sensible heat and latent heat).

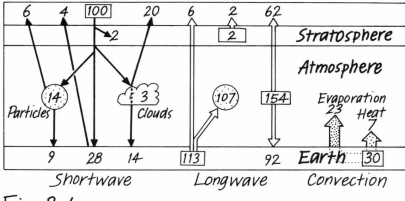

Fig. 9-6

All energy flows in **Fig. 9-6** are expressed as multiples of the 100 entering units of solar radiation, so that, for example, the 14 units of solar radiation absorbed by atmospheric particles is 14% of the extraterrestrial solar flux. All energy *sources are designated by rectangles* — such as the 113 units of longwave radiation departing outward from the earth's surface, which are then divided into the 107 units[4] absorbed by the atmosphere and the 6 units that pass directly to space in the atmospheric radiation window (see the discussion of **Fig. 4-13**). **Table 9-2** demonstrates, for the subsystem of the earth's surface, that the energy flows are in equilibrium: inflows equal outflows. Similar balances can be demonstrated for each of the other three subsystems: space, the stratosphere, and the atmosphere.

We could have a lengthy discussion of the many details in **Fig. 9-6**, but I want to mention only a few. First, the fact that 28 out of 100 solar units reach the surface directly, while only 6 of 113 longwave units reach space directly, is the heart of the idea of *differential transmissivity*, which in turn explains the greenhouse effect[3].

Second, all the rain and snow on Earth, through a whole year, with evaporation moving in the opposite direction, has the net result of transferring only 23 latent heat energy units convectively from the surface into the atmosphere. Only 7 sensible heat energy units are transferred from the surface to the atmosphere — a mere half of the number of units in a seemingly less important pathway: solar energy absorbed by atmospheric particles.

Table 9-2   The annual average energy balance for Earth's surface

Inflow:				Outflow:			
	Direct solar	=	28.		Longwave -		
	Solar scattered -				to space	=	6.
	from particles	=	9.		to the atmosphere	=	107.
	through clouds	=	14.		Convection -		
	Longwave from the				by evaporation	=	23.
	atmosphere	=	92.		sensible transfer	=	7.
Totals			143.				143.

Note: these estimated values are from Lowry (1978).

Third, since there is no convective linkage between the earth-atmosphere system and space, only shortwave and longwave radiant arrows pass, in the figure, through the outer layer of the stratosphere. Fourth, the energy involved in the biological processes of the ecosphere is proportionally so small as to be neglected in discussions, such as this, of geophysical energy flows. That biological energy will be considered presently in connection with **Fig. 9-7**.

Finally, while the network of pathways is actually more complex than shown in **Fig. 9-6**, and while the numbers on the pathways are very different at particular places and times on Earth, the figure demonstrates the basic structural complexity of the system for internal distribution of the energy of our planetary home — Mother Earth.

What of the internal distribution of biological energy within the larger geophysical system? We have already noted that it is a very small fraction of the geophysical energy being distributed. How small a fraction? Roughly one percent — some times and some places as much as several percent, but usually less than one percent overall. **Fig. 9-7** sketches the distribution system for this biological energy — the energy circulating in the ecosphere.

In the upper lefthand corner of **Fig. 9-7** solar energy arrives at Earth's surface, and part of it is absorbed by plants. The part not absorbed does other, physical work, as discussed in the explanation of **Fig. 4-39**, and ends up as environmental heat. Some of the energy absorbed by plants enters into the biological processes of the ecosphere — it enters the dotted rectangle — and the rest also ends up as heat. *Most of the energy arriving follows the non-*

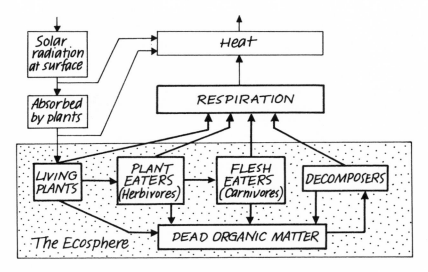

Fig. 9-7

biological pathways as heat energy. Only one percent or so becomes biochemical energy in the ecosphere — one half of one percent of the energy entering and circulating in **Fig. 9-6** (Lowry, 1978).

Within the ecosphere, plants capture and fix the solar energy and convert it, through photosynthesis, into biochemical bonds within their own structures (see **Fig. 6-21**). The energy thus fixed by *plants* — called the **primary producers** — is further distributed into (a) the energy of life maintenance (respiration, about 60%), (b) the energy in plant tissue eaten by *herbivores* (about 6%), and (c), the energy in tissues of plants that die unconsumed (about 34%).

Plant eaters are called herbivores — **primary consumers** in more technical terms. They come in many forms, from leaf-chewing insects to seed-eating birds and to large animals, such as elephants. By far the largest share of plant-to-herbivore energy transfer goes to insects and birds. Each of the boxes within the ecosphere of **Fig. 9-7** represents what is called a **trophic level**. Only 6% of the energy outflow from the plant trophic level goes to the next, herbivore trophic level[5]. The animals that eat flesh are called *carnivores*, and are the **secondary consumers**[B]. Carni-

---

(B) Animals that eat both plants and flesh are called *omnivores*, but they are rare in the whole scheme of things — specialists — and are ignored in this simplified flow chart. Human beings are omnivore specialists, though we usually think of ourselves as too important to be ignored.

vores also come in many sizes, from beetles to birds and snakes, through the family of dogs, and on to the large cats — tigers and lions. Only about 3% of the herbivore energy transfer is to carnivores, with 88% going to respiration, leaving a mere 9% to become dead organic matter uneaten by other animals.

There is no separate pathway shown in **Fig. 9-7** to represent living carnivores being consumed by larger living carnivores. Of course, such pathways exist in nature, but to show a complete **food web** of all the pathways would be prohibitive and beyond our needs. On the average, 88% of the energy leaving the carnivore level goes to respiration, and only 12% to dead organic materials.

Clearly, most energy flowing out of a trophic level is for life maintenance, and only a small fraction goes to feeding other organisms. The energy in the deaths of plants and animals, that becomes dead organic materials (DOM), sustains the lives of the fungi and bacteria together called the *decomposers*, also called the *reducers*. Whatever trophic level energy may reach — and only a tiny fraction of plant energy ever becomes carnivore energy — it all returns eventually to the environment, completing the cycle, as heat.

The GAIA Hypothesis — our planet's Master Plan

We have been considering Mother Earth, in greater or lesser detail, *as she is* — in relationships to our sun, to her sister planets, and to the energy that flows from the sun, through her geophysical and biological systems, and then back to space. We know these relationships have not always been as they are now. For one thing, we have already noted that our atmosphere was once carbon dioxide-rich while now it is oxygen-rich. We also know that the distributions of continents, of climates, and of the multitude and variety of living things have not always been as they are.

In recent years there has arisen, among eminently reputable scientists, a fascinating sequence of investigations and assertions that the progression from whatever was to what is on Earth has not been haphazard and random, but quite according to the principles of a self-regulating and self-stabilizing system. Together, these assertions are called the **Gaia hypothesis**[6].

In brief, the hypothesis views the earth and its biological components, taken together, to be a self-regulating system that, once formed and in place (i.e. once "born"), behaves like a living organism with the capacity to maintain its environmental, geochemical, and climatic steady state favorable to the maintenance and perpetuation of its life. That is, all the processes and systems we discuss in this book work together as one giant organism to assure maintenance and self-preservation for the organism as a whole, though parts of it may become ill or even die.

Fig. 9-8 presents a mini-model of the idea of the self-preserving planetary life system (Lovelock, 1986). It is a model of *Daisyworld* — a cloudless planet whose principal life form consists of white daisies — trillions of them, as far as the eye can see. Daisyworld responds to external changes — potentially disruptive changes — in a manner that maintains the healthy existence of daisies as a species, though some individuals may not survive the impact.

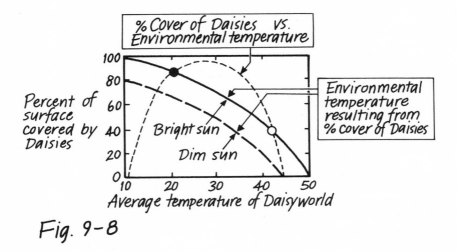

Fig. 9-8

The key to the biophysical feedback on Daisyworld is the fact that the daisies are more reflective — have a larger albedo — than the dark surface beneath them. *When daisies become less abundant in an area, the local temperature rises because more sunlight is absorbed by the darker surface exposed*, and when they become more abundant the local temperature falls. As on Earth, the planetary mean temperature is a weighted average of all the local temperatures. For a given rate of insolation, this relationship is represented in **Fig. 9-8** by the solid arc from the upper left to the lower right.

If, for an example of a potentially disruptive external change, the sun dims, the relationship shifts to that of the dashed arc beneath the solid one. The dashed arc may be read in either of two ways: (i) for a given abundance of daisies, the planet's temperature will be lower with a dim sun; or (ii) to maintain a given temperature with a dim sun, the surface must be darker — a lower abundance of daisies.

As in **Fig. 6-13a** for small organisms generally, daisies have an abundance response to environmental temperature such that they have an optimum temperature (near 28 degrees), and lower and upper thresholds (at 10 and 44 degrees) shown by the dotted curve. An unchanging, equilibrium temperature of the planet will be maintained where the arc and the curve intersect. For a bright sun, that equilibrium temperature on Daisyworld may be at either 20 degrees with 85% abundance (the black dot) or at 42 degrees with 40% abundance (the white dot). What is the difference between the black dot and the white dot?

Take the black dot first. If the sun dims and the albedo does not respond — that is if "life is not present" (abundance/albedo stays at 85%) — the dimming will (in **Fig. 9-8**) result in a drop in temperature from 20 to less than 10 degrees. If the sun dims and the albedo does respond — that is if "life is present" (albedo decreases to 72%) — the dimming will result in a drop in temperature from 20 to only 16 degrees. The presence of life opposes the effect of the changing solar output: *negative feedback — stability.*

Now take the white dot, which also represents a "solution." If the sun dims and the albedo does not respond — that is if "life is not present" (abundance/albedo stays at 40%) — the dimming will result in a drop in temperature from 42 to about 34 degrees (8 degrees). If the sun dims and the albedo does respond — that is if "life is present" — the dimming will result in a temperature fall, which will result in an increase in the abundance/albedo. The temperature will keep falling and the albedo will keep increasing until the optimum temperature for daisies is reached at about 28 degrees (a 14 degree fall): *positive feedback — instability.*

And that's not all. After the albedo/abundance has, in this second case, increased to its maximum value, further cooling with the dimming sun will result now in a decrease in the albedo/abundance. The fall in temperature will continue until it reaches the "stable solution" at the new location of the black dot at (16 degrees and 72%). The same kind of discussion would hold true if the sun brightened rather than dimmed.

Conclusion: on the one hand, as long as the average planetary temperature of Daisyworld is and remains below the optimum temperature for daisies — 28 degrees — life in the form of daisies tends to maintain stability of the whole system by opposing the effects of external changes. If, on the other hand, the average planetary temperature of Daisyworld is somehow above the optimum temperature for daisies, an external change can result either in a zero abundance (extinction) of daisies or a change to the same stable conditions sought in the former case[7].

Being a *hypothesis*, Gaia is merely a statement of how a system works — a statement that seems to fit the observable facts. It is a statement about a system much more complex, of course, than Daisyworld. Being a *hypothesis*, it has not been "proved" or established beyond doubt, but is a proposal: a statement against which to compare new facts and against which arguments can be devised. In the scientific method as applied to the natural sciences, "proof" actually consists of the sustained inability to devise valid counter-arguments — *the inability to disprove*. In that sense, the Gaia hypothesis is a proposal. It cannot (it seems to me) be experimentally tested, except by "thought experiments" such as have played an important role in the development of other branches of science.

These thought experiments are beginning to appear in scientific meetings and journals. Professor Budyko argues, for example[8], that the resupply of carbon dioxide to the environment from volcanoes will not, in the long run, be sufficient to maintain abundant photosynthesis (see **Fig. 6-21**). Furthermore, an insufficient resupply of CO_2, he says, may not maintain the strength of our planetary greenhouse enough to prevent development of an ice-covered earth.

Who knows if the hypothesis can be disproved? To the extent that Gaia is true, we may think of it as our planet's Master Plan: maintaining our system in its intended, successful form without dictating each detail. In that game of environmental design everyone is indeed a winner.

Notes

(1) The details of **Fig. 9-3** must be based on calculations, and in the calculations certain things must be assumed. For one thing, it is assumed that the surface on which the sunlight falls is *horizontal* — parallel to Earth's surface. The assumption of "no atmosphere" makes calculations pretty simple, since matters of the atmosphere's cloudiness and dirtiness are irrelevant. For the case of "clear skies at the surface" the principal assumptions, made for the mathematical calculations, are that (refer to **Fig. 4-7**) the surface is at sea level and that the atmosphere's cloudiness and dirtiness are completely described by the statement that "the direct beam insolation at the surface equals 70% of the extraterrestrial insolation when the sun is directly overhead." That is, in technical terms, *the zenith path transmissivity is 0.7.*

(2) This thinning out takes place according to the **inverse square law**, which says that the density of the energy in a shell *decreases* in proportion to the square of the distance from the sun.

(3) Any planetary atmosphere has a *greenhouse effect*, but some are stronger than others. This effect is the radiant retention of heat within a planet-atmosphere system once it has entered the system as sunlight. More particularly, it is due to the fact that inflowing sunlight finds it easier to enter the system than long-wave radiation finds it to exit — transmissivity larger in short-waves than in longwaves (see the discussion of **Fig. 4-13**). The particular difference in these two transmissivities — the **differential transmissivity** — depends on the composition of a planet's atmosphere. Earth's greenhouse effect, with a lot of oxygen and a little bit of carbon dioxide, is now weaker than it was in the early history of the planet, when there was (like Venus today) a lot of carbon dioxide and a little bit of oxygen. See the discussion of the Gaia hypothesis concerning the reason for the reversals of oxygen and carbon dioxide.

(4) If you are paying close attention, you are probably wondering why there are more radiant units leaving Earth's surface — 113 units — than there are arriving from the sun — 100 units. In fact, there are a very large 154 units being radiated away from Earth's atmosphere — 62 to space and 92 back to Earth. The brief explanation is that the 100 units are being received on an area equal to Earth's cross-sectional area, while the 113 units are being emitted by an area 4 times as large — the whole surface of the planet (**Fig. 9-5**). The more complex explanation involves the many pathways shown in **Fig. 9-6** and the fact, discussed above, that the earth's surface emits the number of units it takes — it has the necessary temperature — to produce a planetary energy equilibrium with our particular atmospheric density and composition.

(5) These percentages are from Lowry (1978) and are for the plants and animals living on land — the terrestrial environment — as opposed to the marine environment of Earth's oceans. As with the energy flows in **Fig. 9-6**, those in **Fig. 9-7** are only averages for the whole system, for a whole year. Short-term local transfers may be quite different, but the transfers between trophic levels seldom exceed 10% and most of that goes to respiration.

(6) The philosopher whose name is most associated with the Gaia hypothesis is James Lovelock, a British scientist of considerable competence in the geophysical sciences in general, and experience in the international space program in particular. "Gaia" translates from Greek approximately as "Mother Earth." Lovelock writes (1979, 1986) that "There is no shortage of planetary ailments to identify with, from the psychosociological drama of Orwell's nightmare vision to the dismal prospect

of nuclear winters and acid rain. As in hypochondria, the real problem is not that these global maladies are unreal, but the uncertainty over whether the present symptoms are prodromal of disaster or whether they are no more than the growing pains of the world." He proposes that the Gaia hypothesis can be the antidote for this global hypochondria — a basis for the practice of planetary medicine.

(7) As Lovelock points out (Lovelock, 1986, and Watson and Love-lock, 1983), very few assumptions are made — and those not different from observed behavior, such as in **Fig. 6-13a** — in the model of Daisyworld. It is not necessary to postulate fore-sight or purpose on the part of the daisies. The interesting thing is that the daisies could as well be black, as long as their color is different from that of the underlying bare ground sur-face, and the feedback effects will still result. I leave the demon-stration of that as an exercise for the reader.

(8) Professor M. I. Budyko is an eminent Soviet climatologist, and his counterarguments to Gaia appear in a recent book — *The Evolution of the Biosphere* — which was reviewed in the *Bulletin of the American Meteorological Society* (September 1987, page 1148).

Peptalk

(P₂₁) The fact that the graph on a logarithmic scale is a straight line is equivalent to the statement that *the ratio of pressures at the top and bottom of an atmospheric layer is the same for any layer of the same thickness, no matter how high or low the layer is in the atmosphere.*

10. What is it like THERE, and why?
The climates and life zones of Earth

In **Chapter 9** we studied Mother Earth as an ecosystem, operating as a whole and interacting with extraterrestrial streams of energy. In this chapter we will examine how that ecosystem — earth, air, land, water, and biota — operates internally, as a huge system distributing energy and water across the surface of the planet.

The ultimate patterns of distribution that satisfy all the laws of physics form a mosaic of climates and life zones on Earth. Since those laws of physics are not random, the patterns of distribution that satisfy all the laws are not random either. Climate and life zones are found across the surface of the earth in quite predictable and recognizable patterns.

As professional designers and planners you will do well to understand these patterns, not only to gain insights into why things are as they are in the region you are familiar with, but also to be able to anticipate differences in other regions of the world in which you may be operating at some other time.

To the extent it is appropriate in each case, we will continue to follow the pattern of examining "what" about the climate and life zones as they are observed, and then dissecting "why" they are as they are. But first, it will help to make clear the differences between the concepts of **climate** and **weather**.

A common quip among atmospheric scientists is that "climate is what you expect, but weather is what you get." That really catches most of the meaning. Climate consists of the overall, repeating, slow-to-change patterns of the various weather elements, on any one of many scales of time and of space. A proper description of the climate of a place will consist of information (see **Chapter 8** on the matter of data analysis) about (a) means (averages), (b) variations around the means, and (c) sequences of events. It is a description — in words, tables, graphs, or equations — about what is expected of atmospheric behavior at a certain place, or in a certain region, during a certain period of time. In a photographic metaphor, *a description of climate is a time exposure of weather.*

In the photographic metaphor, *weather is a snapshot of the behavior of the atmosphere.* Weather, then, consists of the individual behavioral patterns and events in the atmosphere that, together, make up the sequences, the variations, and the averages of climate.

The General Circulation of Earth's Atmosphere

In **Fig. 10-1** you will see the overall view of the large-scale patterns of flow in Earth's atmosphere. On the sphere itself I have sketched the belts of barometric pressure — high and low — circling the globe. In a cross section of the atmosphere, much exaggerated in the vertical[A], you see the giant circulation cells associated with the pressure belts. Within that cross section, rising air and the regions of precipitation — dotted clusters — are associated, while between them are regions of descending air.

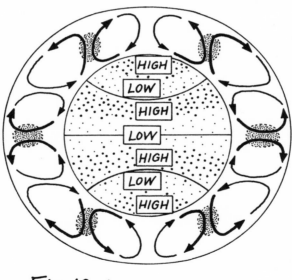

Fig. 10-1

(A) As is shown so clearly in photographs of Earth from space, and confirmed in the discussion of **Fig. 9-4**, the true thickness of Earth's atmosphere is only a small fraction of the radius of the planet.

As you study **Fig. 10-1** a bit more closely, you will notice that in each hemisphere there are three circulation cells. This pattern is called **the three cell model of the general circulation of Earth's atmosphere**. Further consideration shows that, in three dimensions, there are three giant doughnut-shaped circulation patterns in each hemisphere — six in all — and that neighboring cells are circulating in opposite directions. Where they connect with each other, the cells produce rising or descending motion, as we just noted. Our next task is to dissect the general circulation and try to explain it.

Global Scale Patterns of Pressure and Motion

The strategy I am going to use in explaining the general circulation is to liken it to *the design of a complex system*. As designers and planners, you know that that task involves making the final design conform to many individual requirements, while making the final design achieve certain overall goals and objectives. Compromises are made along the way, so that each requirement may not be met in a "pure" manner. The requirements are indeed met — the laws of physics are obeyed — but perhaps in a subtle and disguised manner. My presentation to you, then, will consist of the imaginary process of designing, step by step, the earth's general circulation.

In designing the general circulation I propose to present to you, in sequence, *seven requirements* — laws of physics — pausing en route to present *three accomodations* (compromises) that demonstrate how the often conflicting requirements are met.

Requirement #1. The atmosphere will have to move so as to redistribute excess heat from the low, equatorial latitudes toward the high, polar latitudes in both hemispheres. The associated motion, as seen in our discussion of **Fig. 4-29**, will be the fluid's required *response to differences in heating rate* between the equatorial and the polar regions. If that is all the atmosphere did, its flow would take on the form of one giant circulation cell in each hemisphere, as shown in **Fig. 10-2**. Straight north-south flow like that is called **meridional flow** — along the meridians, or longitude circles. But, of course, that is not all the atmosphere does. Although the redistribution of heat is its primary function, it must meet other requirements and obey other laws.

Requirement #2. As we discussed in connection with **Fig. 4-29**, *the horizontal parts of the flow in a circulation cell must be across surfaces of equal pressure*, — they must respond to pressure differences. The left part of **Fig. 10-3** repeats, in more detailed form, the information in **Fig. 4-29d**. It shows the circula-

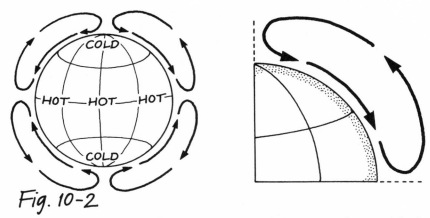

Fig. 10-2

tion between a hot and a cold column of the atmosphere, with the surfaces of equal pressure closer together in cold, dense air and farther apart in hot air.

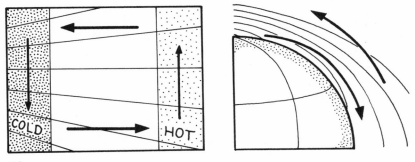

Fig. 10-3

To the right in **Fig. 10-3**, this same circulation is transposed to the huge global, meridional flow of **Fig. 10-2**, with the surfaces of equal pressure added. In this pattern, *the surface pressures are lower near the equator than they are near the poles*, with the opposite true in the upper atmosphere. The pattern in each hemisphere will take on the form shown in **Fig. 10-4**, which will be examined again later in connection with other circulations.

So far, so good. Our model atmosphere is redistributing heat energy meridionally, with flow lines essentially parallel to the earth's surface, and is exhibiting

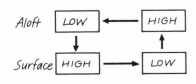

Fig. 10-4

the required pressure pattern associated with — neither purely a cause nor purely a result of — the flow pattern.

A problem: the Coriolis Effect. There is a small problem, which arises as we examine the phenomenon of the Coriolis Effect. According to Sir Issac Newton's First Law of Motion, when air is in motion, at a constant speed and along a straight line, its tendency is to continue that constant flow in a straight line unless some outside force acts on it to make it speed up, slow down, or make its path veer from the straight line.

In our system a "straight line" is just that: *not* a curved line parallel to the surface of a rotating sphere, but truly straight in the cosmic coordinates of space. In order to make use of Newton's physics, atmospheric scientists invoke *an imaginary "force"* — the **Coriolis force** — in order to allow motions like those in **Fig. 10-3** to be considered "straight" — parallel to a curving surface[1].

We will consider more of the Coriolis force, which shows up in flows as the **Coriolis effect**, when we study wind flows in relation to pressures, in a later section. For now we will note only the result that, because of the force, *winds which would otherwise flow directly from high to low pressure actually turn toward the right in the northern hemisphere and toward the left in the southern hemisphere.* In fact, when air flows are not acted on by friction near the earth's surface, they tend toward flowing *perpendicular* to the lines that run directly from regions of high pressure to regions of low pressure. In the northern hemisphere this flow takes place with the lower pressures to the left (i.e. to the left as you fly with the wind), and in the southern hemisphere, with the lower pressures to the right. We will come back later to the way that works.

Requirement #3. If the pressure patterns shown in **Fig. 10-4** were the only thing our model winds had to respond to — lower near the equator at the surface, and lower near the poles aloft — the Coriolis force would produce flow patterns like the ones in the upper two panels of **Fig. 10-5**: parallel to the latitude circles — both at the surface and aloft — so that lower pressure is to the left in the northern hemisphere and to the right in the southern hemisphere. But this requirement of *flow in accord with the rule of the Coriolis force* is in conflict with the first two requirements of flow, as a circulation cell, in response to temperature differences.

The First accomodation. The lower two panels of **Fig. 10-5** show the model atmosphere's first accomodation to the conflicting requirements: flow is partially in response to the first and second requirements and partially in response to the third requirement. The effects of both requirements — carrying heat poleward aloft

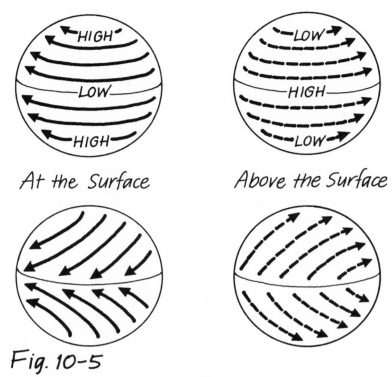

Fig. 10-5

and motion in accord with the Coriolis requirement — can be seen in the compromise pattern. The flow at the surface in the northern hemisphere is both toward the equator and toward the west with lower pressure on the left. The flows aloft and in the southern hemisphere, likewise, show the same compromise.

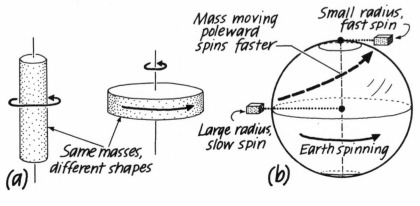

Fig. 10-6

Requirement #4. As air aloft moves from the equator toward either pole, in the first accomodation, seen in the bottom panels of **Fig. 10-5**, it is confronted with the requirement to *obey the principle of the conservation of angular momentum*. **Fig. 10-6** shows the principle in action. It is that a mass spinning around an axis must spin faster if the radius of the spin is shorter (and slower if it is longer). Said another way, it is:

(Radius of spin) x (Angular rate of spin) = (Constant).

This is the same principle which makes a figure skater spin faster after he enters the spin with arms and legs extended, and then pulls them inward to his body. It is the same principle which makes a rock tied to a string spin faster as the string is shortened. It is the same principle which is used in a potter's wheel and an automobile's flywheel: conserving angular momentum, a large mass shaped to have a large radius will spin (except for friction) at a constant rate.

The mass of air moving toward the pole, and remaining at about the same (pressure) altitude will be spinning on a shorter and shorter radius: its distance along a line *perpendicular to the earth's axis*. If the model atmosphere were to respond purely to this principle, it would — in the extreme case — have to spin infinitely fast as the radius approached zero in the polar regions.

The Second accomodation. The atmosphere's accomodation to the fourth requirement of the conservation of angular momentum is the formation of the three celled structure of the general circulation. Not being able to move all the way to the pole, where it would have to spin around the axis at too high a rate, *the air in the cell based at the equator goes no farther than about 30 degrees from the equator* in either hemisphere. This latitude is determined on our planet by a complex combination of factors, such as the density and the viscocity of air and the rate of spin of the earth on its own axis. **Fig. 10-7** shows (for the northern hemisphere) this truncated cell known as the **Hadley cell**. Other features are also shown — windflow, rising and descending air, zones of pressure and precipitation — that form as a result of the "domino effect" caused by the formation of the Hadley cell reaching only to about 30 degrees latitude. This accomodation agrees exactly with the cross section of the atmosphere in **Fig. 10-1**, so that we have already designed a system that looks a lot like the real atmosphere.

The three-celled circulation we have designed will work well for a uniform, spinning earth without seasons — on an axis with no tilt. But we must add features to our design that account for several kinds of departure from a smooth, featureless surface on an earth spinning in a constant geometric relationship with the stream of energy from the sun.

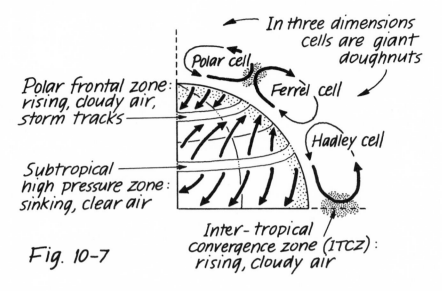

Fig. 10-7

Requirement #5. As the earth moves around the sun and the seasons change, as in **Fig. 9-2**, the latitude receiving the midday sun directly changes, day-by-day, from 23.5° South to 23.5° North between December and June. This latitude can be thought of as a *moving thermal equator* of the planet, where the airstreams from the hemispheres converge in the cloudy, rainy ITCZ (see **Fig. 10-7**). The ITCZ "follows the sun." **Fig. 10-8** shows how our model's three-cell pattern and its belts of pressure, rain, and sun will also adjust to this following of the sun. For one thing, the polar cell nearer the summer sun (in either hemisphere) shrinks so the Ferrel cell and the Polar cell come together at a higher latitude in summer, and a lower latitude in winter.

Requirement #6. The earth's surface is not smooth and uniform. It has alternating land and water masses around most latitude circles, as well as differences in altitudes on the land masses. Our model atmosphere's general circulation must accomodate to that *variation in roughness*.

Requirement #7. Finally, the same alternation of land and ocean imposes a seventh requirement on our system to *include monsoonal flows* on-land in summer and toward the oceans in winter (see **Fig. 4-30**). And along with this requirement goes the requirement that, compared with the oceans, land masses will have

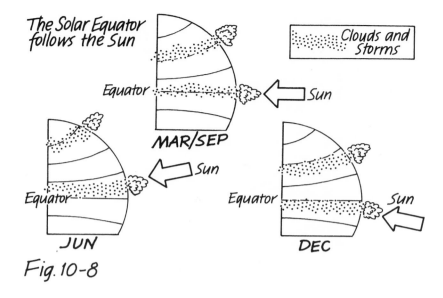

Fig. 10-8

lower pressures in the warm season and higher pressures in the cold season.

The Third accomodation. The system we design will, when accomodating to all these requirements, look something like **Fig. 10-9**, in which the belts of pressure that are continuous around the globe in **Figs. 10-1** and **10-7** are interrupted here because of the alternation of land and water.

So much for the design of our system. How does it compare with the real atmosphere? **Fig. 10-10** shows the global patterns of surface pressure and wind in the extreme seasons: January and July. I will leave it to you to examine the details, but the interrupted pressure belts, and the alternations, between seasons, of high and low pressure between continent and ocean are strikingly evident in the real system.

As an additional check on reality, **Fig. 10-11** shows the global patterns of annual precipitation. Again, I will leave it to you to examine details, but the general belts of higher and lower precipitation amounts are evident at the latitudes — roughly the equator, 30° and 45-50° — where we expected them. The interruptions in these belts are due, in the real atmosphere, mainly to major north-south mountain ranges, rather than simply to the alternations of land and ocean.

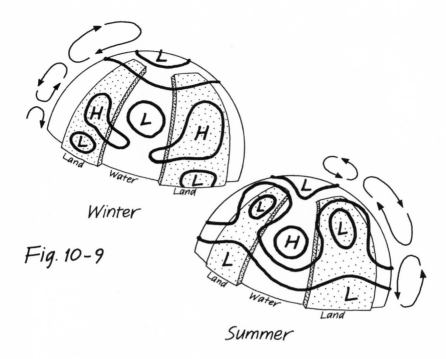

Winter

Fig. 10-9

Summer

These figures and remarks concerning departures of climatic patterns from what one would expect on a uniform earth bring up the subject of **climatic controls**. The term refers to the factors that are the basic determinants of the climate in a region and at a particular place. Here is a list of climatic controls.

a) *Latitude:* it determines most of the climatic components concerning insolation and the seasons.

b) *Altitude:* it constitutes a secondary control on barometric pressure, and therefore on temperature, and on insolation (see **Fig. 4-7**).

c) *Relationships to terrain features:* depending upon the the geographical orientations of the nearest major mountain ranges, and on the position of a place with respect to those ranges, and with respect to local hills and valleys, the climate can depart markedly from what it would be like at that position on an earth without terrain features. Many of these departures were discussed in **Chapter 3** regarding windflow and cloud patterns in uneven terrain.

d) *Relationships to oceans and shorelines:* air temperatures and humidities are greatly modified by the passage of an airstream over oceanic waters upwind of a place.

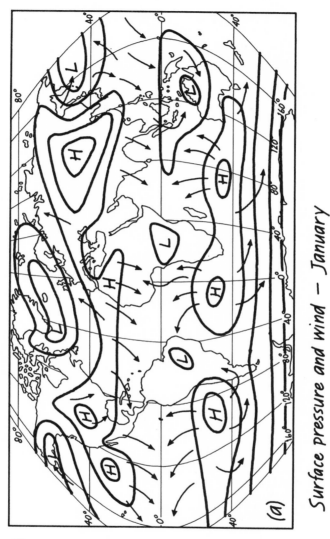

Surface pressure and wind — January

Fig. 10-10

Surface pressure and wind — June

(b)

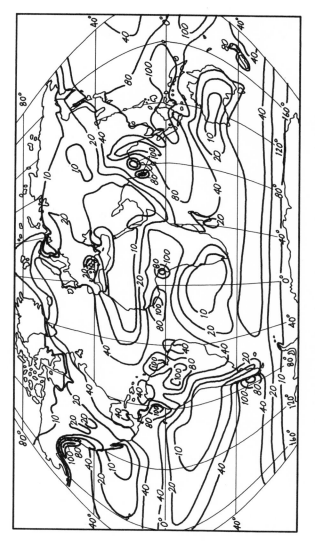

Average Annual Precipitation (in inches)

Fig. 10-11

Because oceanic air passing over a location may not have travelled directly from the nearest shoreline, but rather along a less direct trajectory, it is the distance of a place from a shoreline, measured along the path of the airflow, as well as the characteristics of the offshore waters, that is a climatic control. The longer the path, the less the air will have the same characteristics, of temperature and moisture content, as air lying next to the water upstream.

Air flowing over northern Arizona in summer has often come from the Gulf of Mexico, by way of Texas and New Mexico, rather than from the nearest shores of the Pacific Ocean. Also in summer, air flowing over New York has often come from the Gulf of Mexico, by way of Georgia and the Carolinas, rather than from the North Atlantic nearby (see **Fig. 4-30**).

The first three of the controls on the list have already been suggested in our discussions. We will take another look at the fourth control — shorelines — presently.

Global Scale Patterns of Seasonal Temperature and Precipitation

As just mentioned, the characteristics of oceanic air, which are in turn determined largely by the properties of underlying oceanic currents, have a major influence on climate, even at places far inland on the continents. Following the plan to describe "what" and then "why", **Fig. 10-12** shows the names and locations — as well as the relative temperatures — of the major oceanic currents of the world. The figure labels the currents "warm" and "cool." Relative to what? *Relative to the temperatures one would otherwise expect for that latitude.* We will see presently how that applies to the formation of climate.

Up to this point, we have not discussed, except with respect to pressure and winds, the changes in climate between seasons. **Fig. 10-13** shows the average temperatures for January and July around the world. These temperatures, of course, are the averages not only of days and nights, but also of all the different kinds of weather at each place. They are "what you expect" rather than "what you get."

Broadly, the patterns in **Fig. 10-13** are what one would expect based on our discussions up to this point. The major features are (a) the decline in temperature poleward from the equator, in any season; (b) the movement of the thermal equator — the highest temperatures — from south of the equator in January to north in

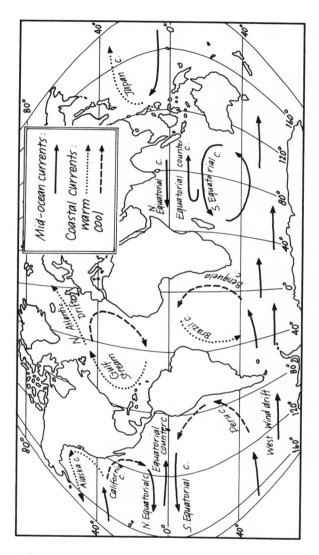

Fig. 10-12

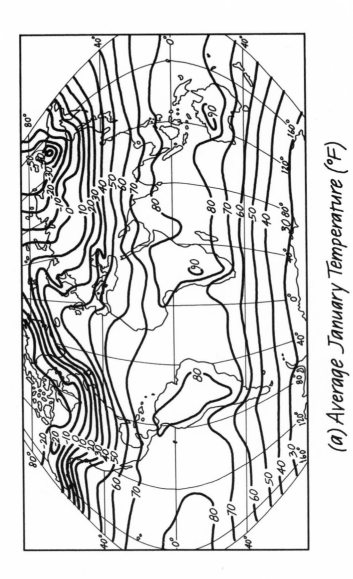

(a) Average January Temperature (°F)

Fig. 10-13

(b) Average July Temperature (°F)

July; and (c) the warmth of land relative to water at the same latitude in summer, and the reverse in winter. There are, however, other departures from a simple pattern of latitudinal belts of temperature, which we will examine as we discuss "why."

Fig. 10-14 shows sketches, for winter and summer, of the ocean currents and average temperatures on imaginary and simplified continents in the northern hemisphere. First, consider the ocean currents in light of the wind patterns in **Fig. 10-7**. Near the equator, easterly winds (from the east) drive the surface waters westward between the continents; and at midlatitudes, the westerly winds (the "prevailing westerlies") drive the surface waters eastward between the continents. Incidentally, the same wind effects and directions apply in the southern hemisphere, although there are distinct differences owing to the fact that scarcely any land mass lies south of 40°S latitude. The currents must diverge, of course, when they reach a shoreline, so the pattern shown is formed. The large circular, oceanic whirls, called **gyres**, are clearly associated with the three atmospheric cells in **Fig. 10-7**, and follow the same seasonal movements north and south between the continents.

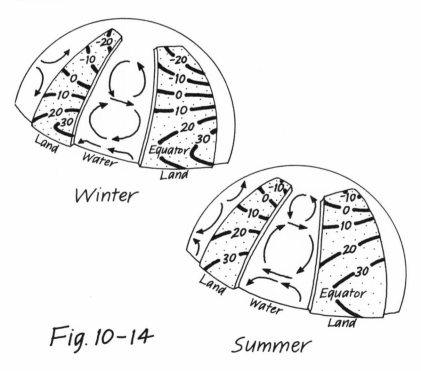

Fig. 10-14

Winter

Summer

On land, the lines are *isotherms* — lines of equal average temperature — like the ones in **Fig. 10-13**. These isotherms are approximately in degrees Celsius, but just consider them to be in conveniently sized numbers for purposes of this explanation. Again for purposes of this explanation, the temperature patterns are keyed to the fact that the "0-degree isotherm" in winter and the "10-degree isotherm" in summer lie east-west across the continents at the same latitude as the ocean current driven by the midlatitude westerlies.

Departures of the temperature patterns from a simple east-west orientation are in response to the relative temperatures of the offshore ocean currents. Currents converging toward midlatitude — cold from the pole and warm from the equator — cause the isotherms to converge on east coasts. The isotherms diverge on west coasts. The whole temperature pattern shifts in response to the shifts in the thermal equator.

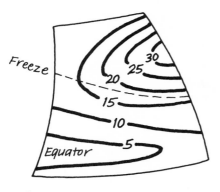

Summer minus Winter

Fig. 10-15

Fig. 10-15 adds another seasonal dimension to this model of temperature climate. In the figure, the lines are of *equal difference between summer mean and winter mean temperature*. The dashed line is the farthest equatorward position of the mean freezing isotherm: zero degrees. Clearly, the most equable temperature climates are near the equator, especially on the west coast. The least equable are on the northeastern coasts.

Comparison of these simplified patterns with the real ones in **Fig. 10-13** reveal distinct similarities. The causes of the shapes of these patterns are the ones I have suggested.

Going on from seasonal temperatures to seasonal precipitation patterns, **Fig. 10-16** presents another simplified, explanatory model of this aspect of climate. In the figure, precipitation types are described in terms that will become clearer in the latter sections of this chapter: drizzly, frontal, and convective. Before those explanations, here is a catalog of the types and their locations in **Fig. 10-16**.

a) "Drizzly" precipitation falls at low rates, spead over considerable time periods, such as a week. It occurs where isotherms diverge because of diverging oceanic currents: on middle west coasts.

b) "Frontal" precipitation falls at high rates, in intermediate time periods, such as a day. It occurs where isotherms converge because of converging oceanic currents: on middle east coasts.

c) "Convective" precipitation falls at high rates, in brief time periods, such as an hour. It occurs equatorward of the frontal type, where mean temperatures are above about 25 degrees.

As we will see presently, the precipitation-free desert regions are found between these three precipitation types because, in those desert regions, there is no sustained mechanism for producing precipitation (see **Chapter 4** concerning clouds and precipitation).

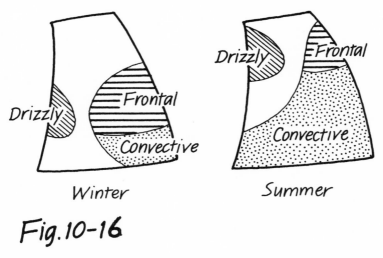

Fig. 10-16

I have presented no global maps of seasonal precipitation patterns, but less specialized textbooks on climatology sometimes contain such maps[2].

Global Scale Patterns of Vegetation

To complete our consideration of the "climates" of Earth — we will consider "weather" presently — let us examine the life zones on the planet. **Fig. 10-17** shows the major vegetation types of Earth; and, as is clear from **Chapter 6**, animal life forms and plant forms are closely correlated.

Fig. 10-17 shows "what" and **Fig. 10-18** suggests "why." It divides the imaginary continent into regions according to the sea-

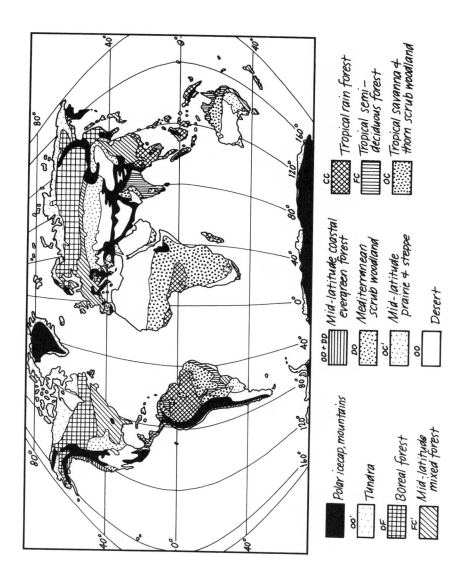

Fig. 10-17

sonal precipitation patterns — **Fig. 10-16** — that each experiences, with the freeze line added.

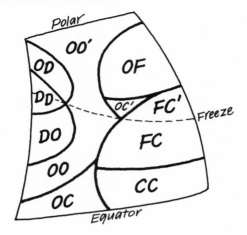

Fig. 10-18

Here is the catalog of these regions, together with the kinds of vegetation associated with them, as included in the key of **Fig. 10-17.**

Symbol	Precipitation		Vegetation type
	Winter	Summer	
DD	Drizzly	Drizzly	Western conifer forest: fir and hemlock
DO	Drizzly	Dry	Scrub oak
OD	Dry	Drizzly	Western conifer forest: fir and hemlock
OO	Dry (No freeze)	Dry	Desert
OO'	Dry (with freeze)	Dry	Tundra
OF	Dry	Frontal	Boreal forest: spruce
FC	Frontal (No freeze)	Convective	Semi-tropical forest: pines and palms
FC'	Frontal (with freeze)	Convective	Deciduous forest
OC	Dry (No freeze)	Convective	Semi-deciduous tropical forest
OC'	Dry (with freeze)	Convective	Grasslands
CC	Convective	Convective	Tropical rain forest

The concordance of this catalog with **Figs. 10-17** and **10-18** is quite clear upon closer inspection. With this explanation we have completed our examination of the climate and life zones of Earth. Now we begin consideration of the global patterns of "weather."

Storms and Weather Changes: Pressure and motion in Mid-latitudes

The individual events and patterns of atmospheric behavior called "weather" occur on various time and space scales, and in recognizable groups with similar characteristics. You will recognize what I mean when I say, for example, that the spectrum of weather events extends from a small scale, short-lived event such as a tornado to a huge storm system stretching from the Rocky Mountains to the Great Lakes and maintaining its identity for three or four days. In the remainder of this chapter we will consider several groups of events and patterns, each one as a concept that can also be considered as a separate atmospheric feature. We will consider atmospheric *waves, air masses,* and *storms.* While in one sense separable, they are connected by the phenomena of atmospheric *motion* and *energy exchange.*

Begin by noting the range of energetic sizes of these phenomena, the energy content in each expressed as a multiple of the solar energy that arrives on Earth each day. Some of them have already been discussed. The rest will be considered presently.

Solar energy received each day	1
Melting of an average winter's snow during the spring	1/10
A monsoon circulation between ocean and continent	1/100
Use of energy by all mankind in a year	1/100
A midlatitude cyclone	1/1,000
A tropical cyclone	1/10,000
Kinetic energy of motion in Earth's general circulation	1/100,000
The first H Bomb	1/100,000
A squall line containing thunderstorms and perhaps tornados	1/1,000,000
A thunderstorm	1/100,000,000
The first A Bomb	1/100,000,000
The daily output of Boulder Dam	1/100,000,000
A typical local rain shower	1/10,000,000,000
A tornado	1/100,000,000,000
Lighting New York City for one night	1/100,000,000,000

The point of this table is that the energy content of perfectly ordinary weather events is huge — much more than is in many man-made events we think of as involving huge amounts of energy. Upon consideration, it is not surprising, to judge by their similar sizes and appearances, that an A Bomb and a thunderstorm are comparable in their energy content. The fact that the bomb's lifetime is so much less makes it seem more energetic. Though they involve the same amounts of energy, a tornado seems more energetic than the lighting of New York City because the release is both sudden and concentrated[B].

We discussed the basic cause of atmospheric motion in **Chapter 4**, but we need now to consider the rules governing the speed and direction of motion. First, a few definitions, as depicted in **Fig. 10-19**.

1) An **isobar** is a line drawn on a surface of *constant altitude* (such as sea level). The barometric pressure is the same everywhere along an isobar.

2) The **pressure gradient** can be measured by the number of isobars crossed by a line of a given length perpendicular to the isobars. Said another way, it is the amount of pressure change per unit distance on a surface of constant altitude (such as sea level).

3) The **pressure gradient force (PGF)** is the force acting on a parcel of air due to the difference in pressure from the high-

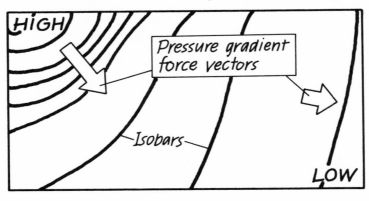

Fig. 10-19

(B) This idea is summed up in the definition of "**power**." It is the time rate of energy use or conversion — (energy/time). Thus, if a given amount of energy is released more rapidly in one phenomenon than in another, the first is said to be more *powerful* than the second. The more precise statements above, then, are that an A Bomb is more powerful (rather than energetic) than a thunderstorm, and a tornado is more powerful than the lighting of New York City.

pressure side of the parcel to the low-pressure side. It is proportional to the pressure gradient.

4) A **pressure gradient force vector** is an arrow whose direction is perpendicular to the pattern of isobars and whose length is proportional to the **pressure gradient force**.

Earlier in this chapter we mentioned the Coriolis Force (CF), noting it is an imaginary force used to account for the fact that atmospheric (and also oceanic) motion takes place on a spinning sphere — a rotating coordinate system. To understand the basic rules governing motion, first consider what happens when an air parcel is set in motion and is then acted upon by two forces: the pressure gradient force and the Coriolis Force[C]. We have already noted that the PGF acts perpendicular to the isobars — along the pressure gradient. Earlier we noted that the Coriolis Force acts to one side of the direction of motion — to the right in the northern hemisphere. In fact, we define the CF as acting perpendicular to the direction of motion.

Fig. 10-20a shows the state of affairs when these two forces are *equal but opposite in direction*[P22]. The direction of motion when

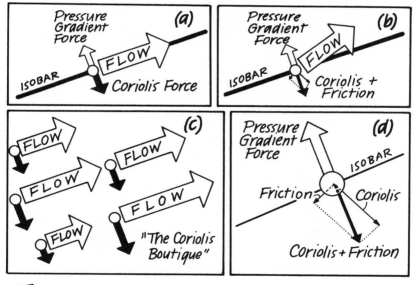

Fig. 10-20

(C) For this discussion we can ignore two other, smaller forces acting on the parcel, both of them vertical: gravity and buoyancy. It is the rules of horizontal motion we are concerned with here. We will add in the (horizontal) frictional force presently.

all these requirements are met is *parallel to the isobar.* This **balanced motion** results because, whatever other speed and direction it may have had, it settles down to being the combination of speed and direction that is associated with the correct CF — the one that exactly balances the PGF. It is as if (pardon the simile) the parcel, first observing the local strength and direction of the local PGF, goes to the "Coriolis Boutique" (**Fig. 10-20c**) and selects the correct CF to balance the PGF, and then assumes the flow speed and direction associated with that correct CF. All the CFs in the boutique are perpendicular to, and proportional to, their associated flow vectors. The CF is defined that way. They differ only in their sizes, or magnitudes$^{(P_{22})}$.

This balanced flow, with the CF equal and opposite to the PGF, is called **geostrophic flow**, and it is the kind of flow mentioned in the discussion of **Fig. 10-5**: parallel to the isobars with the low pressure to the left in the northern hemisphere. **Figs. 10-20b** and **10-20d** show what happens when a frictional force (FF) as added to the mix. In **Fig. 10-20d** the sum of the FF and the CF exactly balances the PGF. The flow is *balanced,* but (see **Fig. 10-20b**) it is no longer parallel to the isobars — it crosses them toward lower pressure. The air parcel has selected another CF at the boutique: the one that balances the PGF and the FF combined. In general, the nearer the flow is to the earth's surface, the greater is the FF. The greater the FF the larger the angle at which the flow crosses the isobars[D].

With these rules governing the direction of motion, you can understand why the winds, *near the earth's surface,* flow outward from a high pressure area — a HIGH (H) — and into a low pressure area — a LOW (L) — across the isobars, which usually lie in a somewhat circular pattern on a map. A HIGH is called and **anticyclone**, while a LOW is called a **cyclone**. **Fig. 10-21** shows the patterns of flow in both hemispheres.

Now we need to add to our ideas, about flow patterns near the earth's surface, an understanding of flow patterns aloft — well aloft, up near the level where about half the atmosphere's mass lies below (see **Fig. 9-4**). I am going to break my own rule, again, about putting "what" before "why" in order to give you the basic concepts and connections involved in several kinds of maps we will examine in the remainder of this chapter.

(D) When the isobars are curved we must consider other forces — centrifugal forces — and balanced flow is then called **gradient flow**. I think we will leave well enough alone and stop with geostrophic and frictional flow. Any reader who wants to know more about this aspect of flow can look at any good textbook on dynamic meteorology, for example Byers (1974).

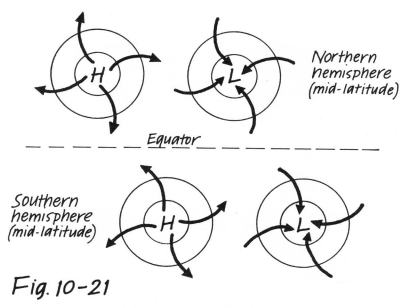

Northern
hemisphere
(mid-latitude)

Equator

Southern
hemisphere
(mid-latitude)

Fig. 10-21

First, you should understand the connections between the three-dimensional, wavelike patterns of pressure surfaces in the atmosphere and the two basic kinds of maps that meteorologists use to depict those surfaces. **Fig. 10-22** shows these connections for what is actually an air wave, having many of the same characteristics as a water wave. In discussing how air columns of different heat contents produce different cross-sectional patterns for *surfaces of constant pressure* (see **Figs. 4-29** and **10-3**) we noted that these surfaces are farther apart in warmer air, and closer together in colder air.

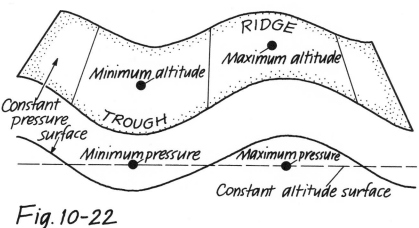

RIDGE

Maximum altitude

Minimum altitude

Constant
pressure
surface

TROUGH

Minimum pressure   Maximum pressure

Constant altitude surface

Fig. 10-22

Based on your experiences with water waves, you will not be surprised to learn that there are waves in air — also fluid. The wave on the *surface of constant pressure* in **Fig. 10-22** exhibits its lowest altitudes in **troughs**, and its highest on **ridges**. In the cross-sectional sketch at the bottom of the figure, there are varying pressures along the *surface of constant altitude*. The minimum pressure on that surface (remember: pressures are always less at higher altitudes) is associated with the location of the trough, and the maximum pressure with that of the ridge. That is, a map of the surface of constant pressure and a map of the surface of constant altitude will both show "highs" and "lows" in the same locations, although the first map would contain lines of equal altitude, and the second lines of equal pressure — isobars.

In meteorological practice, the upper air is described with maps of surfaces of constant pressure — they are like topographic maps, with lines of equal altitude, showing ridges and troughs (valleys). The only map of a surface of constant altitude used in practice is that of the sea level pressure — showing HIGHS and LOWS on maps with isobars.

While "highs" and "lows" have the same locations on the two kinds of map in **Fig. 10-22**, they may not when many pressure surfaces are involved. **Fig. 10-23** shows a cross-section of the atmosphere — it is perhaps 20,000 feet thick — with many surfaces of constant pressure aloft, and the single, sea level surface of constant altitude. In **Fig. 10-23a** the columns of cold and warm air are both vertical, while in **Fig. 10-23b** they are not. You can quickly see that there are limitless possible patterns on the two kinds of maps, depending on the temperature differences and the orientations of air columns.

Constant pressure surfaces

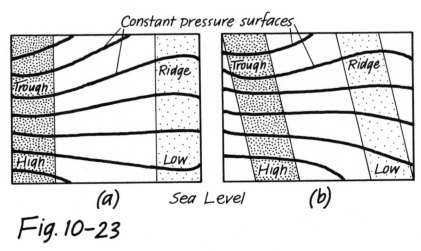

(a) Sea Level (b)

Fig. 10-23

In **Fig. 10-23** the patterns are of *thermal waves* — waves that appear because of the differences in heat content of neighboring air columns. Their patterns will move and change as the locations of heating and cooling change on the earth's surface, and the air columns reflect these changes. **Fig. 10-24** shows cross-sections of *gravity waves* — waves formed by some mechanical impulse applied "upstream" on the pressure surface. Gravity waves are like the waves on a pond that form and move from the point where a falling rock hits the surface. Gravity waves in the atmosphere, like those in water, move downstream from the impulse that generated them.

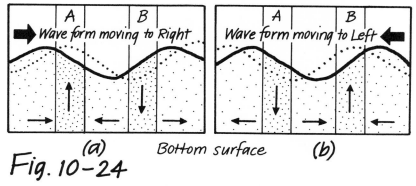

Fig. 10-24

The locations of the ridges and troughs during the next time step are shown as dotted lines in **Fig. 10-24**. Also shown are the combinations of horizontal and vertical air motions associated with these moving wave forms. For example, as a ridge approaches from just upstream (column A in **Fig. 10-24a**), air flows into the column overhead to expand the column vertically.

Clearly, with changes in heating and cooling patterns and the locations and timing of various impulses, and with the presence of some thermal waves and some gravity waves, maps of the real atmosphere — either kind of map — can become very complex. **Fig. 10-25** shows a typical map of the height of the 500 millibar (constant pressure) surface over North America, while **Fig. 10-26** shows a typical map of the sea level (constant altitude) pressure. On the 500mb map, ridges and troughs are shown with dashed lines, while on the sea level pressure map, HIGHS and LOWS are shown with "closed" isobar patterns — like topographic hilltops and depressions[E].

(E) If you are alert, you are wondering how we can map the sea level pressure over the Rocky Mountains — or any place above sea level, Denver for example. **Fig. 10-23** gives the clue: assuming a fictitious air column between Denver and sea level, and assuming it has the same heat content as that of the air measured at Denver, we can calculate how close together the pressure surfaces in that column would be, and thus (an estimate of) the sea level pressure.

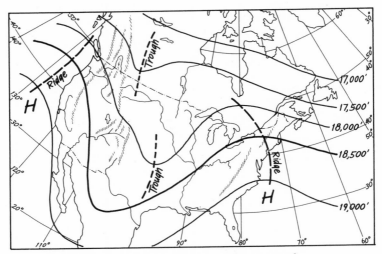

The 500 millibar map is for a constant pressure surface.

Fig. 10-25

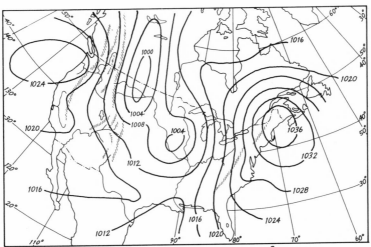

The Sea Level map is for a constant altitude surface.

Fig. 10-26

Storms and Weather Changes:
the Analogy of the Mountain Stream

We have discussed winds in relation to pressure patterns, waves of air, and recognition of these features on two kinds of map. I want to put them all together now in a consideration of how the real atmosphere moves and changes, as a system, over shorter periods of time — the time scale of weather, as opposed to the time scale of climate. First, we will consider the system in the mid-latitudes and then in the tropics. I will make my presentations only *in the orientation of the northern hemisphere,* understanding that some readers may have to translate these presentations (turn them upside down) for the southern hemisphere.

In **Fig. 10-7**, the mid-latitude storm belt lies along the Polar Frontal Zone, between the Ferrel cell and the Polar cell, at latitudes generally between 40 and 55 degrees. Here airstreams of contrasting properties converge and rise into cloudy, rainy updrafts. **Fig. 10-27** is a sketch — a snapshot — of part of that storm belt. You will recognize three centers of low pressure (cyclones) with air flowing around and into them.

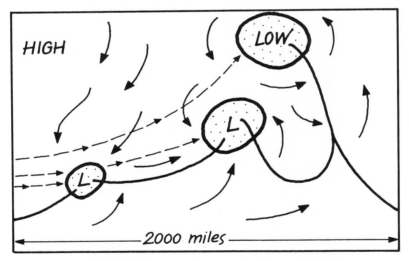

Fig. 10-27

An instructive way to think about the flows and patterns on a map such as **Fig. 10-27** is to imagine you are looking down on the surface of a fast-moving mountain stream. There are several similarities — after all, they are both fluids in motion — that you can recognize from your own experience watching the way water moves. Here are some of those similarities.

1) HIGHS in the atmosphere are like the local domes on the water surface: they are generally stationary and their locations are largely determined by the topography of the stream bottom.

2) There are "quiet pools", near the shorelines, whose locations are largely determined by the shape of the shoreline. These pools will be related to the idea of "air masses" presently.

3) Eddying vortices are formed at stationary, preferred locations.

4) The eddying vortices are formed, at those locations, over a sequence of similar intervals of time.

5) The eddying vortices all move downstream along very similar trajectories.

6) The eddying vortices all pass through similar stages of a "life cycle." In the atmosphere each one is a "travelling mid-latitude cyclone" — also known as a LOW or a "storm system."

7) Each moving vortex exhibits three kinds of motion:

    a) the center itself moves downstream;

    b) there is cyclonic inflow — rotation — around the center; and

    c) the center slowly changes its shape as it moves downstream.

In **Fig. 10-27**, the three storm centers have been generated upstream to the left, they have moved along the dashed lines, and they are increasingly mature from left to right. We will study the stages in the life cycle of a travelling LOW presently.

Earlier discussions make clear that there are connections — though sometimes they are not simple — between the forms and behavior of the atmosphere near the surface and aloft. I did not mention it before, but **Figs. 10-25** and **10-26** are drawn for the same time. Your closer examination will disclose there are distinct connections between the locations of the troughs and the lows, and between the locations of the ridges and the highs.

If you are going to keep track of weather and maps in real time, you ought to be aware of what is called the atmosphere's **Index Cycle**. It is depicted in **Fig. 10-28**, and it involves the slow changes in the shapes of the meanderings of the upper air streams along the Polar Frontal Zone. Generally speaking, the streamlines of arrows in **Fig. 10-28** are closely associated with the **Jet Stream**, about which you have probably heard: a narrow band of fast-moving air, usually near the 500mb constant pressure surface.

Forecasters and climatologists often speak of the mostly east-west flows in **Fig. 10-28a** as "high index" and the mostly north-south flows in **Fig. 10-28d** as "low index." Since the mid-latitude atmosphere seems to go through a cycle — ($a > b > c > d > a$, and so on) — over a period of time such as three to five weeks, the name Index Cycle was used to refer to this tendency to repeat the sequence.

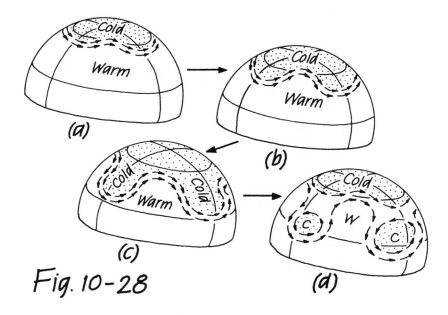

Fig. 10-28

Being aware of the Index Cycle, you should also be aware that the path of a storm center across the map — the dashed lines in **Fig. 10-27** — has a tendency to follow beneath the pattern of the jet stream. This apparent control of the movement of storm centers at the surface by the shape of the flows aloft is called "steering", but I emphasize it is *only a tendency*. The point of mentioning the Index Cycle is to make clear that the *rate of east-west movement* of a storm center across the map, as in **Fig. 10-27**, is faster or slower according to whether the upper air pattern is high index or low index. It is less obviously so for a mountain stream, but the tendencies are present there as well.

Storms and Weather Changes: Air Masses

In your own experience with discussions of weather, you have very likely encountered the term "air mass." The term has a very precise meaning to atmospheric scientists, and the concept it refers to is important for understanding the dynamics of storms in mid-latitudes. An **air mass** is a large volume of air exhibiting, level-for-level, the same characteristics of heat (temperature) and moisture (humidity) over large horizontal distances. What does "level-for-level" mean? **Fig. 10-29** holds the answer. In the figure, there are sketches of three imaginary air masses. They are (conceptually) like three huge blobs of fluid resting (and moving) on an impermeable surface — like water drops on canvas.

Although the sketches suggest all three air masses have similar spatial dimensions — typically a thousand miles horizontally and two or three miles vertically — you see that each one is unique in its layered structure. But each layer — exhibiting some particular combination of heat and moisture — reaches over large horizontal distances. Thus, an air mass is not itself spatially homogeneous; it is an individual layer within the air mass that is spatially homogeneous.

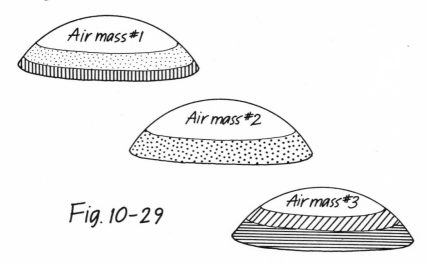

Fig. 10-29

How do air masses form? And where? They form within the atmosphere's equivalent of "quiet pools" along the shorelines of the mid-latitude stream — the Polar Frontal Zone. The volume of air remains in place for several weeks, and slowly takes on its unique characteristics by exchanging heat and moisture (by convection) with the underlying surface and heat (by radiation) with outer space. The characteristics of the air mass — especially in the lower layers — more and more match those of the underlying surface. Eventually, following the dynamic changes in the hemispheric flows — suggested by **Fig. 10-28** — an air mass will move out from its place of origin (called its **source region**) and into the moving stream, much as water in a mountain stream[F].

The system usually used for naming air masses employs the two major characteristics of temperature and moisture, although these

(F) I do not want to leave the impression that the unique layering of air masses is a random process. The fact is, as you can tell since you understand how they form, each source region tends to produce very similar — though never identical — air masses year after year. Over warm oceans, air masses are increasingly cool and dry upward from the surface. Over cold land surfaces, air masses are increasingly warm and moist upward from the surface; and so on.

characteristics are slightly disguised within the names. Recalling that there is a direct connection between the characteristics of the air mass and those of the underlying surface in its source region, it is natural that instead of "hot, or warm" and "cool, or cold" we use "Equatorial, or Tropical" and "Polar, or Arctic" to signify heat content. Likewise, we use "Maritime" and "Continental" rather than "moist" and "dry." Thus, the system yields the following set of names and symbols for air masses. In everyday meteorological practice, the terms "Equatorial" and "Arctic" are seldom used.

Moisture Content	Heat Content			
	Equatorial (Hot)	Tropical (Warm)	Polar (Cool)	Arctic (Cold)
Maritime (Moist)	mE	mT	mP	mA
Continental (Dry)	cE	cT	cP	cA

Fig. 10-30 shows the usual seasonal locations of the various source regions on Earth, and the boundaries between these source regions. The heavy boundary lines between the regions are frontal zones whose importance will become clearer in the next section. The information in **Fig. 10-30** is closely correlated with the three-cell model of **Fig. 10-7**. You will recognize already that the Tropical Fronts are closely associated with the ITCZ (**Fig. 10-7**) and that the midlatitude fronts are closely associated with boundaries between the Ferrel cells and the Polar cells — the Polar Frontal Zones of both hemispheres. You will recognize also that the entire pattern of source regions and fronts moves with the thermal equator, as in **Fig. 10-8**.

As will become evident in the next section, it is along the boundaries between the air masses having the greatest contrasts across them — cP and mT — that the most intense storms of midlatitudes are born and live out their lives. In the Northern Hemisphere, these boundaries are found in the northeastern quadrants of North America and Asia, and in the northwestern quadrant of Europe.

Storms and Weather Changes: The Mid-latitude Wave Cyclone

As we have just noted, it is along the air mass boundaries having the greatest contrasts across them that the most intense storms of midlatitudes are born and live out their lives. These storms are called **midlatitude wave cyclones** (MWC), and the general model for their anatomy and life history was worked out in detail about 50 years ago by Scandinavian meteorologists, who studied particu-

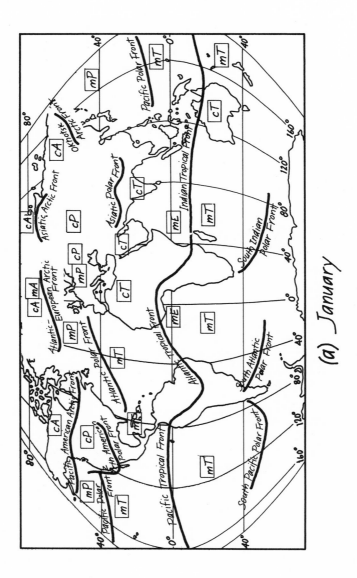

(a) January

Fig. 10-30

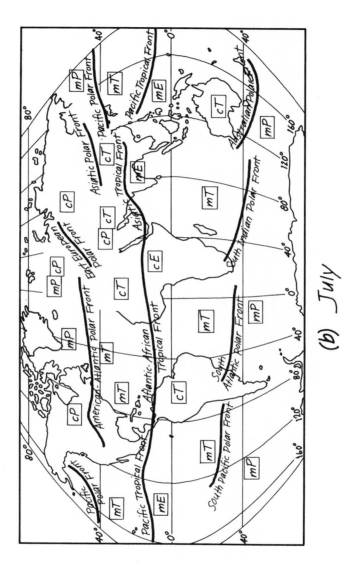

(b) July

lar examples in the northwestern quadrant of Europe. This section presents those details in a form I think will be most useful and instructive to designers and planners. As with any model, this one represents the "ideal" case. Reality seldom agrees exactly with the model, but the model helps one analyze an otherwise confusing set of observations. In the section, because of the nature of the subject, I will reverse my usual sequence and present the model — the theory — before discussing actual examples and the ways the theory differs with reality.

Fig. 10-31 presents mainly the life history of a typical MWC, while **Fig. 10-32** presents greater detail of its anatomy in its "age of early maturity" — the time when the storm has its greatest intensity. In both figures, the east-west dimension is approximately 1000 miles. To connect the life history in **Fig. 10-31** most directly with what we have already considered, I will begin by saying you can see the same stages of development of a storm in **Fig. 10-27** where we considered the idea of the mountain stream analogy. In **Fig. 10-31** the life history of the storm unfolds from top to bottom.

In the lefthand column the story is shown in the form you would see it on a weather map — the same form as in **Fig. 10-27**. In the righthand column the sequence is enhanced with three-dimensional block diagrams. As the storm matures and changes structure, it also moves across the surface of the earth — over distances typically two to four thousand miles in the course of its lifetime.

The storm forms on one of the boundaries between contrasting air masses, shown in **Fig. 10-30**. As the storm matures, the single boundary assumes two or three identities, called *fronts*. The **cold front** — drawn by meteorologists on constant altitude (sea level) maps as a line with sharp barbs pointing downwind — *is the moving air mass boundary along which colder air is replacing warmer air as the two air masses move*. The **warm front** — drawn as a line with rounded barbs pointing downwind — *is the moving air mass boundary along which warmer air is replacing colder air as the two air masses move*. Note that this two-dimensional line is the constant altitude surface's intersection with a three-dimensional surface between the air masses.

As the storm matures — as the huge swirling vortex "winds up" and the warmer air mass is pinched off and separated from the center of the storm — the cold front catches up with the warm front, and the map projection becomes the **occluded front** where the cold and the warm fronts are one above the other. It is drawn as a line with both the sharp and the rounded barbs, both pointing downwind. The particular temperature structures of the original air

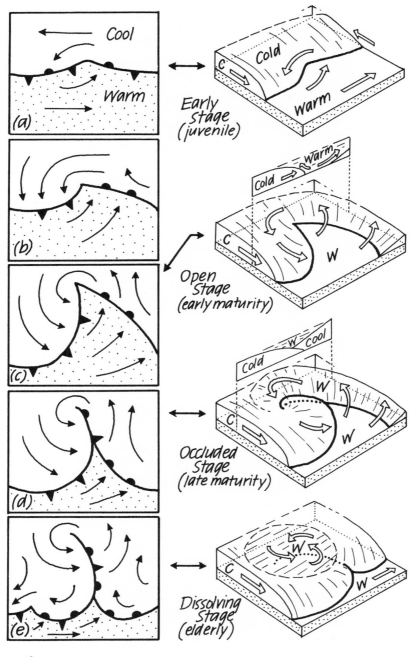

Fig. 10-31

masses determine which front is above the other, so there may be several different kinds of occluded front, which we will not need to discuss further.

If you look carefully at **Fig. 10-31** you will see there are places and times when *the air mass boundary is not moving across the map,* even though there is air motion (wind) in the system. These times and places are found in the very early and in the very late stages, and, along the line, neither air mass is replacing the other. This kind of boundary — called a **stationary front** — is drawn with sharp and rounded barbs on opposite sides of the front.

In early maturity, the MWC is at its most intense. **Fig. 10-32** shows more detail about the anatomy of this stage. As you study this figure, occasionally look also at the same stage in **Fig. 10-31** where there is a three-dimensional sketch to supplement the two cross-sectional drawings in **Fig. 10-32**. In three dimensions you can see that *along the cold front, colder air is undercutting warmer air, while along the warm front, warm air is overriding colder air.* This undercutting and overriding provide the basis for

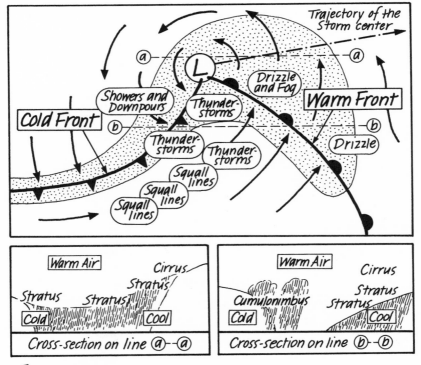

Fig. 10-32

the cloud types and locations associated with the storm. The clouds associated with the storm are found mainly in the shaded areas in **Fig. 10-32**. You will recall from **Chapters 3** and **4** that abrupt, local lifting of moist air produces cumulus clouds, while a gradual, regional lifting produces stratus clouds. As you would expect, then, cumulus clouds tend to form in a relatively narrow band associated with the cold front, while stratus clouds tend to form in a relatively wide band associated with the warm front. Noting the storm's trajectory in **Fig. 10-32**, you can see that the higher altitude stratus-like clouds associated with the warm front usually precede the storm's arrival — they are the first sign at a place ahead of the storm that the storm will soon arrive.

Other features of the storm's anatomy are distributed outward from the center of lowest pressure marked with the "L" for Low. Toward the south from the center (toward the north in the southern hemisphere) lies the **warm sector** — the warm, moist air mass feeding energy into the developing storm. Recall from **Chapter 4** that both the higher temperature and the greater moisture content of the tropical air mass (mT) represent energy content. The large contrasts in energy content over short distances, as we discussed in **Chapter 4** regarding the basic cause of wind, produce vigorous motion. The vigorous motion, in turn, produces clouds and precipitation, which are forms of energy conversion; and the storm develops and feeds on the energy of the air masses, especially that of the mT air.

As we discussed in relation to the precipitation types in **Fig. 10-16**, drizzle and fog are associated with the stratus clouds of the warm front, while showers and downpours break out in the cumulus clouds near the cold front. With the right conditions of stratification for temperature and moisture in the warmer air, what might otherwise be only cumulus clouds along the cold front, yielding just showers, will grow into thunderstorms producing downpours. For reasons beyond our scope to explain, thunderstorms can, in some storms, form in warm air overriding the warm front. If the local lifting associated with the cold front is more than ordinarily abrupt, and the vertical structure of the air in the warm sector is right, **squall lines** develop along the cold front and run ahead of it into the warm sector. These squall lines act just like fast-moving mini-cold fronts, complete with the **thunderstorms** usually associated with a vigorous cold front.

Although they are not necessarily associated spatially with squall lines, **tornadoes** most often occur somewhere within a warm sector that is of the right structure to support the formation of squall lines. A very violent witch's brew full of **severe storms**. Very shortly, in other sections, we will consider severe storms in more detail — thunderstorms and tornadoes of mid-latitudes, and the tropical cyclones nearer the equator[G].

Now we need to place the MWC within a spatial context. **Figs. 10-33** and **10-34** are sketches of the atmosphere over North America on April 1, 1971. The date is not important, except it suggests a time of year when contrasts over the continent are large. What is important is that the upper air, 500mb, constant pressure map of **Fig. 10-33** and the sea level, constant altitude map of **Fig. 10-34** are for the same time, and you now know how to interpret these maps. At the top of each figure is a small, generalized sketch showing windflows within the system, and connecting the information with previous discussions of the mountain stream and the MWC (**Figs. 10-21, 10-25, 10-26, 10-27, 10-28, 10-31,** and **10-32**).

Perhaps the main features of these two maps are that (a) from them you can get a better feel for the spatial dimensions of the MWC, and that (b) there is a definite, though loose, correlation between the sizes and locations of pressure features on the two maps: the MWC (L) at sea level is slightly east of the deep trough (T) at 500mb, and the two sea level highs (H) are matched by the two ridges (R) straddling the trough. You have already encountered — in **Fig. 10-23b** — the idea of a sea level Low being east of an upper level trough.

By now, you should have a good feel for the various aspects of the mid-latitude storminess found in the Polar Frontal Zone of the three-cell atmosphere of Earth. I will leave this subject by noting, as I did earlier, that the "model" of the MWC we have just discussed works much better in some places and times than others. Generally speaking, the model works best (a) over large, uniformly featureless mid-latitude regions — such as oceans, and the central portions of the wide continents of North America and Eurasia — and (b) in the spring and fall seasons when contrasting air masses are most likely to encounter each other. Mountainous regions distort and complicate the workings of the model, as do complex shorelines such as those surrounding the Mediterranean Sea. Finally, on the historical side, American meteorologists working in the Midwest added, to those of the Scandinavians, most of the ideas in the model about thunderstorms, squall lines, and tornadoes.

(G) While the discussion of severe storms has been introduced as part of the consideration of the MWC, the fact is that most thunderstorms on Earth occur in the tropics and have no relationship to warm fronts and cold fronts, and much less relationship to time of the year than in mid-latitudes. Tornadoes, on the other hand, are almost exclusively a mid-latitude phenomenon; in fact, they occur almost nowhere else than in North America. Finally, the severe storms of the tropics are called Tropical Cyclones, and they are quite seasonal. Roughly speaking, tornadoes that occur over water are called **waterspouts**, but they are much less seldom severe or destructive.

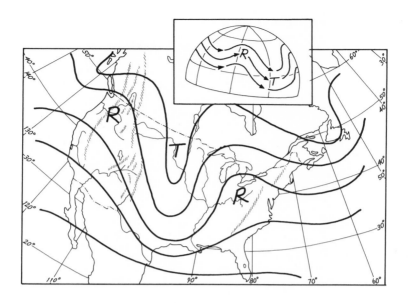

Fig. 10-33

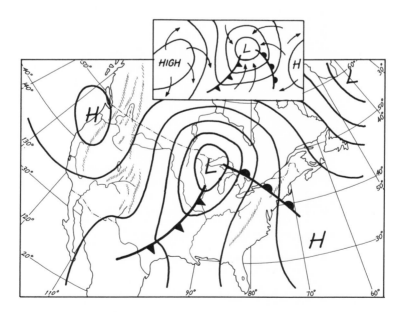

Fig. 10-34

Severe Storms: Winter

Generally speaking, the severe storms of winter take the forms of coldwaves, blizzards with heavy snow, and ice storms. Coldwaves have few design implications, but a recent summary of the climatology of coldwaves in the central United States will give you basic data on these storms (Wendland, 1987). Blizzards, with their heavy snow loads, have important design implications, but the local variability of loads — that is, snow accumulations at places other than official stations — is too great for me to be able to provide useful data of a generalized nature. Such data as there are on frequency of various snowdepths can be found in the publications described in **Appendix B**[(H)].

Baldwin (1973) provides an overview of the geography of ice storms in the United States. I reproduce it here since the information is less readily available elsewhere. **Fig. 10-35** provides average annual numbers of days with glaze and/or freezing rain, though design implications doubtless involve icing thicknesses as well. As with snow depths, local variability makes any generalized statements about ice loads beyond the scope of this book.

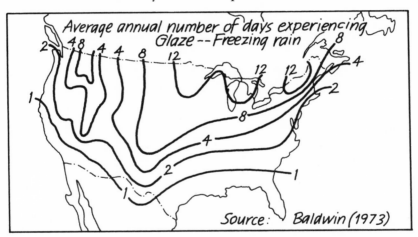

Average annual number of days experiencing Glaze -- Freezing rain

Source: Baldwin (1973)

Fig. 10-35

Severe Storms: Thunderstorms

Fig. 10-36 is a summary reminder of the discussion in **Chapter 4** about cumulus life cycles that reach the mature stages of a

(H) For most states in the United States, detailed data on various weather occurrences, such as snow depths, are available through the Agricultural Experiment Station of the state's University.

severe storm — thunder and even hail. In that discussion, we saw that any cumulus cloud will try to reach this mature stage, but, as in any life cycle, the number of special requirements increases each time the next stage is entered. That is, there are many more cumulus births than there are cumuli reaching the thunderstorm stage, and even fewer reach the hail stage. Thunderstorms, seen that way, are relatively rare events, and hailstorms much rarer still.

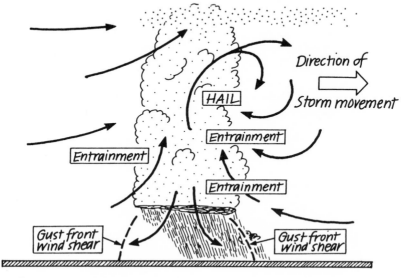

Fig. 10-36

Understanding how severe thunder and hail storms work does not tell us where and when they form — where the very special set of conditions arises. **Fig. 10-37** is a rudimentary answer for the United States, though it does not give information on when the storms occur. As noted earlier, thunderstorms are quite common in the tropics.

Knowing the requirements for thunderstorms from **Chapter 4**, we can easily grasp why Florida, the Southeast, and the central Rockies experience the most thunderstorms. Warm, moist air from the Gulf of Mexico flows onto the continent in North America's annual monsoon (see **Fig. 4-30a**). A generous supply of moisture, and solar heating of the surface to yield a sustained updraft provide two of the major requirements for mature cumulus growth. What is more, the Florida peninsula is especially well suited because the air flows across all the shorelines (see **Fig. 3-24**) and the airstreams converge, and then rise, in central Florida to provide an even more reliable updraft.

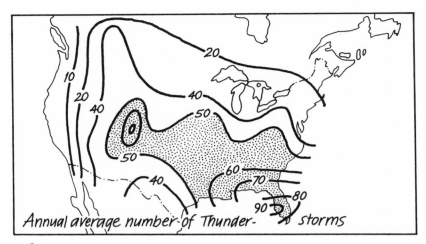

Annual average number of Thunder-storms

Fig. 10-37

The central Rockies may seem well removed from warm, moist air, but as noted earlier in the discussion of climate controls, air flowing over the central Rockies in summer has most often come from the Gulf of Mexico, by way of Texas and New Mexico. The added effects of abrupt increases in elevation to provide lift and updrafts makes for frequent and reliable thunderstorms and hailstorms. As you have probably seen already, the regions of high frequency of thunderstorms are just exactly suggested by the region named "convective" in **Fig. 10-16**.

In Florida and the Rockies, according to the records, there are enough thunderstorms to provide one nearly every day for three summer months in an average year. A map of hailstorm frequency shows a distinct maximum — about 8 days per year — in the central Rockies, but none at all in Florida and the south Atlantic Coast.

Severe Storms: Tornadoes

The name of nature's most violent storm on land comes from the Spanish verb *tornar:* to turn or twist. *Physically,* each of these storms is a very narrow vortex of air rotating at a very high speed. The direction of the spin is usually (but not always) "cyclonic" — counterclockwise in the northern hemisphere. This motion obeys the *principle of the conservation of angular momentum* (see **Fig. 10-6**), so that, for example, if a radius of 5 km is associated with a speed of 40 km/hr, then a speed of 200 km/hr must be associated with a radius of 1 km. The National Weather Service estimates tornadoes typically have radii of only 50 yards — 0.045 km — with rotational winds typically 500 km/hr.

It is certain that speeds within a tornado can, for short periods at least, exceed the speed of sound — about 1200 km/hr — since only that could account for the fact that wheat straws have been driven into tree trunks, like nails, by a tornado. This inward rush of fast moving air can occur only if the rate of removal of air at the top of the vortex tube is great enough to maintain a pressure difference of 50-100 millibars between the low pressure core and the air just outside — a very large pressure gradient indeed!

Tornadoes form in the same environments — they are creatures almost exclusively of North America — that produce squall lines: warm sectors of mid-latitude wave cyclones in early maturity, where air masses of large contrast converge. The likelihood of a tornado outbreak increases when the jet stream (see **Fig. 10-28**) occurs simultaneously above an especially vigorous stream of mT air flowing into the warm sector of a midwestern MWC. It seems that another condition for initiation of a tornado is a sudden and nearly explosive upward motion — a sudden breaking out into a very unstable air mass above — an upside-down version of what happens when we remove the cork from an inverted bottle full of water. The updraft in the center of a funnel is typically 300 km/hr.

As we noted in discussing thunder and hailstorms, they are rare events. So it is with tornadoes, since the conditions must be just so. Tornadoes occur almost exclusively in association with severe thunderstorms, even the waterspouts over oceans. The National Weather Service has estimated that there are, on the average, about 10,000 *severe* thunderstorms each year in the United States, of which only about 700 will spawn a tornado, and of these only about 20 will be truly violent and potentially lethal. As noted earlier, despite its power, a tornado typically involves only 1/1000 as much total energy as a thunderstorm.

Visually, the tornado has several aspects. The "inner funnel" forms when air, rushing to the low pressure core of the tube, condenses and forms a cloud. This cloud is shaped like a dangling rope. Though the bottom seems to descend to the earth, it is actually the case that the air in the cloud is rushing upward, and the place of lowest condensation is occurring closer and closer to the ground as the general environment is cooled.

If the low pressure core of the vortex extends to the ground, the insides of buildings, mobile homes, and automobiles will suddenly be at much higher pressures; and these structures will explode. The debris from these explosions is then lifted into the rising air, forming — visually — an "outer funnel."

Careful examination of the aftermath of many tornadoes reveals that debris is not lying all in one direction on the ground. It is often the case that debris lies in all directions — like jackstraws. This is the case because tornadoes often come in families — several funnels within a kilometer or so of one another, each rotating around its own core, the funnels rotating around each other, and the whole family moving across the ground. That kind of complex, triple motion produces very complex trajectories and patterns of debris and damage; but the path of the storm center — the center of the family — is nearly always *from the southwest toward the northeast.*

Table 10-1 presents various annual statistics concerning tornadoes, organized according to their location in the various geographical regions. Within a region, states are listed from northwest to southeast. Careful examination of these statistics reveals several interesting aspects of tornado occurrence and societal impact.

(a) As shown also in **Fig. 10-38**, there is a marked difference in reported annual frequency, depending on which sample of years is examined.

(b) When frequencies are expressed *per unit area,* there is a definite band of high-frequency states — called "Tornado Alley" — from the Gulf of Mexico up the Mississippi valley.

(c) The region of greatest probability of tornadoes is roughly the same as that for Thunderstorms, including two states outside "Tornado Alley": Florida and Colorado.

(d) Although the states with the largest tornado-related death rates *per unit area* — Mississippi and Indiana — are in "Tornado Alley", they are not the states with the largest tornado frequencies. Furthermore, Oklahoma and Florida, two states with high frequencies *per unit area,* have relatively low death rates. This is some-

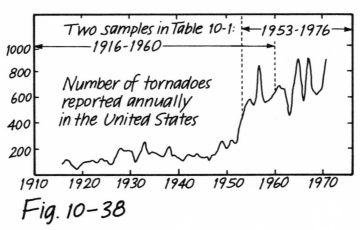

Fig. 10-38

-342-

times attributed to a vague factor called "public awareness", and to the types of housing involved. An historical "accident", such as a rash of many severe storms in the years of record just before warning systems were organized, could also account for some of the differences in this respect.

TABLE 10-1 OCCURRENCES OF TORNADOES IN THE UNITED STATES

State	Annual Number of Tornadoes[a] (1953-1976) (24 Years)		Annual Number of Tornadoes[b] (1916-1960) (45 Years)	Annual Number of Tornado Days[c] (1953-1971)	Annual Number of Deaths from Tornadoes[a] (1953-1976)
New England					
(Six States)	12.1	(1.8)[d]	3.2	9.9	4. (0.6)[d]
North Atlantic					
New York	3.3	(0.7)	0.8	2.5	0. -
Pennsylvania	6.6	(1.5)	3.2	4.4	0. -
New Jersey	1.6	(2.1)	0.7	1.2	0. -
Maryland	2.3	(2.2)	1.5	1.7	0. -
Delaware	0.8	(4.3)	0.2	0.6	0. -
West Virginia	1.9	(0.8)	0.3	1.4	0. -
Virginia	5.3	(1.3)	1.9	3.3	1. (0.1)
North Central					
Minnesota	15.5	(1.9)	5.0	9.7	3. (0.3)
Wisconsin	17.3	(3.1)	5.1	10.0	2. (0.4)
Michigan	14.7	(2.5)	4.1	6.6	9. (1.6)
Iowa	25.6	(4.6)[f]	14.5	11.6	2. (0.3)
Illinois	28.3	(5.0)[f]	8.4	11.2	5. (0.9)
Indiana	23.0	(6.4)[f]	6.6	10.5	8. (2.3)
Ohio	13.5	(3.3)	3.9	7.1	6. (1.5)
South Atlantic					
North Carolina	10.8	(2.1)	3.0	5.4	1. (0.2)
South Carolina	9.2	(2.9)	3.6	6.3	1. (0.3)
Georgia	21.3	(3.6)	6.6	11.8	3. (0.5)
Florida	36.7	(6.3)	7.2	19.8	2. (0.3)
South Central					
Missouri	29.8	(4.3)[f]	11.2	14.6	5. (0.7)
Kentucky	7.9	(1.9)	1.8	4.2	4. (1.0)
Arkansas	18.4	(3.5)[f]	10.8	9.1	5. (0.7)
Tennessee	11.2	(2.6)	4.6	5.2	3. (0.7)
Gulf Coast					
Texas	115.6	(4.3)[f]	25.8	45.7	13. (0.5)
Louisiana	19.0	(3.9)[f]	6.6	10.2	3. (0.7)
Mississippi	22.1	(4.6)[f]	7.	10.5	13. (2.7)
Alabama	19.6	(3.8)	7.5	9.3	7. (1.5)
Great Plains					
North Dakota	14.8	(2.1)	3.5	8.1	1. (0.1)
South Dakota	22.7	(2.9)	4.6	11.6	0. -
Nebraska	34.0	(4.4)	11.9	16.6	2. (0.3)
Kansas	46.8	(5.7)[f]	25.9	22.7	6. (0.7)
Oklahoma	55.3	(7.9)[e,f]	22.8	24.5	7. (1.0)

Mountain West					
Montana	3.5	(0.2)	1.7	2.9	0. -
Idaho	1.4	(0.2)	0.4	1.1	0. -
Wyoming	6.8	(0.7)	2.0	4.9	0. -
Nevada	0.60	(0.1)	0.02	0.4	0. -
Utah	1.3	(0.2)	0.2	1.3	0. -
Colorado	14.5	(1.4)	3.8	9.8	0. -
Arizona	3.8	(0.3)	0.6	2.7	0. -
New Mexico	8.5	(0.7)	2.2	6.6	0. -
Pacific Coast					
Alaska	0.04	(0.0)	0.02	0.04	0. -
Washington	1.0	(0.1)	0.2	0.9	0. -
Oregon	1.0	(0.1)	0.2	0.7	0. -
California	2.8	(0.2)	0.8	1.9	0. -
Hawaii	0.5	(0.1)	0.04	0.4	0. -

(a) Lutgens and Tarbuck (1979). These data are updated by two years in Lutgens and Tarbuck (1982)
(b) Critchfield (1974)
(c) Baldwin (1973)
(d) Data in this column are (per 10,000 square miles — 100 miles x 100 miles)
(e) This figure exceeds 16 in *Central* Oklahoma
(f) This state considered to be in "Tornado Alley."

As noted above, there is a marked difference in reported annual frequency, depending on which sample of years is examined. **Fig. 10-38** makes this point dramatically, and the discussion of The Builder near the end of **Chapter 7** emphasizes your need to understand it. In the figure, it appears as if the number of tornadoes occurring in the United States increased almost explosively — at least tripling in the early 1950s. While dramatic changes of climate can and do occur, it is also quite possible that the increase was in the density and efficiency of the reporting network. The point for designers and planners, of course, is that you should be as aware as possible of all readily available summaries of climatological data, rather than simply accepting uncritically the first one you encounter. The secondary point is that your atmospheric consultant is probably your best source of information on the status of such summaries.

As noted above, **Fig. 10-37** does not give information on *when* thunderstorms occur. **Fig. 10-39** demonstrates the distinct seasonal variability of the mid-latitude severe storms in the United States. The conditions already described for the growth of cumuli into mid-latitude thunderstorms, and for the outbreaks of tornadoes, are most prevalent in the late spring and early summer in both hemispheres. Within the seasonality of **Fig. 10-39** there is an annual cycle in which the tornado season begins in late winter in the central Gulf Coast states, and then moves up Tornado Alley to the north and west during March and April. By the peak months of May and June, the center of tornado activity is in the southern Great Plains, the upper Midwest, and the Great Lakes.

The data in **Fig. 10-39** are from Baldwin (1973), and the very same trends appear in the data for a longer sample: 1953-78 (Lutgens and Tarbuck, 1982, page 251).

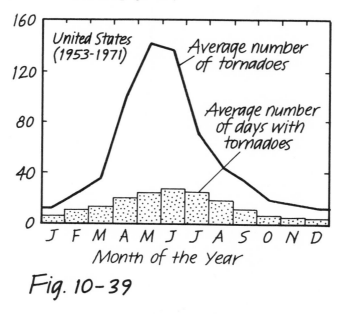

Fig. 10-39

Severe Storms: Tropical Cyclones

Our considerations now turn from the mid-latitudes to the tropics. As we have noted, while thunderstorms are creatures of both the mid-latitudes and the tropics, tornadoes and hail are found mostly in the mid-latitudes. Severe storms of the tropics are called **Tropical Cyclones**[I]. These storms, like tornadoes, are so reliably and massively destructive that no ordinary *design* can withstand their direct onslaught, though there are *planning* principles that can be invoked as responses.

Tropical Cyclones go by various names in different parts of the world. Here, and in **Fig. 10-40**, is a partial catalog of the most widely used names. The storms occur in many places not listed in this catalog, but the list serves to tell you that these storms,

---

(I) To a meteorologist, *cyclonic flow* is any flow — at any altitude or latitude — that is counterclockwise in the Northern Hemisphere (and the opposite south of the Equator). To a meteorologist, a *cyclone* is a complete storm system with the general size of 500-1000 miles across, although the term is often used in the rural American midwest to mean the same thing as a tornado.

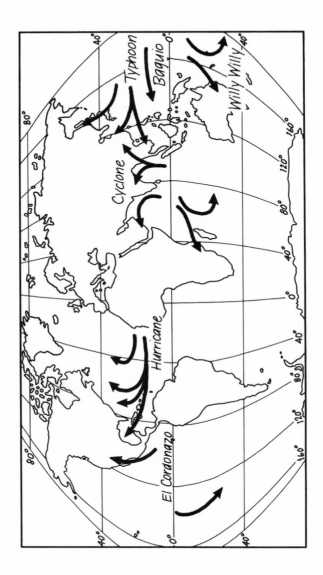

Fig. 10-40

although with different names, are all really the same thing, meteorologically speaking. The generalized storm tracks for the storms are also shown in **Fig. 10-40**. Because of the requirements for formation that we are about to discuss, these storms occur mostly in the quiet seasons at the end of summer when the local surface waters are at their highest temperatures.

Region	Local Name for Tropical Cyclone
North Atlantic and Caribbean	Hurricane
Mexican West Coast	El Cordonazo ("The Whip")
Northwestern Pacific	Typhoon
The Philippines	Baguio (A city on northern Luzon Island)
Australasia	Willy Willy
Bay of Bengal	Cyclone (see Footnote (I))

Following our pattern of what-then-why, **Fig. 10-41** shows the general anatomy of a mature Tropical Cyclone. The cyclonic flow looks familiar from our considerations of the MWC, but here *the storm system is somewhat smaller, there are no contrasting air masses and no fronts*. The storm is almost exactly circular, and its symmetry is also circular: any cross section through the central **eye**, regardless of direction, will look about the same. The clouds and winds are in "bands" whose shape and motion spiral inward toward the eye. The eye is a quiet, cloudless area — about 20 km in diameter — in the core of the storm, and it is separated from the violent parts of the storm by the **eye wall**: a cylindrical zone of transition. As in a MWC, the outflow at the top of the storm is anti-cyclonic — in the opposite direction from that at the surface — so that the whole system resembles a giant doughnut, with the eye for a hole, except that there is a spiraling motion making it different from the simple doughnut patterns in **Fig. 3-26**.

As in the tornado, the motion within a tropical storm must obey the *principle of the conservation of angular momentum* (see **Fig. 10-6**), but since the central radius is larger in the Tropical Cyclone than in the tornado, the maximum wind speeds are not so great as in a tornado. Wind speeds greater than 75 mph (120 km/hr) are designated as *hurricane force winds* — 12 on the Beaufort Scale (see **Chapter 3**) — but these speeds occur at the outer boundaries of the moving Tropical Cyclone. The wind speeds near the eye are more typically near 320 km/hr. Broadly speaking, if winds reach Force 12 at a station, the assumption is usually made that the station will be on or near the track of the storm, which is usually approaching at a speed of 40-45 km/hr.

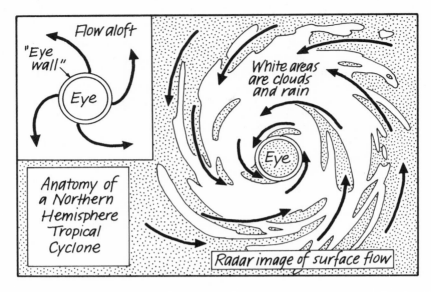

Fig. 10-41

Fig. 10-42 sets the scene for our discussion of the formation and life history of Tropical Cyclones. Recall from Fig. 10-14 that "cool" ocean currents flow equatorward on the eastern shores of the major ocean basins. As these currents reach very low latitudes, they turn westward, becoming parallel currents — one in each hemisphere. This westward turning off the coast of West Africa is shown in the figure as an example — the example leading to the Hurricane, most familiar to North Americans.

Between the cool currents and the warm, tropical air above there is a shallow, cool layer of air, formed by contact with the cool waters. As we shall see in **Chapter 11**, this condition of cool-beneath-warm is called *stable,* and is a condition in which the boundary between the cool air below and the warm air above maintains its identity under ordinary circumstances: very little mixing takes place between layers. In fact, this boundary between air layers is so stable in these parts of the world that we can think of it almost as a huge sheet of air — almost like a fabric sheet — stretching thousands of miles across the tropical ocean. The boundary is much more substantial and impenetrable in the eastern parts of the oceans, but because of gradual warming of the waters and gradual mixing between cool and warm air layers farther westward — "downstream" — the boundary becomes less substantial.

As the boundary becomes less substantial, the depth of the cool layer also becomes greater, as shown in the sketch of **Fig. 10-42**.

Now imagine that some major impulse — the equatorward edge of a MWC, for example — sets up a westward moving wave train on the stable boundary surface, much like the wave trains described in **Fig. 10-24**. As the wave train moves westward, it sets up patterns of vertical and horizontal motions, and patterns of cloud, also shown in the sketch of **Fig. 10-42**. As we have noted, the columns of air just ahead of — downstream from — the wave crests experience organized rising motions, with tall clouds and the accompanying release of latent heat into the upper layers of air. It is usually in these circumstances that tropical cyclones are born.

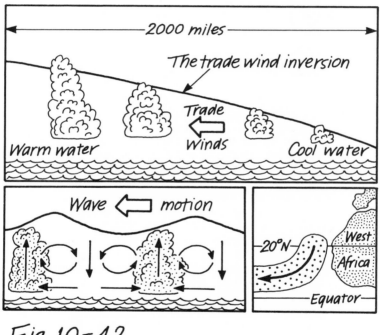

Fig. 10-42

If rising, cloudy air in the tropics were all that is needed to form a tropical cyclone, there would be many times more of them. Although we know that the latent heat released in the rising tropical air is the fundamental source of energy that drives a tropical cyclone on to maturity, other factors must be acting cooperatively — as in the cases of thunderstorms and tornadoes — before rising air in a tropical wave train becomes a severe storm. The two principal additional factors are *very warm surface waters, and a source of spinning energy* for the young storm system. Researchers have concluded that large expanses of open ocean with water temperatures exceeding 27 °C are essential for maturity. These high temperatures are necessry to assure that the vapor content of the air (remember, this content rises with temperature as in **Fig. 4-22**) is sufficient for the required rate of energy release in condensation. For one thing, that probably accounts for the absence of tropical cyclones in the South Atlantic, which only rarely has these high temperatures.

As for the source of spinning energy — called **vorticity** — researchers believe it lies in the cyclonic upper air flows, above the young storm system, that are in turn part of a mid-latitude storm system that lies poleward of the young storm just at the critical time it is forming. We didn't mention it as part of the discussion of the Coriolis force, but the strength of the CF — the length of the CF vector in the "Coriolis Boutique" (see **Fig. 10-20**) — increases toward either pole, but it is *zero at the Equator.*

From this you see again the long list of special requirements needed for the maturing of a severe storm. Without the CF in the oceanic regions near the Equator, there is no local source of vorticity, so a mid-latitude source must be tapped — piggybacked — for the storm to grow. For this to happen, in cooperation with the presence of a wave train and warm waters, the mid-latitude source must be at just the right location poleward of the wave train. In addition to being a source of vorticity, the mid-latitude system — though perhaps a thousand miles away — circulates in such a way as to help remove the air rising in the middle of the tropical storm, as there must be a removal of air at the top of a tornado. All parts of the atmosphere interact — sometimes enhancing and sometimes inhibiting what happens elsewhere.

The impacts of tropical cyclones on the built environment result not so much from wind force as from the pounding and flooding of near-coastal areas by giant surges of water pushed into huge waves by the winds at sea. When the storm center "runs out of warm water" — when it moves over a cooler surface — it begins to die. Movement over rough, friction-producing land adds to the rate at which the storm dies. Thus, tropical cyclones tend to reach their maximum intensities at the westward ends of their cross-ocean tracks (see **Fig. 10-40**) where they also tend to "recurve" northward and then eastward, moving into the mountain stream of the mid-latitude westerlies.

Weather Changes in The Tropics

All this discussion of severe storms in the tropics might leave you thinking weather there is mostly violent. Of course, the familiar pictorials of the lovely, calm tropical sunsets — with towering clouds on the horizon — really do describe the most frequent weather conditions of the tropics. As we will see presently, *weather changes in the tropics are mostly diurnal — between day and night.*

If you stop and think about the contents of **Fig 10-8**, describing the seasonal movement of the Thermal Equator, you will quickly see that *tropical seasons are "wet and dry"* rather than "warm and cold" as they are in mid-latitudes. More particularly, the length and strength of the dry season, relative to that of the wet season, increases the farther one goes from the Equator toward the Tropics of Cancer and Capricorn. At the Equator there is virtually a continuous wet season; and the dry seasons at slightly higher latitudes occur in the months of the mid-latitude winter of that same hemisphere. For example, the dry season in Central America is centered on February and March, while that in Northern Australia is centered on June and July.

Tropical weather changes in the form of cyclones are seasonal, as we have noted. This kind of weather is restricted to the tropical regions 10-15 degress from the Equator, where there are distinct wet and dry seasons. Mainly because of the fact that the CF is zero at the Equator — in the rainy tropics — cyclones don't reach the rainy tropics. *In the rainy tropics, weather changes are mostly diurnal.* The diurnal temperature range is typically 10 °C, while the annual temperature range — the range of monthly mean temperatures (see **Chapter 8**) — is typically only 2 °C. And the rainfall, almost exclusively from thunderstorms, occurs with a distinct maximum probabilty in mid-afternoon.

The afternoon storms are the principal weather of the rainy tropics — their dying stage provides the tall clouds on the horizon in the sunset color pictorials. The storms are a result of the local sea breeze circulation (see **Fig. 3-24**) — a sort of daily monsoon. In the saturated air, slight lifting by even minor terrain features can determine the preferred locations of the storms and enhance their production of rain. The convergence of sea breeze streams over the centers of modest sized islands produces lifting effects similar to those of terrain features. The storms are so reliable that "western" hostesses in the rainy tropics send invitations for parties to begin "after the shower."

Notes

(1) Actually, it was a 19th century French mathematician, G. G. Coriolis, who made the first complete theoretical study of the effect (McDonald, 1952). In mathematical terms, the imaginary force accounts for motion on a *rotating coordinate system* rather than simply over a curved surface. For example, if the earth were rotating in the opposite direction on its axis, the Coriolis force would act in the opposite direction, and winds would turn left in the northern hemisphere.

(2) See, for example, pages 186-188 in the excellent text by Lutgens and Tarbuck (1982).

Peptalk

(P₂₂) All of a sudden, we are using ideas about *vectors*. It is a natural technique for dealing with wind velocity, which, you recall from **Footnote (L)** in **Chapter 8**, includes information on both speed and direction. For our purposes, there are two ways to look at the use of vectors, in particular the *addition of vectors.*

First, as in the sequence of wind vectors of **Chapter 8**, we consider a group of vectors as each representing *the same thing at different times*, in this case wind velocity at a certain place and different times. Adding the vectors to estimate trajectories, we placed the vector arrows end-to-end and noted where they went (on a map) and where they ended with respect to the point where they began. We noted that *any sequence of addition* ended at the same place — in the particular example of **Chapter 8**, at the starting point. Such a sequence would not have to end at the starting point. If it did not, the vector from the starting point to the ending point would be called the **resultant vector**.

Second, as in the consideration of forces in this chapter, we consider a group of vectors as each representing *a different thing at the same time,* in this case different forces acting on a certain air parcel at one time. If the vector arrows are added end-to-end and they return to the starting point, the forces are said to be **balanced**, or in balance. In such a case the parcel will continue its motion at the same velocity (speed and direction) according to Newton's First Law of Motion. If the vectors are not balanced — they end at a different point — the velocity of the parcel will change (it will be *accelerated*) in speed, direction, or both.

11. What are we DOING here?
Air Pollution and other losing bets

Introduction

One of the assumptions we have been making throughout most of this book — without putting it into words — has been that designers and planners have, in some sense, *control over the environments* they are designing, planning, and managing. That assumption, quite clearly, has been understood in the section called **Environmental Design — Chapters 7** and **8**. If you stop and think about it, however, you see it has not been understood in this section called **Supplements — Chapters 9** and **10**. We have been considering, in this section, the larger scale processes and their results observed in Earth's atmosphere, not necessarily because we think they are manageable processes, but because we recognize they are as they are and have to be lived with by any designer, planner, or manager.

In a larger sense, most things we address in this section called **Supplements** are beyond our control. The subject of this final chapter — air pollution — fits somewhere between the controllable and the uncontrollable. Before we begin to consider a catalog of concepts concerning air pollution control, let me remind you, in a philosophical way, of the uncontrollable aspects of air pollution that will lie in the background of all our discussions in this chapter. I want to do this because I have encountered too many designers and planners who think of any pollution problem as one whose solution is just a matter of making a design and preparing a report. My reminder is in the form of a quotation from Buckminster Fuller (1978, page 77).

Typical of the subsidiary problems within the whole human survival problem (whose ramifications now go beyond the prerogatives of *planners* and must be solved) is the problem of pollution in general — pollution not only of our air and water but also of the information stored in our brains. We will soon have to rename our planet "Poluto." In respect to our planet's life sustaining atmosphere we find that, yes, we do have technically feasible ways of precipitating the fumes, and after this we say, "But it costs too much."

The larger problem Fuller alludes too, of course, is the one centered on the need to sell the plan to those who will pay for it and benefit from it. That problem is not a new one to designers and planners, of course. You, probably more than I, know what a client is and how difficult he can be at times. But somehow, again in my experience, problems concerning air pollution seem to many designers and planners to be on space and time scales that tend to make them forget there is a client. Perhaps that is because problems concerning air pollution "go beyond the horizon" and involve so many people — clients — not readily identifiable with the problem and the plan.

To finish this point in consideration of the economics of pollution, let me explain why I have included, in the title of this chapter, the idea that *air pollution is a losing bet.* I put it that way because, it seems to me, these designers and planners of whom Fuller and I speak, and most of their clients, have made a bet that Earth's atmosphere is so huge and so efficient at diluting and distributing whatever we wish to dump into it, that dumping can be done pretty much without cost. That, we are beginning to see all too clearly, is a losing bet.

Be that as it may, it's time to present for your consideration a catalog of concepts concerning air pollution control. **Fig. 11-1** shows a standard paradigm, used through the years by most air pollution control specialists, to present a basic framework for

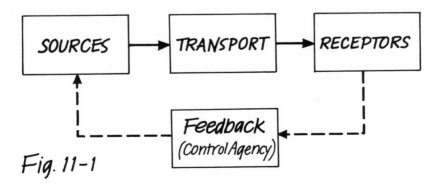

Fig. 11-1

studying the subject. Although the diagram looks similar to those for systems in **Chapter 2**, in fact it does not describe the flow of energy or mass, but only of concepts. It is of value mainly to provide a framework — an outline — of the major areas of concern.

Following the next section in this chapter, the presentation is organized according to the three major areas: sources, transport, and receptor effects. Then following is a section in which our concern is with the feedback and management associated with air pollution control concepts. In passing, note that this diagram could as well describe nearly any form of environmental pollution: water, air, or land. Note also that, for air pollution in particular, all three of the major areas are responsive to changes and differences in weather and climate.

The kinds and emission rates of air pollutants respond to weather (and climate): fuels for space heating and cooling, for one example, are weather-dependent. The transport of air pollutants from any source depends on the state and dynamics of the atmosphere. The effects of air pollutants on receptors is also dependent on weather (and climate): the wetness of building stone, for one example, affects the rate of erosion by airborne acids.

In the diagram the feedback loop is from the receptor phase to the source phase — it is **reactive** feedback — because most existing control agencies work that way. It could as well be from the source phase to the receptor phase — **proactive** feedforward — but agencies have not customarily done much in the preventive mode. That deficiency could, it seems to me, be the point of entry for designers and planners.

Airfill: the atmospheric garbage dump

Perhaps you have heard about the term "**airshed**" as an atmospheric counterpart to the hydrologic term "**watershed**." Both terms refer to a semi-isolated portion of a geophysical sub-system — of the atmosphere in the former case and of the hydrosphere in the latter — that has characteristics of size, catchment (inflow) rate, and release (outflow) rate from a known outlet, or drain. I have even heard of the term "milkshed", used by geographers to suggest the area from which dairy products flow into an urban area within the shed.

I suggest there is also an atmospheric counterpart to the solid waste term "**landfill**." That term is "**airfill**" — the volume of atmosphere into which, as noted earlier, we systematically inject waste products with the expectation (the bet) that those products will deteriorate and become harmless during their residence period in the fill.

Fig. 11-2 catches several of the aspects of the concepts of air-shed and airfill. Perhaps the best way to appreciate the system shown is to conduct another imaginary experiment — or even a real one — using your own domestic water basin or tub as the laboratory. The figure provides the principle truth contained in the concepts — the familiar mass balance from **Chapter 4**: *as long as inflow rate exceeds outflow rate, there is an increase in storage.* Here, unlike **Chapter 4**, the upper limit of storage capacity is defined. It is the size of the basin. Here, also unlike **Chapter 4**, the upper limit of the outflow rate is defined. It is the size of the drain. Revised, the principle truth becomes: *as long as inflow rate exceeds the maximum drain size, the basin will overflow, and the only unknown factor is "when."*

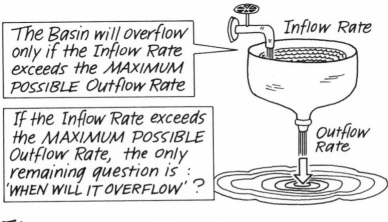

The Basin will overflow only if the Inflow Rate exceeds the MAXIMUM POSSIBLE Outflow Rate

If the Inflow Rate exceeds the MAXIMUM POSSIBLE Outflow Rate, the only remaining question is : 'WHEN WILL IT OVERFLOW' ?

Inflow Rate

Outflow Rate

Fig. 11-2

Notice these important results of your experimnent:

a) as long as the inflow rate is less than the outflow rate, the basin will remain empty;

b) as long as the inflow rate is never greater than the outflow rate, the inflow rate can have any value, and *need not be equal to zero;*

c) as long as the inflow rate equals the outflow rate, the volume stored in the basin will remain constant at the value it had at the moment the inflow rate and the outflow rate became equal; and

d) if the inflow rate ever exceeds the outflow rate, it matters not how many sources there are nor which is the largest or the smallest, the basin will overflow, and the only unknown factor is "when".

These homely experimental results provide several lessons for the strategies of air pollution control:

1) the central operational problem for planning and management is to define the maximum drain size;

2) once the maximum drain size is defined, it also defines the maximum allowable emission rate from all sources; so that

3) those of the public (see **Chapter 7**) and those environmentalists (see **Chapter 8**) who cry "*Zero emissions*" are being unnecessarily restrictive, and are probably demonstrating their lack of understanding of the fundamental processes involved; and

4) as noted in **Chapter 2**, the flow rates and the storage volume need have no particular size correlation.

Building on these homely results requires that we examine the notion of "the maximum drain size" in the context of this chapter. In the context of the airfill, the drain size is correlated with the rate of removal of pollutants by all means, or processes. In fact, pollutants in significant masses are seldom removed artificially — by man-made means — once they enter the air. Removal, therefore, is mostly by natural processes, such as (a) *conversion* to other compounds by air chemistry, (b) rainout and washout in *wet removal,* and (c) fallout and impaction by *dry removal.* These three routes constitute *the atmosphere's self-cleansing.*

Each of these routes has a maximum possible rate (which varys with weather and climatic conditions) and their *sum* is directly correlated with the "the maximum drain size." Incidentally, conversion to another compound may change only the nature of the air pollution problem if the new compound is also troublesome. And removal of a compound may only change the problem from one of air pollution to one of water or land pollution. But we are discussing only air pollution. Here is a suggestion of why the drain size changes with changing weather and climate.

The rate of . . .	. . . varys with these atmospheric properties . . .
Conversion Wet removal[a]	air temperature, humidity, insolation
rainout	storminess: cloud type and rainfall rate
washout	storminess: cloud type, rainfall rate, humidity beneath the clouds
Dry removal[b]	
fallout	wind speed and turbulence
impaction	wind speed

(a) **Rainout** is the capture of particles and gases by raindrops within a cloud, while **washout** is the capture of particles and gases by raindrops falling beneath a cloud.

(b) Solid (particulate) pollutants **fall out** by gravity, obviously according to their size, shape, and mass. They can be removed by **impaction** when they strike an object and are then "stuck" to it.

It should be apparent, even from this simplified list of weather elements, that the drain size, and therefore the allowable emission rate, is a changing thing. Only the concept remains static. Be that as it may, *the critical research problems associated with our increasing understanding of air pollution and its control lie in the area of discovering the natural removal processes and rates associated with each of the important pollutants.*

Fig. 11-3 sets the stage for a further examination of the matter of volume and storage as they relate to the drain size in Earth's air-fills. The maximum storage capacity of the basin in **Fig. 11-2** has its equivalent, in **Fig. 11-3**, in the volume of air labelled "lower layer," which will vary in size according to varying atmospheric conditions — it is the size of the airfill.

The reason we need to know the size of the airfill is that (Total mass/Volume) is the Concentration of the pollutant in the airfill[A]. Once we quantify the maximum allowable concentration (usually for reasons associated with the responses of receptors) and the size of the airfill, we know the total allowable mass. These factors would really be irrelevant to the idea of the maximum drain size except for the fact that — unlike our simple experiment — on the local scale *the volume of the airfill changes with changing weather.* For example, if the weather changed so that, as from **Fig. 11-3a** to **Fig. 11-3b**, the volume would become smaller with no change in the total mass contained, the concentration, to which

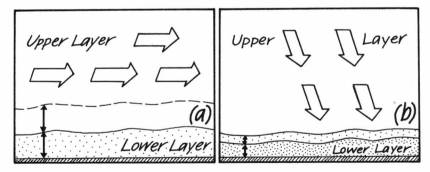

Fig. 11-3

(A) You will recognize this discussion is simplistic, but we need to clarify certain fundamental ideas that are best introduced simplistically. For example, (Total mass/Volume) is the Concentration of the pollutant, really, only if the mass of the pollutant is uniformly distributed within the volume. For another example, the pollutants mentioned in discussion of the drain size must usually be considered separately, rather than *en masse,* as they are implicitly considered in that discussion.

most receptors respond, would increase markedly. The point of this paragraph is that *the relationships among the volume, mass, and flow rates* — one of which is related to the drain size — *are dynamic because the atmosphere is dynamic.*

It is not at all fanciful to say that the size of the storage volume — the local airfill — changes constantly. It changes from night to day as the sun warms and expands the lower layer, from one day to the next as storm systems come and go, and from one season to the next as they become wetter, drier, and more or less sunny and stormy. You have already examined, earlier in this book, all the processes by which these changes are constantly taking place.

In **Fig. 11-3a**, the vertical motion is mostly associated with the *daily* expansion and contraction of the lower layer. On the *weekly* time scale, the upper air streams, when they are stormy, assist in the expansion. These circumstances usually exist, in mid-latitudes at least, in the transition seasons of spring and autumn.

In **Fig. 11-3b**, the vertical motion is associated with massive downward pushing on the lower layer — the airfill — by subsiding upper air. This is called **subsidence** by meteorologists, and it is usually associated with either the midsummer or the midwinter seasons (depending upon geographical location) when the regional atmosphere is relatively calm and free of storminess. Clearly, it is in these times that the basin and the drain size are smallest. These "worst case" conditions, unfortunately, tend to persist for most of an entire *season,* without the daily and weekly relief of other seasons.

If the inflow (emission) rates of local sources of pollutants are held constant while the volume of the airfill and the drain size are changing, is it any wonder that the concentrations of pollutants increase and decrease, changing on all the time scales from daily to seasonally?

From this last question arise two basic strategies for air pollution control: (i) implement a *dynamic system* to manage the emission rates in response to the changing volumes and outflow rates, or (ii) implement a *static system* by setting emission rates so as not to exceed the allowable rates under "worst case" local conditions. In passing, note that the skills of an expert atmospheric consultant are required for the first, dynamic system; and that, for the non-expandable storage volume of the whole of Earth's atmosphere, only the second, static system will work in the final analysis. What is more, society must soon begin to realize that, when the wind blows, pollutants may be diluted and transported — giving the impression of a form of "removal" on the local scale — but on the regional and global scales, they are not removed at all. They just become someone else's problem. No, *the bet that the maximum drain size can be stretched is a losing bet.* Buckminster Fuller is correct.

To close the discussion of the dynamics of the airfill, we need to understand a concept that is absolutely central to much of atmospheric ecology, and certainly to the subject of air pollution control. That concept is called **stability**, and it explains why some air masses exhibit vigorous vertical motion and, with it, larger airfills, while other air masses exhibit the opposite. The concept of stability resides in the relationships between air temperature, on the one hand, and altitude — which is to say pressure — on the other hand. Accordingly, each of the four panels of **Fig. 11-4** is plotted on coordinates of temperature increasing from left to right, and altitude increasing from bottom to top.

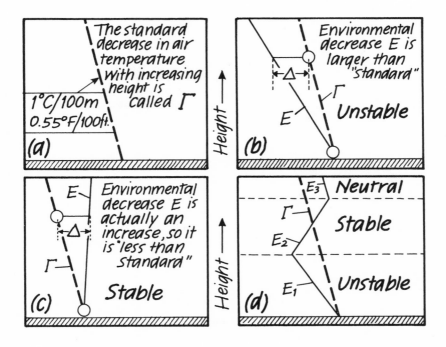

Fig. 11-4

Fig. 11-4a shows that the standard relationship between temperature and altitude is that there is a *1 deg C decrease in air temperature for every 100 meters of increase in altitude.* This translates into a 0.55 deg F decrease for every 100 feet. It is "standard" because it is the relationship that will eventually develop if a volume of air is well mixed, and then allowed to sit quietly without the addition or subtraction of any heat energy.

This ratio of (temperature/altitude) is called a **lapse rate**, and the standard lapse rate is called the **dry adiabatic lapse rate**[P23]. Meteorologists give it the symbol Gamma: Γ. It has that particular value because Earth has a particular acceleration of gravity (that is, a particular mass) and it has an atmosphere of a particular density. Other planets have different standard lapse rates.

Now I need to note the difference between a *process lapse rate* and an *environmental lapse rate*. The process lapse rate is the rate of change of temperature that a parcel of air will undergo if it, for some reason, is forced to change its altitude/pressure — either an increase or a decrease of altitude/pressure. In this discussion the only process lapse rate we will be concerned about is Γ. In **Figs. 11-4b** and **11-4c** the small balloons experience the process lapse rate Γ as they change altitude.

An environmental lapse rate, E, is the *observed relationship* between temperature and altitude/pressure in an air mass. From the discussion of the formation of air masses in **Chapter 10** you can see how different air masses can have quite different lapse rates — different even from one layer of air to another in the same air mass. In **Figs. 11-4b** and **11-4c** you can see what happens when an air parcel moves vertically within two different air columns: one with a value of E larger than Γ, and one with a value of E smaller than Γ[1]. In **Fig. 11-4b** an air parcel moving from the surface upward into the environmental air mass will change its temperature according to Γ, and will find itself, upon arrival aloft, *warmer than its new surroundings by an amount* Δ. It will be *positively buoyant,* that is, and will tend to continue its rising motion. With a little thought about it, you can see that the *downward* motion of an air parcel in this environment would make the parcel *cooler* than its environment Thus, this positive environmental lapse rate *enhances vertical motion.* It is called **unstable**[1].

In **Fig. 11-4c** an air parcel forced to move upward will find itself cooler than its new surroundings by an amount Δ. Vertical motion is *suppressed,* and the lapse rate is **stable**. When E equals Γ, the lapse rate is **neutral**, and vertical motion is neither enhanced nor suppressed. In **Fig. 11-4d** an air mass is layered, each layer having a different value of E.

In summary we have these combinations, in which the additional term **inversion** is introduced: a stable lapse rate in which temperature *increases* with increasing altitude/pressure.

Lapse rate	Value	Any vertical motion — up or down — is ...
Unstable	Positive	enhanced (**Fig. 11-4b**)
Neutral	Equal to Γ	neither enhanced nor suppressed
Stable	Positive	moderately suppressed
	Zero (isothermal)	suppressed
	Negative (INVERSION)	strongly suppressed (**Fig. 11-4c**)

The middle layer of air in **Fig. 11-4d** is *stable,* and in fact it exhibits a thermal inversion, just defined. "The more negative the lapse rate," meterologists say, "the stronger is the inversion." The stronger the inversion, the more difficult it is for a rising air parcel to penetrate the inversion layer, which, as noted in the discussion of **Fig. 4-32**, often limits the growth and the lifetimes of cumulus clouds. It is this same condition that is shown in **Fig. 10-42** and labelled as the Trade Wind Inversion.

In the context of air pollution, the circumstances depicted in **Fig. 11-3** often include an intermediate inversion layer separating the upper layer and the lower layer — the airfill. It is usually the case, for reasons far beyond our discussion here, that a very strong inversion layer lies at the base of the subsiding upper layer and forms an impenetrable "lid" on the pollutant-choked airfill. This is the set of conditions that persists for weeks over coastal Southern California in the worst-case air pollution season there[B].

A more detailed discussion of lapse rates and vertical motion is not needed for our purposes, and may be found in any good meteorology textbook[2]. All you need to know now is that vertical motion and stability — the thermal structure of an air column — are not only correlated, but also *interactive:* one can change the other, and *vice versa.*

Atmospheric pollutants: their sources

A discussion of the kinds of air pollutants known, and their sources, could fill a whole book[3]. In this brief presentation I will mention only the most basic notions about classifying sources and about interactions between the atmosphere and a stream of pollutants entering the atmosphere.

(B) Although the airfill over coastal Southern California is seasonally very small, as discussed, and although the amount of pollution dumped there is not negligible, in fact the amount of pollutant is much less than over other regions with larger airfills. What makes California's pollution so bad is that the automobile effluent is chemically **converted** (see "atmospheric self-cleansing" above), under the bright sun, to a more harmful chemical than was emitted.

Although we usually think of a "factory chimney" or an automobile exhaust pipe when the words "source of air pollution" are mentioned, there are several other configurations of sources. The system of classifying these configurations is usually taken to be six-fold, as follows. Sometimes an additional characteristic is used in the classification: elevated *versus* ground level, but you'll get the idea from the table.

Spatial	Time	Examples
Point	Instantaneous	An explosion yielding pollutants.
	Continuous	A factory chimney, usually called a "stack" in air pollution parlance.
Line	Instantaneous	One pass from a pesticide spray aircraft.
	Continuous	A major highway with continuous heavy traffic.
Area	Instantaneous	A large-scale demolition project (not a common configuration).
	Continuous	An industrial area, urban area, or forest fire.

Fig. 11-5 demonstrates several basic points about pollutants near their source, no matter what the configuration, in this case a *continuous point source.* The sketch is a familiar scene, stylized in a particular way to make these points. The stylization is in the form of the assumption that the effluent comes into the atmosphere as a continuous sequence of individual puffs. Each puff contains the same amount of pollutant and the stack emits the same number of puffs in each unit of time (called the **source strength**), but there are two wind speeds carrying the "plumes" downwind. One puff fills a larger volume in the case of the faster wind speed; and since one puff contains the same amount of pollutant in both cases, the pollutant concentration is less for the faster wind speed.

Summarizing the points of **Fig. 11-5** in the form of a mathematical expression, we have:

$$\textbf{Concentration} \quad \text{is proportional to} \quad \frac{\textbf{Source strength}}{\textbf{Wind speed}}$$

As we will see shortly, there needs to be another factor in the equation to take account of the fact that, as suggested in **Figs. 11-5, 11-6** and **11-7**, the plume widens downwind as the pollutant mass diffuses away from the **plume centerline**.

The various conditions of *stability,* discussed in the last section, enter in **Fig. 11-6**, where the plume takes a different shape for

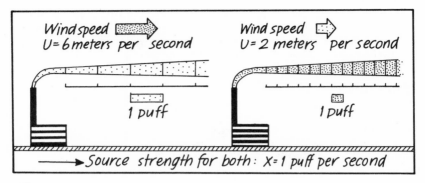

Fig. 11-5

each set of conditions. In particular, I have shown each puff as an elongated box, as in **Fig. 11-5**. In the last panel of **Fig. 11-6** the puffs are blended to look more like a plume you have seen. Finally, the source, which is above the ground surface a distance equal to the **stack height**$^{(P_{23})}$, lies on the boundary between two air layers, each of which may have its own lapse rate and stability.

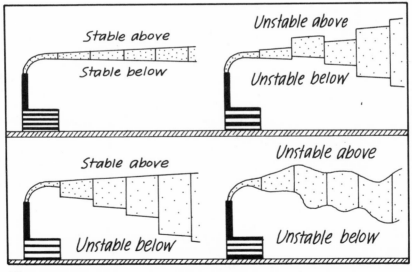

Fig. 11-6

Without pursuing these ideas farther, I am certain you can see how an observer can diagnose the stability conditions of the lower atmosphere by observing plumes as they take various forms, knowing that stability suppresses vertical motion and instability enhances it. These ideas are, perhaps, most clearly seen in a plume from a continuous point source, but the ideas themselves are at work in a pollutant mass from any source configuration.

Atmospheric pollutants: their transport and dispersion

Fig. 11-7 shows several additional ideas about plumes as they move downwind and are further deformed by diffusion. Incidentally, I could as well have written "Dispersion" rather than "Transport" in the first box of **Fig. 11-1**, but *diffusion* refers to a process on the scales of molecular motion and of individual turbulent eddies (see the discussion of **Fig. 3-15**), while dispersion and transport refer to the results of all motion on all scales.

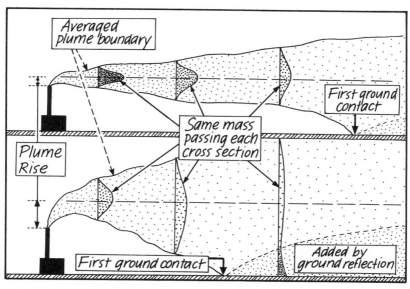

Fig. 11-7

In **Fig. 11-7** there are two plumes, the upper one being dispersed in an air mass slightly more stable than that in the lower panel. Each has a different **plume rise**, which is the extra height of the plume centerline above the stack height. The additional ideas illustrated here are:

a) the concentration of pollutants — their spatial density — within any vertical cross-section of the plume is greatest near the centerline, and departs from that maximum in the familiar form of a bell-shaped curve, associated with random processes (see **Fig. 3-13b**);

b) the total pollutant mass passing through any vertical cross-section is essentially constant, so that, geometrically, the areas under all of the bell-shaped curves are the same, and are equal to the source strength;

c) the dowstream widening of the plume, due to diffusion, is more rapid in unstable air than in stable air; so that

d) the first ground contact with the plume takes place closer to the source in the unstable plume; and

e) downstream from the first ground contact, the reflection of unabsorbed pollutant from the ground surface, back into the plume, modifies the shape of the cross-section of concentration, to produce an asymmetric bulge just above the ground line.

You can see now that our basic equation for pollutant concentration at some point in space downwind — a so-called **plume model** — must look like this:

$$\textbf{Concentration equals } \frac{\textbf{(Source strength)}}{\textbf{(Wind speed)}} \times \textbf{(Diffusion factor)}$$

As you can also see, the "Diffusion factor" gets smaller away from the centerline, either vertically, or horizontally — cross-wind. What is more, this factor gets smaller faster, away from the centerline, in the stable air mass. Thus, the diffusion factor has, built into it, the modeller's information about the atmosphere's local stability. Typical results from a plume model are shown in **Fig. 11-8**, which reinforces the several ideas just discussed.

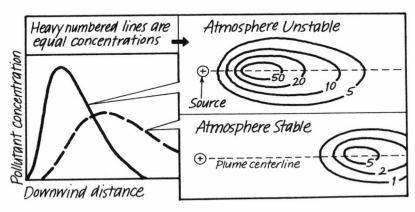

Fig. 11-8

The notion of **time averaging** is the last one we need to mention in discussing this aspect of air pollution. A plume model, with its diffusion factor, produces results such as those in **Fig. 11-8**, showing values — isolines — of time-averaged values of concentration. Values of the *instantaneous concentrations* in the downwind region — with a *zero time average* — would probably look quite different from those of the averages. Visually, this statement is equivalent to saying that the instantaneous plume in the last panel of **Fig. 11-6** is a *snapshot* of the plume, while the plumes in **Fig. 11-7** are *time exposures,* as are the results in **Fig. 11-8**. As shown in **Fig. 11-9**, the longer is the averaging time, the wider is the average plume: the smaller is its *average* concentration at any point in the plume. The diffusion factor has, built into it, the modeller's information about averaging time.

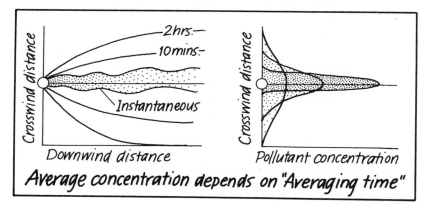

Average concentration depends on "Averaging time"

Fig. 11-9

Figs. 11-6 and **11-9** each show instantaneous plumes — snapshots. After the discussions just completed, you can see intuitively that the "blobs" of pollutant *farthest from the centerline* in such a snapshot arrived at that place mainly because, at the instant they left the source, there was a momentary, turbulent windflow toward that direction — a turbulent component added to the constant, average wind velocity (see the discussion of **Fig. 3-15**). Those blobs were given an initial "shove" toward that side of the plume. That is, the instantaneous plume shows *streamlines,* not *trajectories* (see the discussion of **Fig. 4-30**).

If you stop and think about it, you see that these plume models are quite *static:* they deal with a rather constant set of circumstances. The ground surface is a flat plane, the wind speed and direction do not change, nor do the emission rate and the stability and amount of turbulence. These models are useful for generalized planning, but they do not handle the complex situations of the real world at all[4].

Complexity in the real world arises in different, unique combinations of meteorological circumstances combined with unique geographical circumstances. **Fig. 11-10a** suggests what might happen to plumes from different sources located within the sea breeze, or lake breeze, regime discussed in connection with **Fig. 3-24. Fig. 11-10b** suggests what might happen to plumes from different sources located within the urban circulation regime discussed in connection with **Fig. 5-16.** With local complexities involving major changes of stability within short distances, the simple, static models just don't do the job.

Fig. 11-11 extends the discussion of **Fig. 3-27**, on windflow in rough terrain, to a context of the dispersion of pollutants. Each situation in the real world is unique, and the complexity of site factors can become daunting, but these sketches at least suggest the ways that information presented earlier in this book can be adapted to analysis of various siting problems in air pollution planning and control.

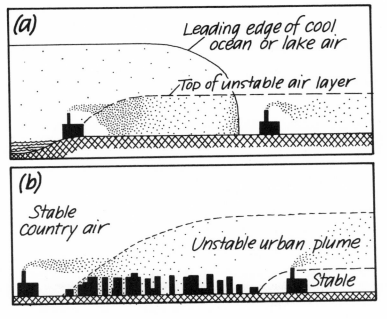

Fig. 11-10

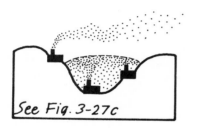

See Fig. 3-27c

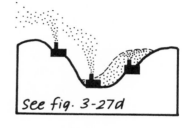

See fig. 3-27d

Plume behavior
depends on topographic
position and inter-
actions of sun, wind
and terrain.

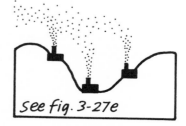

See fig. 3-27e

Fig. 11-11

Atmospheric pollutants: their impacts on receptors

In this section, of all the many effects of air pollutants on receptors that could be discussed[3,5] I have selected only two: visual range[6] and one aspect of human morbidity and mortality.

Common experience attests to the effect of pollutants on how far we can see through the atmosphere. The phrase "how far we can see" requires a bit of clarification. As suggested in **Fig. 11-12**, what is actually meant is more like "at what distance can we discern an object in contrast with its visual background?" That restatement of the question provides the basis for analyzing the physics of the effects of pollutants on visual range.

When we "see" a distant object — called a "target" in the technical literature on this subject — our eyes are actually reacting to light energy reaching them *from the direction of the target*. Some of that energy travels the full path length from target to eye — that energy is *reflected* from the target — but some of the energy reaching the eye has its point of entry into the sight path partway between. In an analog to direct and diffuse solar energy (see **Fig. 4-7**), we can call the target-reflected energy "direct" and that scattered into the sight path from the environment "indirect."

The farther the eye is from the target, the more chances — the more scattering molecules and particles — there are for (i) light energy to be scattered *out of* the sight path, (ii) light energy to be absorbed along the sight path, and (iii) light energy to be scattered *into* the sight path. The first two of these three represent a *depletion of the direct beam* of light energy, while the third represents

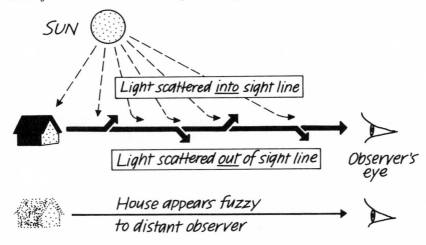

Fig. 11-12

an *augmentation of the indirect.* Even in the cleanest air, scattering takes place by molecules of air. Pollutant molecules only add to this "natural" depletion and augmentation. In a mathematical sense, we can think of the process this way:

$$\text{Visual range} \quad \text{proportional to} \quad \frac{\textbf{(Direct reaching the eye)}}{\textbf{(Indirect reaching the eye)}}$$

The first two effects of depletion reduce the numerator, and the last effect increases the denominator: the fraction gets smaller with increasing distance.

Obviously, not only distance reduces the visual range. The **pollutant loading** — the spatial average concentration of pollutants along the sight path — also reduces the visual range. What is more, it is not just the mass loading of pollutants, for the three effects are related to these characteristics of the pollution particles: (a) size, (b) shape, (c) chemical makeup of the materials, and (d) color. This scattering/absorption phenomenon is indeed complex — even counter-intuitive at times. For example, *the same mass loading* (total mass of particles per unit volume of air) of sea salt and industrial smoke usually have quite different effects on visual range: you can't see as far in "good clean salt air" as you can in "dirty city air" *of the same mass loading.*

Fig. 11-13 combines these two effects of distance and pollutant loading into one statement about visual range. The statement is very simple in appearance, and though the physics is complex, the simple statement allows us to understand several very useful ideas:

(Concentration) × (Distance) = Constant

in which the "distance" is the meteorological range[6] and the "concentration" is the average pollutant loading along the sight path.

The curve connecting the dots representing "observations" in **Fig. 11-13** is called a hyperbola — for one value of the "constant" in the equation. The value of the constant is dependent mainly on the properties of the pollutant particles and of the detector — in this case the human eye. These are the several very useful ideas:

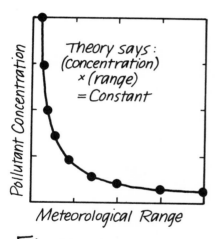

*Theory says:
(Concentration)
× (range)
= Constant*

Pollutant Concentration

Meteorological Range

Fig. 11-13

a) the distance goes down as the air gets dirtier (that's obvious), but the change takes place in a very particular way;

b) in relatively clean air, a small change in concentration makes a large difference in visual range; while

c) in relatively dirty air, a large change in concentration makes a small difference in visual range.

The second idea explains why the outcry from a complaining public — about visibility, at least — can be expected to be greatest in the earliest stages of the deterioration of air quality. The third idea explains why it is so difficult for a new air pollution control program, beginning operations in an already-fouled environment, to yield results that are noticeable by and satisfying to the general public.

The human body responds in many ways to various forms of air pollution. Pulmonary function — breathing — is perhaps the most studied and the most insidious. **Fig. 11-14** provides several insights about these effects. The shapes of the curves may not be exactly according to reality, but the ideas are.

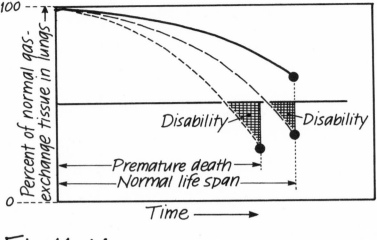

Fig. 11-14

The upper, solid curve says that, in the course of the normal life of a healthy human being, the structural and physiological status of lung tissue gradually deteriorates. At death, lung function is well above a threshold — the horizontal line — separating normal health from disability. Lifelong exposure to mild air pollution — the dashed curve in the middle — does not shorten the life span, but it produces disability — morbidity — in the later years. Life-long exposure to severe air pollution — the dotted curve at the bottom — shortens the life span and it produces disability.

These observations are probably not particularly novel for you, but they will serve to remind us that, even at the end of a very successful air pollution control program, there will still be many "walking wounded" whose lives were irreversibly affected before the cleanup. This is so because, in the sense of **Fig. 11-14**, their lives, once on the "dirty" track, are already foreshortened in their youth, even if they are allowed to move, later in life, to the "clean" track. This change of tracks, notice, means that their lives move *parallel to* these tracks: moving from the dirty to the clean means only that you begin to move *parallel to the clean,* not on it. If you don't live your whole life on the solid (clean) line, you can never get back to it. You can only make an improvement. The damage is irreversible.

Air Pollution and the Planner

I have mentioned the Gaussian plume models, and the more complex dynamic models for air pollution dispersion. They are usually beyond the needs and skills of designers and planners themselves, though, of course, these models are available to you through a competent atmospheric consultant. There is a concept, called the **box model**, that can be within your skills as a designer or a planner, and that can be used here to illustrate several useful insights about the subject. **Fig. 11-15** sets the mathematical stage[P24] for consideration of these box models. It shows a volume of atmosphere defined by its *width* (Y) across the direction of mean windflow, its *height* (h) above the ground surface, and its downwind *length* (X) measured along the mean windflow. There is a sub-volume that is the same except its length is (dX). Thus, the volume of the smaller is:

Sub-volume = (h) (Y) (dX).

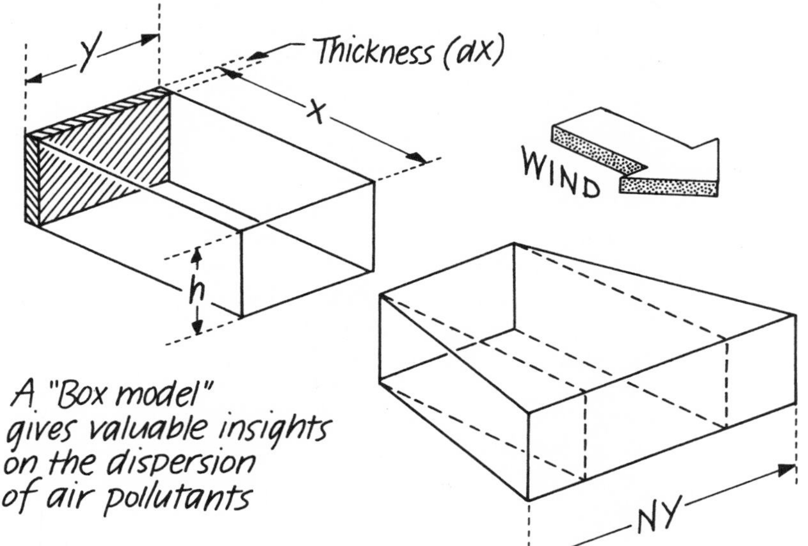

A "Box model" gives valuable insights on the dispersion of air pollutants

Fig. 11-15

What are these box dimensions in the real world? (Y) is the width of an area source, such as an upwind industrial area, while (h) is the height above the surface of an inversion layer — a lid. It is the thickness of the lower, unstable layer — the airfill — in **Figs. 11-3** and **11-4**. The downwind distance (X) is the distance from the source, whose length is (dX), at which the analyst wishes to estimate the concentration.

The *source strength* for the box system is (Q) — measured in units of [mass/(time·area)] — and the pollutants enter the box from the floor area of the sub-volume: (Y)(dX). We assume that the mass of effluent released from this floor area per unit of time — (Q)(Y)(dX) — is mixed instantaneously through the sub-volume, which tells you at once that *the length of the sub-volume (dX) is the length of wind travel in whatever unit of time is chosen* — like a minute or an hour. We get the concentration (C) of pollutant within the sub-volume immediately by dividing the mass by the volume — the definition of concentration:

$$\text{Concentration (C)} = \frac{\text{Mass}}{\text{Volume}} = \frac{(Q)(Y)(dX)}{(h)(Y)(dX)} = \frac{(Q)}{(h)}$$

as long as the wind has blown only the distance (dX) during that time unit. If, instead, the wind speed has (V) distance units (not shown in the figure) during the time unit, then the adjusted concentration in the sub-volume must be (C) = (Q)/(hV).

The sub-volume we have just analyzed moves downwind through the box system, still containing the mass injected from the area (Y)(dX). By the time it has moved a total distance (X) it has had added to it the mass contributions of all the other subvolumes along the way. There are (X) of those sub-volumes, and now in the larger box:

$$\text{Concentration (C)} = \frac{(QX)}{(hV)}$$

As one additional complication (added only to suggest the kinds of complications that could be added) we will say that, while the box does not increase in height — the inversion remains in place — it will widen by diffusion beneath the lid. More particularly, it widens at the rate given by a multiple, *greater than one,* of (N). Finally we have, in the widening box:

$$\text{Concentration (C)} = \frac{(QX)}{(hNV)}$$

Lining up boxes from the west coast to the east coast of the United States, each with a different source strength, will yield a result such as that in **Fig. 11-16**. In that transect, it is clear that the contribution of each city passed is proportional to its source strength, and that, when the cities are closer together, the atmosphere doesn't get a chance to clean itself before the next city is reached. The air, quite simply, gets dirtier downwind of a megalopolis, *even though* (as from local control programs) *the source strengths are not excessive.*

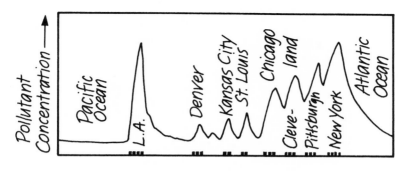

Fig. 11-16

In a very complex use of box models, far beyond our concern as to the model itself, **Fig. 11-17** is a marvellous example of the utility of this concept. The model yielding **Fig. 11-17** consisted of a huge array of boxes, covering the whole of northwestern Europe, and stacked in several layers above the surface. With a large computer, the analyst (Reiquam, 1970) estimated the patterns of sulfur concentration in each of the boxes, using information on the source strengths for sulfur in the region, and using information about the average, or typical, wintertime windflow patterns lasting a week or so at a time in the region. The results agreed well with observations.

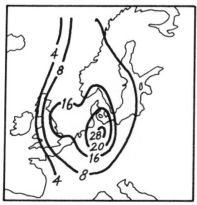
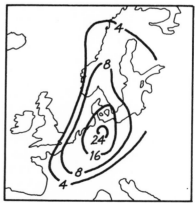

With sources in U.K.   Without sources in U.K.

Fig. 11-17

In **Fig. 11-17a** all the known sources were included, while in **Fig. 11-17b** *the sources in the United Kingdom were ignored* — "turned off" — in an experiment that could never be conducted in real life. The differences between the two maps, it is asserted, are due to the contributions from the UK. Such "what if" games, based solely on the concept of the box model, can be of considerable use in regional planning.

Insights for planning may be obtained also from the concept of the Gaussian plume models, without most of the elaborate calculations describing the complete, three-dimensional model of the plume. One need consider only a horizontal slice through the plume — one altitude. **Fig. 11-18** suggests how these insights come about. In the figure, the source at (A) produces a plume, whose trace on the horizontal slice is shown with the heavy line in the left panel of the figure. In that same panel, the mirror image of the plume — the "backward plume" — points upwind from a receptor at (B).

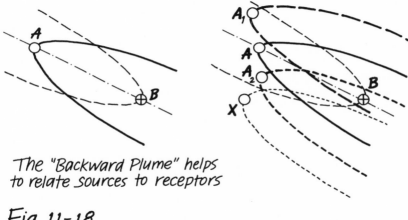

The "Backward Plume" helps to relate sources to receptors

Fig. 11-18

The insights come from the understanding that *as long as (A) is anywhere within the backward plume from (B), the source at (A) will have an effect at (B). Otherwise, it will have no effect.* As usual, remember, these factual-sounding results are no better than the models themselves; but insights can be important without being entirely precise.

In the right panel of **Fig. 11-18** we see the same receptor at (B) and a new set of hypothetical sources: A, A_1, A_2, and X. The first three have effects at (B), but the source at (X) does not. That statement, of course, holds approximately true only as long as the meteorological conditions produce the plume depicted. The statement, therefore, holds true a certain fraction of the time. Similar statements can be made, with a large computer and long weather records, for all the fractions of time at (B).

These considerations suggest several kinds of application for planners:

a) predicting concentrations downstream, at an *existing* receptor, from a known set of sources;

b) predicting concentrations downstream, at a *planned* receptor, from a known set of sources;

c) estimating, under specified atmospheric conditions, the relative pollutant contributions at an *existing* receptor, from each one of a known set of sources;

d) estimating, under specified atmospheric conditions, the most likely pollutant source, from among a known set of sources, of an *observed impact at an existing receptor*;

e) delineating a preempted zone around an existing receptor, within which sources should not be located; and

f) delineating a preempted zone around an existing source, within which receptors should not be located.

Again, these techniques are no better than the models themselves; but they can be important without being entirely precise.

While not in the realm of planning, but rather as a part of "management", simple correlation techniques can answer questions having to do with enforcement or adjustment of a control program. **Fig. 11-19** displays several results using the general concept of the *wind rose* (see **Chapter 8**). In **Fig. 11-19a** the frequency is plotted of the number of times *the wind direction at the receptor was from each octant **and** the concentrations of Pollutants #1 and #2 exceeded certain thresholds.* Frequencies for different thresholds can be used to elaborate the wind rose as one would do with different wind speeds from each direction (see Appendix E of Stern *et al,* 1973). It seems that the primary source of #1 lies north of the receptor, while sources of #2 lie in several directions.

In **Fig. 11-19b** there are four receptors, each with a wind rose, and the evidence suggests the source lies in the middle of their region. Examples of such "detective work" to produce "evidence" for use in regulation and litigation are numerous.

Before I leave the subject of the meteorological connections between sources and receptors, I want to remind you that *the source of a pollutant may not always lie in the direction from which the wind was blowing at the time the pollutant arrived at the receptor.* **Fig. 11-20** illustrates this point while also suggesting to you the nature of another kind of complex dispersion model — one we will not consider beyond this particular discussion. It is

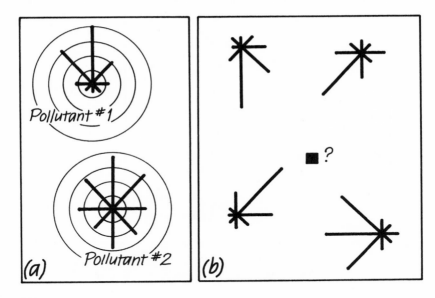

Fig. 11-19

the **Puff model**, and it is in the class of dynamic models that can take account of shifting weather conditions.

In **Fig. 11-20**, the source — **heavy black dot** — is *southwest* of the receptor, T. These relative positions remain the same through

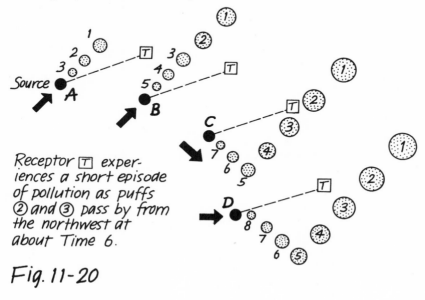

Receptor T experiences a short episode of pollution as puffs ② and ③ pass by from the northwest at about Time 6.

Fig. 11-20

the four steps of the figure, because these two places are station-ary. The puffs of pollutants move around the source and the recep-tor through the four time steps. In the first time step, the source is labelled (A); in the second (B); and so on. The sequence of puffs from the source is labelled, each puff keeping its original serial number as it expands (by diffusion) and moves (by dispersion) across the landscape.

The shifting regional wind velocity is shown by the **heavy arrows** next to the source. You can trace the various changes for yourself, but note the key point of the demonstration: Puffs #2 and #3 pass the receptor T from the northwest, even though the source is located toward the southwest.

It ought to be possible, for planning purposes, to be able to say, generally, what kind of regional air quality problem might exist in the future if the atmosphere behaved as it does now, but pollutant sources were added to the environment. We can speak, therefore, about a *potential* problem which the atmosphere's behavior repre-sents. This is called **air pollution potential** (APP).

One way to approach the assessment of APP is through the cal-culation, through time, of some sort of APP Index number. There are many such indices, and a good place to begin looking into that idea is in Stern *et al* (1973, Chapter 22). A method for assessing the spatial aspects of APP under prescribed conditions is illustrated in **Fig. 11-21**. The analyses are for two regions — each centered

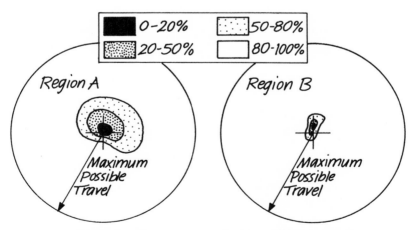

Dispersal of pollutants during episodes of air stagnation.

Fig. 11-21

on a major airport that is the source of the wind data used — and for the particular set of atmospheric conditions that produce increasing concentrations of pollutants in stagnant air: the "worst case" scenarios.

The maps in **Fig. 11-21** were drawn as follows, step by step:

1) from weather records at the airport (or any other nearby site with hourly wind data) select from the historical record (it will usually be about 20 years) a complete sample of episodes in which wind speeds were continuously less than some low threshold value — let us say (X) — for at least a certain number (N) of hours, such as 24: these circumstances to be defined as *stagnation;*

2) for each episode plot the wind vectors in sequence (see **Peptalk 22** in **Chapter 10**), each sequence beginning at the origin of the diagram;

3) on the diagram mark each point where two vectors join, and call these points "nodes;"

4) draw smooth curves around the origin so as to include, in each directional sector, particular percentages of the nodes in that sector;

5) label each one of the smoothed curves with the percentage of nodes included within it; and

6) draw the line to be labelled "100%" as the circle, with center at the origin, and radius equal to (NX) units. This is called the "maximum possible travel" in **Fig. 11-21**.

The resulting map is to be interpreted in this way. The coded legend on the figure gives the probability that any puff emitted from the source at the origin, during a stagnation episode, will be found in that sub-area at the end of the episode. Roughly speaking, the two regions are to be compared, in their relative severity of APP, by the ratio:

$$\text{Relative severity of APP} = \frac{\textbf{Area enclosed by 100\% probability line}}{\textbf{Area enclosed by 50\% probability line}}$$

In the cases shown, Region B has by far the greater APP. The probability lines may have roughly circular patterns, as in Region A, or they may show a preference for certain directions, as at a seacoast or lakeshore, or in a valley location such as Region B.

Other losing bets

I don't know if you have noticed, but several losing bets have been lurking in the discussions of this chapter — some explicit, and some neatly hidden within familiar and usually unquestioned assertions about the way the world (and society) works. In writing this section I have gone back through the chapter and listed several of these bets — or, if you prefer, assumptions. There are probably more, and a trained philosopher would probably deal differently with them; but the ones I have listed will illustrate very nicely the points I want to make.

I have divided the bets into short-term bets and long-term bets. I suspect the short-term bets are really just "fervent hopes," but the long-term bets are surely based on genuine faith.

Among the short-term losing bets — the fervent hopes — too many of us have made, as designers, planners, atmospheric scientists, and ordinary citizens, are these.

1) *A problem converted is a problem solved.* I mentioned earlier that it is tempting to assume that we have solved an air pollution problem by converting it to, say, a water pollution problem or a land pollution problem. That is a very expensive way to solve a problem. To save money, we might as well try to solve the problem of too much air pollution simply by redefining the words "air pollution." Alas, physical and biological facts are not changed by changing human perceptions, although even the history of science and technology is full of examples of problem solvers who pretended this was not so.

2) *Out of sight, out of mind, problem gone.* This bet has been mentioned earlier in the discussion about dispersion of pollutants to someplace else — "over the hill." If this were a winning bet, then breathing carbon monoxide would not kill because it is odorless and colorless. If this were a winning bet, then water tables on the Great Plains and in the arid Southwest would not be falling as water-loving lawns, golf courses, and crops are grown. In a more direct form of example, related to the effects of air pollution on visual range, I want to quote Stern *et al* (1973, page 308):

> The segment of our society most in touch with the real magnitude and time scale of the problem of vision through the atmosphere is the fraternity of airline pilots. The availability of electronic technology to make seeing unnecessary during flight does not reduce their negative reactions to the impact of reduced visibility.
>
> For very few people is seeing through the atmosphere more than an aesthetic matter, but to anyone who has ever really enjoyed a good long view through clear air, the fouling of our air is a source of considerable sorrow and frustration. To all, no matter what their personal history, place of residence, or level of income, the loss of visual range ought to be perceived as "nature trying to tell us something before it is too late."

3) *Society will absorb all external costs without complaint.*
Economists, as I understand it, use the term "externalities" to cate-
gorize those costs that stem from development and use of a
resource and are borne, or exacted, "outside the plant" — in a
location or at a time other than where and when the firm does
business. In a none-too-subtle example, a company is not liable for
the fact that its plant uses the local airfill to dump a gas with a
bad odor, and then the local merchants lose trade because of a
decline in tourism. More subtle examples abound, but the point is
that recent national trends in consumer advocacy and current local
legislative debates are gradually changing the formerly winning bet
to a losing bet, at least in the United States. Elsewhere in the world
— this is my bet — the expansion of the "global village" will
make "freebie externalities" a losing bet as well.

Similarly, too many have made long-term losing bets like these.

1) *Earth's airfill is huge enough and efficient enough at diluting and distributing.* This losing bet was labelled clearly earlier in
this chapter. It seems to me it has in common with the second
short-term bet the notion that Man can freely alternate between
two views as it is convenient to do so: (i) Man is a member only
of a local community — his "tribe" — and (ii) Man is a member of
both local and a global communities — the former by choices of
convenience, and the latter by necessity.

2) *The maximum drain size of Earth's atmosphere can be
stretched.* This losing bet was also labelled clearly earlier in this
chapter. Physical and biological laws, that operate in the same way
anywhere on Earth (and probably everywhere in the universe)
govern the rates at which reactions and processes proceed. If this
were a winning bet then international airline pilots would not have
noted, with dismay, the gradual reduction in visual range
worldwide.

3) *Man can and will find a technological solution — a technological fix — for any environmental problem.* You might argue
that this is really a short-term bet rather than a long-term one,
because, in his heart of hearts, Man knows he cannot always come
up with the technological fix, even though his success rate has
been first rate up until now. But this is a losing bet because of the
interactions of time and economics. If this were a winning bet,
then water rates would not have to be increased in districts that
have learned clever ways to conserve water, the increase needed to
make capital debt payments. As will be noted in the next example,
technological solutions are too often only an expensive way to
"buy time" until other things can happen; and the amount of time
that needs to be bought is seldom evident until long after the so-

called solution is in place. The ultimate irony of the technological fix is caught in a quotation by Kenneth Boulding (Bates, 1969).

> In the West, our desire to conquer nature often means simply that we diminish the probability of small inconveniences at the cost of increasing the probability of very large disasters.

4) *Solving the air pollution problem can be done on the safe and even unstated assumption that the behavior of Earth's atmosphere and the climates of Earth are constant and not responsive to how Man uses the global airfill.* To an atmospheric ecologist, this is obviously a losing bet, but it seems not to have been recognized by enough designers and planners that Man cannot tinker around with trial solutions to problems such as "acid rain" until he finds one that works, confident that the atmosphere will remain constant, like some giant plumbing system, while he tinkers. Indeed, there are two aspects of this kind of blindness that are particularly disturbing.

First, the atmosphere will often be *interactive* with the tinkering: a trial solution, laid aside as ineffective, may already have changed the nature of the system and the problem before the next trial solution is tried, without that fact being recognized. Mentioning this puts me in mind of our discussions of *Phase C* and the *Balance of Nature* in **Chapter 8**.

Second, the atmosphere — indeed the whole of Gaia — has an inevitable *time lag*. The results of one of Man's trial solutions — his tinkering — may not be evident until a long time has passed. Indeed, though I personally doubt it, it is possible we may already be too late to undo the damage of an ozone layer growing too thin, or of a Greenhouse Effect growing too strong.

Mention of the Gaia Hypothesis (**Chapter 9**) in this context — and here we get almost into a metaphysical discussion — requires two observations about it. First, even the most ardent proponents of the validity of the hypothesis say that the system *tends* to maintain and perpetuate itself within limits. They quite freely admit there are limits beyond which changes may not go. Second, Man is an agent for change on Earth like no other has ever been. The metaphysical debate here would have to be about whether the "Master Planner" of Gaia would include, or would not include, the possibility of such a disruptive agent in the plan.

To complete this chapter, I want to pursue this last losing bet by presenting a very brief outline of some of the accepted theories and ideas concerning the *interactions between air pollution and global climatic change*. We know that the burning of both wood and the fossil fuels, coal and petroleum products, yields three major products: (i) water vapor, (ii), smoke particles, and (iii) carbon dioxide. Wood has been burned for millenia as fuel by Man. Its use has been in rough proportion to his numbers. The burning of fossil fuels, on the other hand, has increased more rapidly than Man's numbers[7]. These three products of combustion are most often indicted as causally related to climatic change, so my brief centers on them.

Carbon dioxide and climatic change. It is beyond question that the concentration of carbon dioxide in Earth's atmosphere is increasing — roughly doubling during this century — but the reason is quite open to question. Obviously, CO_2's source strength exceeds its drain size on the global scale, but what are the natural and the anthropogenic source strengths? Environmentalists discussing the Greenhouse Effect usually assert that the anthropogenic burning of fosil fuels is the culprit. Since, philosophically, *pollution must be man-made* rather than natural[8], it follows that the CO_2 increase is a form of pollution.

It is physically inescapable that, *other things being the same,* the greater is the CO_2 concentration the stronger is the Greenhouse Effect. The doubling now under way, it is estimated, will increase the mean surface temperature of the earth and its inner atmosphere by about 2.5 °C. Most experts agree it would be much more than that near the poles and much less near the Equator, but other things may not be equal. The horror stories of melting ice caps and drowning cities may not even turn out to be accurate, not to mention precise. Here are some of the "other things" to think about.

(a) The inner atmosphere would be warmed at the expense of the outer atmosphere, so that the atmosphere as a whole would become *less stable* — warm beneath cool — and would overturn more readily, diluting the pollutants more effectively and forming more sunshine-reflecting clouds: negative feedback.

(b) Although it is true that both the CO_2 concentration and Earth's temperature increased between 1870 and 1945 or so, after that the CO_2 concentration continued to increase but the temperature did not.

(c) We could just as easily explain an increase in the CO_2 concentration by an increase in temperature rather than the other way around. Warmer "garbage" — the ecosystem's dead organic materials (see **Fig. 9-7**) — decays and produces CO_2 faster than cold.

(d) We know that there were major fluctuations in the CO_2 concentration after green plants appeared (**Chapter 9**) but long before Man appeared on Earth.

Water vapor and climatic change. Another product of Man's burning of fossil fuels is water vapor. We don't ordinarily think of water vapor as a pollutant, but it contributes in a minor way to the Greenhouse Effect (as do other gases, we are now learning[C]), so in that context we should. On the other hand, an increasing concentration of water vapor makes possible, with a less stable atmosphere (see the discussion just above), more sunshine-reflecting clouds: negative feedback.

Solid particles and climatic change. Solid particles as *smoke* are a product of Man's burning of fossil fuels, while solid particles as *dust* result from Man's prodigious land-clearing operations. The fact that visual range is deteriorating worldwide shows that the source strength for particulates exceeds its drain size on the global scale, but, again, what are the natural and the anthropogenic source strengths? *Other things being equal,* so the simple argument goes, the particles from the same combustion that produces carbon dioxide will reflect sunlight and counteract the strengthening of the Greenhouse Effect. Comforting, but are *other things really equal?* Is reality that simple? Here are some other things to think about.

(a) If the particles are of the correct size and composition, they might enhance the production of rain by acting as additional condensation nuclei (see **Chapter 4**), thereby cleansing the atmosphere: negative feedback.

(b) The particles both reflect and absorb sunlight, depending upon their color, shape, size, and composition. Thus, particles that more readily absorb than reflect would actually enhance the Greenhouse Effect.

(c) If some particles reflect and some others absorb, their *altitude* above the surface is important in determining whether the atmosphere's stability — tendency for overturning — is increased or decreased. Decreasing stability, as we have just noted, makes for more clouds and for greater dilution of pollutants.

Notes

(1) The relationships among process and environmental lapse rates, and vertical motion, may be likened to what I call "the elevator game." Consider these analogies. *Rule 1* says that for each floor an elevator ascends, its internal temperature must decrease one degree, and for each floor an elevator descends, its internal temperature must increase one degree. That is the standard lapse rate for the elevator. *Rule 2* says that when an

---

(C) Such as methane and the oxides of nitrogen.

elevator stops and the door opens, if its internal temperature is greater than that in the hallway, it must ascend one floor; and if its internal temperature is lower than that in the hallway, it must descend one floor. *Rule 3* says that the elevator must spend the night, with its door open, on the last floor it reached on the previous day.

Now, the temperatures in the hallways can be controlled by thermostats on each floor, so the maintenance engineer who sets the thermostats at any hour can control what happens to the elevator by controlling the environmental lapse rates in the building. The answer to the question "At which floor will the elevator come to rest?" may change, on any given day, at the whim of the maintenance engineer.

(2) A particularly clear explanation is in Oke (1987b, page 51).

(3) As an entry point to that huge literature, try Stern *et al* (1984, Chapter 6 for sources and Chapters 7-10 on effects).

(4) The static plume model you will most likely encounter is called the **Gaussian Plume** model, because it incorporates (in the idea of the bell-shaped curve) a theoretical basis invented by the early 19th Century German physicist K. F. Gauss. Dynamic models for dispersion are being used more and more, but they are very complex, requiring first class computer systems to use. Please remember, though, that *increasing complexity in a mathematical model does not always represent increasing validity,* because assumptions and simplifications are inescapable. Complex models often only disguise these assumptions and simplifications, and leave the uninitiated with a feeling of great deference to authority. I'm sure you've heard "Garbage in, garbage out." These models are no exception.

(5) You may obtain quick access to current technical writings on air pollution control through current issues of the *Journal of the Air Pollution Control Association* — for most technical aspects — and *Atmospheric Environment* — for meteorological aspects.

(6) A clarification of technical terms relating to how far we can see through the atmosphere:
   (a) **Visual range** is, simply, the distance the human eye can distinguish objects *along a specified sight path,* day or night.
   (b) **Visibility** is the average, or *median,* visual range for all sight path directions from the observer (see the statistical discussion of averages in **Chapter 8**). The term was devised for use by the air transport industry. At certain well instrumented airports, information is available to pilots, during low-visibility

periods, concerning the **runway visibility**, which is just the visual range along the runway in use at the time.

(c) **Meteorological range** is the distance, in a specified direction, that the human eye can distinguish a black target from its white visual surroundings under cloudless, sunny conditions with the air layers well mixed. This is, clearly, a distance that exists only in precise physical theory and is seldom used in non-specialist discussions.

(7) The human population has doubled in about 40 years over the most recent decades (today's two percent per year yields a doubling time of about 35 years). Roughly, Man's "production" of coal has had a doubling time of about 30 years, and his "production" of crude oil has had a doubling time of about 10 years over that period (Hubbert, 1969, Figures 1 and 2). I have put "production" in quotation marks because Man's process is actually "extraction" — only the Master Planner *produces* these resources.

(8) It is a philosophical judgment, of course, whether or not Man and his consequences are "natural" or in another category. Assuming the latter, and assuming pre-human ecosystems adapted in a non-destructive way to changes in air quality, the statement that "*pollution must be man-made* rather than natural" seems to follow logically.

Peptalks

(P₂₃) More technical terms. More meteorological vocabulary. They are included mainly because they are what your atmospheric consultant will use to describe these concepts. To avoid later confusion, notice now that a *positive* lapse rate always means a *decrease* in temperature with *increasing* altitude. That's just the way it is defined.

(P₂₄) The mathematics for box models, as presented here, is well within your ability to understand, but even this simple idea can, and has been, made complex. I believe the insights available from following the math are worth it, so take the time to follow through.

12. So what?

Summary of the book and its purposes

Here we are at the end of the journey — the journey intended to begin a long overdue dialogue between designers and planners — your profession — on the one hand, and atmospheric ecologists — my profession — on the other hand. In the first half of the book we have examined and discussed facts and concepts concerning the atmosphere and its interactions with the organisms of Earth's ecosphere. Later, we addressed in several ways the possible interactions between members of your profession and members of mine. It is these inter-professional interactions I want to consider again as we part company in this last chapter.

In **Note 2** of **Chapter 8** I suggested an answer to my own question: "When do you need the consultant?" I suggested the answer is "You need him when you no longer feel honestly comfortable and confident that you are still within the limits of your own technical expertise on the subject of the design component." Further, I suggested that, while you probably knew that answer all along, what you probably have not known is where to find an appropriate atmospheric consultant, and how to understand what he would tell you. After your journey through this book, I trust, you know where to find him and how to listen to him.

What else is there to discuss?

I believe it is worth examining the concepts of **generalist** and **specialist** as they apply to you and to me in this common enterprise of designing and planning changes in our physical — and perhaps even our cultural — environment. In addition, I would like to examine, one last time, that common enterprise itself: **ecological design, planning, and management**.

Take the enterprise first. I am struck by what seems to me the urgency of understanding, and at the same time the reluctance of too many to accept, the idea contained in this passage from a current prospectus:

> Unfortunately, our understanding of the relationships and dependencies of organisms with one another and with their physical environment has not developed as quickly as our capacity to modify them. Anthropogenic changes in the (planetary) environment are a unique feature in that they may well exceed the limits of natural regulation. There are many indications that human-induced changes are substantive enough to affect the survival of other organisms, both directly and indirectly, and to pose a potential threat to mankind itself. — U. S. Committee for an International Geosphere-Biosphere Program (1986)

You and I, as members of our professions, are actively engaged in producing those "anthropogenic changes in the (planetary) environment that are substantive enough to affect the survival of other organisms, both directly and indirectly, and to pose a potential threat to mankind itself." We are not the only players in the game, but we are players, and we need to talk to one another about this common ecological enterprise that has brought about such an important problem. Perhaps this sounds too much like a sermon, or a campaign speech, for your taste. But I feel strongly about these ideas, and I may never get another chance to tell you how strongly I feel.

I propose that our final discussion center on the assertion that *one of the primary causes of the problem of man-made change running ahead of Man's knowledge of the consequences is Man's devotion to the concept of the **professional specialist**.*

In my view of our ecological enterprise, as I said at some length in **Chapter 7**, *you* — the designer/planner — *are the central player:* the facilitator and the integrator. I — the atmospheric consultant — enter the game at your invitation. You are the generalist and I am the specialist — *in this particular enterprise*. But I'll bet you have been in other games in which you are considered a specialist. I know I have been in many games where I was considered the generalist. Which are we? What are generalists and specialists? Is the distinction useful and productive, or is it, as I have asserted, the heart of the problem?

Buckminster Fuller (1978, page 13 *et seq*) makes a fine case for the idea that specialists are too often very bright people who have become slaves of the "Great Pirates" — the Establishment. In return for their willingness to say they can't deal with complex problems because they are specialists, they are set up in a "socially and culturally preferred, highly secure, lifelong position." Fuller perceives that the Great Pirates are the true generalists, who really know about causes and effects, about changes and consequences, and who do most of what they want to do — make the changes they want to make — while taking advantage of *Man's devotion to the concept of the **professional specialist***.

So, finally to my point. I see the informed ecological designer/planner acting as generalist — you — as our primary antidote to the problems caused by the Great Pirates. I see you acting as a generalist — an integrator — as you oversee the man-made changes in the physical environments we all must share. And I believe that the best way for you to be truly an antidote to *the problem of man-made change running ahead of Man's knowledge of the consequences* is to be able to choose and use the professional specialist wisely.

In my experience, the atmospheric specialist has not been well used by members of your profession. Neither have atmospheric specialists held their knowledge out to you in a form in which you can readily recognize its utility, so I have put these ideas of mine about atmospheric ecology down for your consideration. If my hopes are realized, this book will provide you with a vocabulary of facts, concepts, and skills with which to integrate responses to the atmosphere into your designs and plans.

If you are the bright person I think you are, you will understand that what I offer here is simply *my way of seeing the world.* Others in my profession will have their ways of seeing the world. Among the skills you must develop, along with the vocabulary of facts and concepts, is the ability to discern when *not* to try to to be your own specialist, and the ability to tell which specialist — of any field, not just my field — will best help you be that antidote to the problem of changes and unattended consequences.

Another of the skills you must develop is the ability to frame the right questions when you seek a specialist — to insist that you get answers that make sense to you. Specialists, in my experience, have too many answers to questions they hope you will ask, and not enough of the kind you need. That is particularly true of atmospheric specialists, because the absence of dialogue with you has given them no proper basis for finding out what you need to know. So part of your task is to train that atmospheric specialist to understand what you need to know.

You must be the generalist — the activist, *the doer.* Most of what I think I can do is in these pages. I do not envy you your chosen assignment, but you are absolutely essential. All the best.

Appendix A.
Specialized Examples of Energy Balance Systems

Summaries of representative energy balances in *percent of maximum R*$^{(m)}$

Surface	Ref	R*	Time	R	B	H	E	Δ
Dry lake bed	a	600.	Morning	63.	-33.	-30.	0.	-
			Afternoon	75.	-3.	-72.	0.	-
			Night	-13.	13.	0.	0.	-
Coniferous forest								
Fir	a	600.	Morning	83.	§	-35.	-45.	§
			Afternoon	75.	§	-23.	-50.	§
			Night	-10.	§	10.	-5.	§
Pine	a	690.	Morning	88.	§	-49.	-29.	§
			Afternoon	75.	§	-49.	-28.	§
			Night	-3.	§	-1.	0.	§
Shallow pond	a	730.	Morning	96.	-26.	-9.	-34.	-27.
			Afternoon	27.	-3.	-12.	-48.	36.
			Night	-10.	7.	4.	-3.	2.
Soil, glasshouse	a	400.	Morning	80.	-40.	-§	-§	-
			Afternoon	65.	0.	-§	-§	-
			Night	-8.	15.	-§	-§	-
Tropical ocean	a	700.	Morning	77.	-	-§	-§	-63.
			Afternoon	71.	-	-§	-§	-26.
			Night	-4.	-	-§	-§	41.
Melting glacier	a	540.	Morning	89.	-	9.	1.	-99.
			Afternoon	65.	-	11.	1.	-77.
			Night	-9.	-	12.	3.	-6.
Urban canyon:	a							
East facing wall		180.†	Morning	100.†	-	-72.	0.	-28.
			Afternoon	28.	-	-28.	0.	0.
			Night	-17.	-	0.	0.	17.
Floor		420	Morning	38.	-8.	-29.	-1.	
			Afternoon	12.	0.	-9.	-3.	-
			Night	-7.	7.	0.	0.	

Roof level plane		520	Morning	71.	#	-50.	-4.	#
			Afternoon	35.	#	-33.	-8.	#
			Night	-12.	#	4.	-2.	#
Rice crop	b	670.	Morning	73.	§	-13.	-48.	§
			Afternoon	60.	§	-7.	-63.	§
			Night	-4.	§	0.	-3.	§
Maize crop	b	700.	Morning	63.	-4.	-20.	-38.	§
			Afternoon	61.	-3.	-16.	-34.	§
			Night	-14.	§	3.	§	§
Sugar beet crop	b	400.[n]	August	100.	-1.	-14.	-83.	§
Mature sunflower	b	625.	Morning	74.	-6.	-16.	-52.	§
			Afternoon	80.	-12.	-4.	-64.	§
Deciduous forest								
Mixed species	b	419.[n]	July	100.	-1.	-2.	-97.	-
Oak	b	419.[n]	July	100.	-2.	0.	-98.	-
Maple	b	452.[n]	July	100.	-1.	-3.	-96.	-
Canadian prairie	b	600.	Morning	72.	-15.	-32.	-25.	-
June			Afternoon	72.	-15.	-35.	-22.	-

m — Values of the midday net radiation, R^*, are in watts per square meter — W/m^2.

n — All values are means of daylight hours.

a = Oke (1978)

b = Monteith (1976)

§ — these terms are not separated in the published balance.

† — The maximum R on the east-facing wall occurred at mid-morning.

— Some of the stored heat being released is from the floor **(B)** and some from the walls **(Δ)**. They are not separated here.

Appendix B.
Sources of weather and climate data

In this Appendix I have tabulated basic information on the standard sources of weather and climate data available in the United States and Canada, including addresses and telephone numbers of the government agencies from which your own copies can be obtained. Naturally, major academic libraries have these same documents, so I have listed the Library of Congress call numbers under which I have found the documents listed.

As with any reference document, you must use it to know the details of what it has available. I have set out enough information for you to know where to begin seeking certain kinds of data, but my simple tables cannot convey all there is in these documents.

In the American documents, you will quickly discover that there are three kinds of information available from very few stations, even though my notes say that document represents a large network. These three kinds are : (i) solar radiation data, (ii) soil temperature data, and (iii) evaporation data. It is because these are observed at only a few places. In addition, as a general rule, wind information for particular dates is best obtained from the stations themselves.

When a document contains a station index ($\checkmark$) it provides information on the location (county, latitude, longitude, elevation) and on the operation schedule (observer and times of observations).

Each document is listed as being available from a central agency office. Here are the agencies:

NCDC:
National Climate Data Center   1-(704) 259-0682
Federal Building
Asheville, NC 28801

NOAA/USDA:
Weekly Weather and Crop Bulletin
Commerce/Agriculture Crop Service
Room 5844, USDA South Building
Washington, D.C. 20250

NOAA/AISC:
Assessment and Information Services Center — NOAA
Universal Building, Room 512
1825 Connecticut Avenue NW
Washington, D.C. 20235

GPO:
U.S. Government Printing Office   1-(202) 783-3238
Division of Public Documents
Washington, D.C. 20402

AES:
Atmospheric Environment Service/Environment Canada
4905 Dufferin Street
Downsview, Ontario, M3H 5T4

WMO:
World Climate Data Programme (WCDP)
World Meteorological Organization
C.P. 5
CH 1211 Geneva 20, Switzerland

THE UNITED STATES

(CD): Climatological Data by Sections: QC 983 U531 or QC 983.A5

HOW OFTEN IS IT ISSUED?---------- Monthly + Annual Summary
FOR WHICH STATIONS OR
   NETWORK OF STATIONS?--------- One state[a].
   (a) Both the major airport stations for which LCD is available,
      and many stations operated by volunteers. Approximately eight
      stations per county.
WHERE IS IT OBTAINED?------------- NCDC
WHAT DATA DOES IT CONTAIN?

Variable	Hourly	Daily	Monthly	Annual	Other
Temperature					
Air	-	AB	ABCD	ABC	X
Soil	-	X	-	AB	-
Precipitation	-	XB	XC	AC	-
Snow	-	X	X	-	-
Snow depth	-	X	-	-	X
Evaporation	-	X	X	XBC	-
Station index	✔				

X = Total or other appropriate observed value. Combinations of
      observations are:
A = Averages   B = Extremes   C = Departures from Normal
D = Degree Days

Climatological Data, National Summary: QC 983 U533

HOW OFTEN IS IT ISSUED? ---------- Monthly + Annual Summary

FOR WHICH STATIONS OR
    NETWORK OF STATIONS? --------- Essentially the same as LCD

WHERE IS IT OBTAINED? ------------- NCDC

WHAT DATA DOES IT CONTAIN?

Variable	Hourly	Daily	Monthly	Annual	Other
Temperature					
Air	-	-	ABD	A	Map
Precipitation	-	-	AB	X	Map
General weather summaries			X	-	-
Solar radiation	-	-	X	-	-
Storm summaries			X	-	-
Upper air data	-	-	X	-	-

X = Total or other appropriate observed value. Combinations of
    observations are:

A = Averages B = Extremes C = Departures from Normal

D = Degree Days

Hourly Precipitation Data: QC 925.1 U17

HOW OFTEN IS IT ISSUED? ---------- Monthly + Annual Summary

FOR WHICH STATIONS OR
    NETWORK OF STATIONS? --------- One state[c]

    (c) All of the stations for which LCD is available, and many sta-
    tions from which CD is available. Approximately five stations
    per county.

WHERE IS IT OBTAINED? ------------- NCDC

WHAT DATA DOES IT CONTAIN?

Variable	Hourly	Daily	Monthly	Annual	Other
Precipitation	XB	X	X	XB	
Station index	✔				

X = Total or other appropriate observed value. Combinations of
    observations are:

A = Averages B = Extremes C = Departures from Normal

D = Degree Days

(LCD): Local Climatological Data:
QC 984.07 U513 or QC 983.A514

HOW OFTEN IS IT ISSUED?---------- Monthly + Annual Summary
FOR WHICH STATIONS OR
   NETWORK OF STATIONS?--------- One for each major airport[b]
  (b) Each series for one station. Also available from certain obser-
    vatories. About 6-10 locations per state, with New England
    and Maryland/Delaware each considered to be one state.
    Observations were tabulated hourly until 1964, after which
    they have been tabulated for *every third hour* ($\square$ in table)
WHERE IS IT OBTAINED?-------------- NCDC
WHAT DATA DOES IT CONTAIN?

Variable	Hourly	Daily	Monthly	Annual	Other
Temperature					
Air	$\square$X	ABCD	ABCDH	ABCD	B
Precipitation	X	X	XH	X	X
Snow	-	X	XH	X	B
Humidity	$\square$XA	-	A	A	-
Pressure	$\square$XA	A	A	-	-
Sky cover	$\square$XA	-	A	-	-
Storm events	$\square$X	-	A	A	-
Visibility	$\square$XA	-	-	-	-
Wind	$\square$XA	-	AB	B	

X = Total or other appropriate observed value. Combinations of
    observations are:
A = Averages B = Extremes C = Departures from Normal
D = Degree Days H = Historical values

Weekly Weather and Crop Bulletin: QC 983 U98

HOW OFTEN IS IT ISSUED?---------- Weekly
FOR WHICH STATIONS OR NET-
   WORK OF STATIONS?-------------- United States
WHERE IS IT OBTAINED?------------- NOAA/USDA
WHAT DATA DOES IT CONTAIN?
   The Bulletin is intended for use in keeping track of the advance
of the agricultural growing seasons around the nation. It contains
descriptions of general weather highlights of a week's weather
across the country, and status reports on crop development by
state, as well as maps of temperature, precipitation, and snow.
Cumulative status of seasonal soil moisture is also mapped.

Daily Weather Map, Weekly Series —
Document No. 003-019-80001-9

HOW OFTEN IS IT ISSUED?---------- Weekly

FOR WHICH STATIONS OR
    NETWORK OF STATIONS?--------- Rawinsonde stations[d], U.S.
                                   + Canada

(d) Upper air soundings, about 300 miles between stations.
    Observations are taken twice each day, but these maps are
    plotted for only once each day (□ in table)

WHERE IS IT OBTAINED? ------------- GPO

WHAT DATA DOES IT CONTAIN?

Variable	Hourly	Daily	Monthly	Annual	Other
Temperature					
Air	□	B (Map)	-	-	-
Precipitation	□	-	-	-	-
Rates	X (map)	-	-	-	-
Cloud cover	□	-	-	-	-
Pressure	□	-	-	-	-
Wind	□	-	-	-	-

X = Total or other appropriate observed value. Combinations of
    observations are:

A = Averages   B = Extremes   C = Departures from Normal

D = Degree Days

Monthly Climate Impact Assessment: QC 983 C52

HOW OFTEN IS IT ISSUED?---------- Monthly

FOR WHICH STATIONS OR
    NETWORK OF STATIONS?--------- The Industrialized World

WHERE IS IT OBTAINED? ------------- NOAA/AISC

WHAT DATA DOES IT CONTAIN?

    The publication contains various written, tabular, and graphic
summaries of climate-related impacts on energy utilization (mostly
population-weighted heating and cooling indices) during recent
time periods of various lengths, mostly for the United States, but
also for the major subareas of the industrialized world.

CANADA

Monthly Record of Meteorological Observations in Canada: QC 985.A5

HOW OFTEN IS IT ISSUED?---------- Monthly
FOR WHICH STATIONS OR
 NETWORK OF STATIONS?--------- A network similar to that for CD and LCD (see Notes (a) and (b) above).
WHERE IS IT OBTAINED? ------------- AES
WHAT DATA DOES IT CONTAIN?

Variable	Hourly	Daily	Monthly	Annual	Other
Temperature					
Air		B	ACF		Map (C)
Soil		X			
Precipitation		X	X		Map
Rates	X				
Cloud cover	F				
Evaporation		X			
Pressure	F				
Solar radiation		X	X		
Vapor pressure	F				
Visibility	F				
Wind	BF				

X = Total or other appropriate observed value. Combinations of observations are:
A = Averages B = Extremes C = Departures from Normal
D = Degree Days F = Frequency distributions

Monthly Record - Meteorological Observations - Eastern Canada: QC 985.A51

HOW OFTEN IS IT ISSUED?---------- Monthly
FOR WHICH STATIONS OR
  NETWORK OF STATIONS?--------- Similar to CD.
WHERE IS IT OBTAINED? ------------- AES
WHAT DATA DOES IT CONTAIN? Hourly observations are tabu-
  lated for *every sixth hour* (□ in table)

Variable	Hourly	Daily	Monthly	Annual	Other
Temperature					
Air	□	AB	B		
Precipitation	□	X	X		
Rates	X				
Evaporation		X			
Humidity	□				
Pressure	□				
Solar radiation		X			
Wind	□BF				

X = Total or other appropriate observed value. Combinations of
      observations are:
A = Averages   B = Extremes   C = Departures from Normal
D = Degree Days   F = Frequency distributions

Monthly Record - Meteorological Observations - Northern Canada: QC 985.A52

HOW OFTEN IS IT ISSUED?---------- Monthly
FOR WHICH STATIONS OR
  NETWORK OF STATIONS?--------- Similar to CD
WHERE IS IT OBTAINED? ------------- AES
WHAT DATA DOES IT CONTAIN? Same as Monthly Record -
  Eastern Canada

Monthly Record - Meteorological Observations - Western Canada: QC 985.A53

HOW OFTEN IS IT ISSUED?---------- Monthly
FOR WHICH STATIONS OR
  NETWORK OF STATIONS?--------- Similar to CD
WHERE IS IT OBTAINED? ------------- AES
WHAT DATA DOES IT CONTAIN? Same as Monthly Record -
  Eastern Canada

INTERNATIONAL

World Survey of Climatology: QC 981.W63

Harry Van Loon, Editor
National Center for Atmospheric Research
Box 3000, Boulder, Colorado 80307

There are 15 volumes, one for each region or continent, and the world's oceans. The data are long-term averages and normals of many kinds. The data are presented in tables and maps, and the reasons for the patterns are analyzed and explained.

Monthly Bulletin, Climate System Monitoring

HOW OFTEN IS IT ISSUED?---------- Monthly
WHERE IS IT OBTAINED? ------------- WMO[e]
WHAT DATA DOES IT CONTAIN?

Maps and tables depict departures from normal for temperature and precipitation. In addition, maps and text describe drought and crop conditions globally.

(e) The World Climate Programme and the WMO have a wide variety of publications and data sets available. In addition, as an agency of the United Nations, the WMO has information about data available from individual member nations. It is best to inquire directly and state your requirements.

Appendix C.
Three example design problems
with atmospheric components

In this Appendix I present details concerning three of the example design problems discussed at the end of **Chapter 8**. The first has to do with an architectural problem centered on a single building. The second deals with the cooling of urban canyon space with transpiring trees. The third addresses the choice of the most appropriate air conditioning alternatives in certain climatic regions.

In all three discussions I use mathematical symbols to condense the analysis and make it explicit. In the second discussion, on street trees, I make calculations with the mathematical formulations to illustrate the point that some design questions cannot be answered only with words, which are non-quantitative symbols. They must be answered with numbers — quantitative symbols.

The large rehab commercial block.

The building is a large, cubical commercial building in the shape of a "U", with interior skylit court facing east. To the immediate west lies a major topographical ridge — it rises 600 feet within a quarter of a mile of the block. The building was formerly a regional warehouse for a catalog sales company. It is to be remodelled into a trade center, with offices, display areas, and attractive space for receptions and meetings.

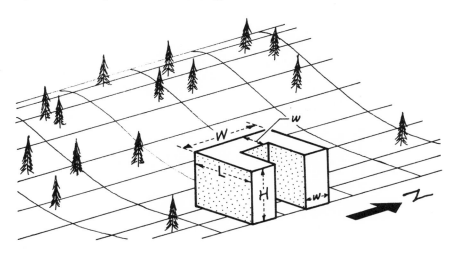

The design question requiring an answer is whether or not to enclose the east-facing court with glass, making it a huge atrium space. On the one hand, the enclosure might significantly increase the annual heating bill by requiring the heating of more space, especially since heat would tend to rise above the level of the occupants. On the other hand it might decrease the annual bill by permitting passive solar heating of the space, and by reducing the total surface area exposed to cold air and winds.

The summer cooling associated with an atrium, and heat flow through the ground line, will not be considered. The principal source of detailed weather data for the area is the published observations for the regional airport 5 miles to the east.

I approach the problem by analyzing the energy flows through the surfaces of the skylit court — the energy balance of the court — with and without the glass, since the flows through other surfaces of the building will be little affected by whether or not the glass is present. It is likely the energy requirements for the building without glass — and therefore the energy flows for the court/atrium *as a fraction of those for the whole building* — could be estimated well from past performance data, despite the differences between past and future uses of the building.

Table C-1 is a catalog, moving westward around the building from the southeast corner, of walls and roofs, together with their dimensions and areas, and their exposure to sunlight. Areas of walls, of course, are (H)x(Running length).

From this catalog we can make some instructive calculations concerning basic proportions of areas before and after the glass enclosure of an atrium space. To make the calculations, we will express *all dimensions as multiples of the height, H.* **Table C-2** is the result.

It is clear at once that, although the enclosure with glass reduces the wall area by 30%, there is a 56% increase in the roofed area. Because the wall area is so much larger than the roofed area, however, there is a 14% reduction in the total surface area in contact with the atmosphere. This is very rudimentary information, and I have included it mainly to support my contention that enclosure with glass will, overall, expose less interactive surface to the atmosphere.

Table C-3 is a catalog, for *mid-morning*, of the surfaces affected by the decision whether or not to use glass. These five surfaces (and the ground surface) enclose the skylit court — the volume whose energy balance is to be estimated under various conditions. The areas and the access to solar energy for each surface will be referred to presently.

Table C-1

Wall	Description	Running length	Affected by glass?	AM	Midday	PM
1S	South facing, south wing	L	No	Yes	Yes	Yes
2	West facing, west wing	W	No	No	No	Yes
1N	North facing, north wing	L	No	No	No	No
3N	East facing, north wing	w	No	Yes	No	No
4S	South facing, courtyard	(L-w)	Yes	Yes	Yes	Yes
5	East facing, courtyard	(W-2w)	Yes	Yes	No	No
4N	North facing, courtyard	(L-w)	Yes	No	No	No
3S	East facing, south wing	w	No	Yes	No	No
6	East facing, glassed atrium	(W-2w)	Yes	Yes	No	No

Roof	Description	Dimensions	Affected by glass?
RS	South wing	(L) x (w)	No
RC	Connecting wing	(w) x (W-2w)	No
RN	North wing	(L) x (w)	No
RA	Atrium	(L-w)x(W-2w)	Yes

Table C-2

Dimension or surface		Without glass	With glass	Ratio (With/Without)
Length	L = 2.00 H			
Major width:	W = 1.50 H			
Minor width:	w = 0.50 H			
Total wall area		10.00 H	7.00 H	0.70
Total roof area		2.25 H	3.50 H	1.56
Total surface area		12.25 H	10.50 H	0.86

Table C-3

------- MID-MORNING HOURS -------

Surface	Description	Area	Sunlit on sun day?
4S	South facing, courtyard	(H) x (L-w)	Yes
4N	North facing, courtyard	(H) x (L-w)	No
5	East facing, courtyard	(H) x (W-2w)	Yes
6	East facing, glassed atrium	(H) x (W-2w)	Yes
RA	Atrium roof	(L-w)x(W-2w)	Yes

Up to this point, any competent designer could have constructed the same tables and derived the same conclusions as I have done. From this point, I believe, the role of an atmospheric consultant becomes clear.

Since there are engineers who have worked out — and made readily available — many useful guidelines for estimating requirements, design alternatives, costs, and payback times for different, generalized architectural problems involving both passive and active solar heating, it might seem the next step would be to turn the problem over to such an engineer. But bear with me. There is still a place for the atmospheric consultant.

Keep in mind that the approach is to estimate the energy balances of the atrium space, without and with glass, under different conditions of time and weather. I trust it is clear the problem is more than simply a matter of adjusting the energy cost experience, already built up for the building, for a different set of surfaces and for uses now requiring human comfort rather than simply warehousing space.

If those two adjustments — for a different set of surfaces and for uses now requiring human comfort rather than simply warehousing space — were all that was required we could simply develop, and then adjust, a curve for the building such as the ones in **Fig. 5-11a**, which say in a slightly different way that the energy requirement is proportional to the number of heating degree days (**HDD**) below the inside (thermostatted) temperature we have called $(T_{inside})^*$. The problem is more complex than that because there is passive solar heating involved in the problem: a glassed atrium would make it really a different building presenting different directional aspects to the weather, not simply a different surface area. The reason more analysis is required is that some of the energy differentials (between unglassed and glassed) are cost-free because they are passive, while some are not. I need to support this contention.

As a general proposition, the energy cost associated with each one of the surfaces in **Table C-3** (wall or roof) of the atrium space, *during one time period, such as an hour,* has several components:

(a) the net energy flux through a unit of area of the surface under a specified set of weather conditions, call it the *weather-specific energy function* (**W**) with dimensions of (energy units per unit area per hour);

(b) the area of the surface (**A**) with dimensions of (area);

(c) the monetary cost of delivering one unit of (heating/cooling) energy, call it the *cost function* (**C**) with dimensions of (cost units per energy unit); noting that the operating cost for passively obtained energy is ($C = 0$); and

(d) the probability that the specified set of weather conditions will be observed at that hour on that date at that location, call it the *probability function* (**P**) with no dimensions, since probability is always just a number between zero and one.

Multiplying these functions yields a product, call it the *weather-specific cost function*, (**X**) = (**P**)(**W**)(**A**)(**C**), with dimensions of (cost units per hour). The way to form this product involves following the logic leading to the results in **Footnote Q, Chapter 4**, and writing down the *net energy flux* — energy per unit area per unit of time — of the surface during the hour, when it is exposed to the specified weather conditions.

We can write down a broad, approximate description of the *net energy flux for a unit area* of one of the surfaces of the atrium:

Net Shortwave ± Net Longwave ± Convection ± Conduction = 0

| + S | ± L | ± H | ± B | = 0 |

or, slightly re-written:

| + S | ± L | ± H | = | ± B |

which is formulated in this way. First note that in this approximation there is no energy flux due to latent heat transfers, **LE**, and negligible heat storage in the surface, Δ, during the period. When the surface is sunlit, the cost-free net shortwave energy (+**S**) enters the equation and is always positive (an energy addition to the atrium system). The cost-free net longwave energy (±**L**) is always present, sometimes an outflow and sometimes an inflow, but its magnitude is responsive to the surface-atmosphere temperature difference. The cost-free convective energy flux (±**H**) is a function of both the wind action on the surface — the wind speed — and the surface-atmosphere temperature difference.

The fourth component in the energy balance equation (±**B**) is special in two ways, so I have set it aside on the right in the last version of the equation. It is special because (i) its value (magnitude and direction of flow through the surface) is *the residual required to balance* the algebraic sum of the other three components, and (ii) it is *the only flow associated with a non-zero cost* (**C**). In terms of the energy balance, then, it is the term involving the "metabolic" heat released inside the building in order to maintain the inside (thermostatted) temperature we have called (T_{inside})*.

The next step after cataloging the atrium surfaces would be to write and use a special computer program to estimate each component of each surface in **Table C-3** under different conditions of time (solar geometry) and weather. The program would require that certain conditions be met, for example that the temperature within the building itself would be $(T_{inside})^*$ at all times, and that the temperature would be $(T_{inside})^*$ inside the glassed atrium. The program would also include the information about how each of the terms of the balance equation is related to weather variables. For example, we know from previous discussion that these dependencies are involved:

Balance component	Depends on —
+ S	Latitude, date, hour, surface color, *sky cover.*
± L	Surface-atmosphere *temperature* difference.
± H	*Wind speed and direction* — wind action directly on the surface — and the surface-atmosphere *temperature* difference.

In quantifying these dependencies, for example, the program would include a much reduced convective component — much reduced wind speed and surface-atmosphere *temperature* difference — within a glass-enclosed atrium as compared with the same surface without glass.

The computer program would have to be used to estimate the ensemble of balances for all the atrium surfaces *taken together.* Because *the surfaces are interactive*, their balances could not be estimated separately. While an engineer can write and use such a program, an atmospheric ecologist could do that too, and would bring other important understandings to the enterprise. Once the ensembles of energy balances are estimated for different weather conditions, the resulting balance components can be used to fill the cells of tables such as those in **Table C-4**. Following that, the non-zero costs could be associated with the residuals — the values of $(\pm B)$ — which could not be done unless these terms were estimated separately by the program.

Before we consider the potential contribution of the atmospheric consultant to solving the planning problem, we need to consider the matter of probability.

What is the probability function? You can construct, from hourly weather observations taken at the regional airport, a *frequency function* (F) showing the number of occurrences of each

particular combination of weather elements[A] for each hour —
Midnight, 1 a.m., 2 a.m., etc. — of the record at that location. You
can estimate the probability (**P**) needed to calculate (**X**) by dividing
the number of days (**F**) on which the particular weather combination occurred *at that hour* by the total number of days in the
record from which you calculated the frequency function (**F**).
There will be one complete **Table C-4** for each surface, and there
will be one value of (**P**) for each of the (32N) lines of that table[A].

With values for components of the *weather-specific cost func-
tion*, (**X**) = (**P**)(**W**)(**A**)(**C**), we would need to associate these with
the probabilities of the various combinations of weather conditions
to get the basis for the economic decision about whether or not to
use glass in the atrium. The basis would be the comparison of the
two sums of products:

Operating cost (No glass) = $\sum \sum$ (X)$_{\text{No glass}}$ doubled-summed
over all weather types, over all times of the day, and over all surfaces, *versus*

Operating cost (With glass) = $\sum \sum$ (X)$_{\text{Glass}}$ double-summed
over all weather types, over all times of the day, and over all
surfaces.

Finally, *what is the contribution of the atmospheric ecologist to
be?* Certainly he can translate his knowledge of microclimates into
(i) the categories of each relevant weather variable to be analyzed[B], and (ii) the probabilities for the area, **P**(X), derived from
the hourly records at the regional airport. But is that all? I think
not.

I have included in the circumstances of this problem the fact
there is a prominent terrain feature — the major ridge to the west
of the building — because (see **Fig. 3-27**) the airflows at the building, and thereby the direct wind action on the surfaces, may often
be quite different from those represented in the hourly records at
the regional airport. Perhaps *the greatest contribution of the atmospheric ecologist* will be to adjust the airport records for direct use
at the site by using his knowledge of these terrain effects on wind
flow.

(A) **Table C-4** uses a scheme for classifying weather conditions that has (32N) possible types: 4 sky conditions, 8 wind conditions, and N temperature conditions
(the size of N is as required — only 2 are illustrated in **Table C-4**: 40 °F and 50 °F).

(B) These categories, or "classes", are those found in **Table C-4**: Sunny, partly
cloudy, and overcast; light wind and brisk; etc.

Table C-4

------ WALL 4S: SOUTH-FACING IN THE ATRIUM, MID-MORNING ------

Weather Conditions	Energy Balance Component							
	Net Shrtwv		Net Longwv		Convection		Conduction	
	Glass	None	Glass	None	Glass	None	Glass	None
Clear - Low Sun								
Temperature = 40 °F								
Wind NW - Light								
- Brisk	Each line of entries in this Table							
Wind NE - Light								
- Brisk	constitutes one set of components of							
Wind SE - Light								
- Brisk	the *weather-specific energy function* (**W**).							
Wind SW - Light								
- Brisk								
Temperature = 50 °F	Combined with the area function (**A**), the							
Wind NW - Light								
- Brisk	probability function (**P**), *with one value of (P)*							
Wind NE - Light								
- Brisk	*for each line of this Table*,							
Wind SE - Light								
- Brisk	and the cost function (**C**), the result is							
Wind SW - Light								
- Brisk	the *weather-specific cost function*, (X)							

NOTES:
This Table is for *one surface*, so —
   a) to complete this Table, repeat for other temperatures under the same time and sky conditions, and then —
   b) for *the same surface*, repeat for other *time conditions*, for example: midday, midafternoon, and nighttime;
and the same set of *sky conditions*, for example:
   Clear - Low Sun; Clear - High Sun; Partly Cloudy; Overcast.
There will be a set of tables like this one for each atrium surface.

Trees to cool urban canyons.

In this example I want to accomplish two things. *First*, I want to follow through on my assertion, in **Chapter 1**, that the idea of street trees cooling urban canyons by evaporation is a myth. *Second*, I want to present a series of mathematical formulae, and calculations with them, to illustrate an important point: *some arguments must be made with mathematics and calculations, because they cannot be made only with words.*

While it is true that mathematical models, and the calculations that come from them, are only approximations to reality; still they offer insights that words alone cannot give. The following numbers and calculations are *representative* in the sense that their magnitudes and relationships agree, generally, with those in nature. *In fact, the numbers are generous in the direction of larger cooling rates.* So here we go.

First, we define our terms, and the "building blocks" with which to make the case that cooling urban canyons with evaporation by street trees is a myth.

The canyon space. We will consider the problem by studying one "canyon unit" — a cubical volume with 10 meters on an edge, as shown in the sketch. The volume enclosed is:

Volume of one Canyon Unit = $\mathbf{V}$ = 10^3 m^3 (1000 cubic meters).

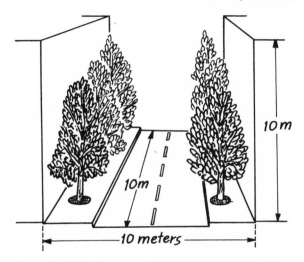

The trees. In each canyon unit there are *six trees*, each tree with a ground area of 5 m^2, and a leaf area index LAI(T) = 5. (See **Chapter 3**) Knowing that, we know that the total leaf area for one tree is (5)x(5 m^2) = 25 m^2, so that the total leaf area for the canyon unit is (6)x(25 m^2)

Total leaf area in one Canyon Unit = $\mathbf{A}$ = 150 m^2.

The air to be cooled. We will use a typical sea-level *density* for air, $\mathbf{d}$, of (1.2)(10^{-3}) gm/cm^3, which is (1.2)(10^3) gm/m^3 — 1.2 kilograms per cubic meter — so that

Total mass of air in one Canyon Unit

= $\mathbf{M}$ = $\mathbf{V}\,\mathbf{d}$ = (10^3) m^3 x (1.2)(10^3) gm/m^3 = (1.2)(10^6) gm.

Having defined our building blocks, we can now make the calculations we need.

Cooling power of the trees. A reasonable midday transpiration rate, expressed as an energy flow rather than as a volume flow (see Lowry, 1969, or Gates, 1980), is T = (0.1) calorie/cm^2-minute, or (10^3) cal/m^2 *of leaf area*-min. The total evaporational energy flow in one canyon unit, then, is

E = AT = (150) m^2 x (10^3) cal/m^2-min = (15)(10^4) cal/min.

The **specific heat of air at constant pressure** is *the number of calories each gram of air holds for each degree of temperature it has*. That number is

c_p = (0.24) cal/gm-deg C.

From that we calculate that the rate of cooling of the air in one canyon unit, due to the transpiration from the six trees, would be

Δ = [E / (c_p M)]

= $\Big($ (15)(10^4) cal/min $\Big)$ ÷ $\Big($ [(0.24) cal/gm-deg] x [(1.2)(10^6) gm]$\Big)$

= (0.52) deg/min, or about *30 degrees C per hour.*

That looks as if the air in the canyon unit would be cooled 30 deg C every hour! Who says it's a myth?

Conclusion. But wait. That's the cooling rate *if the air is stagnant and the transpiration rate is constant through the hour.* Common sense tells us neither of those conditions holds true in our case. First of all, even if the wind were calm, the canyon air would become more and more humid, so the transpiration rate would decrease steadily.

What really happens is that the transpiration rate, and the lower humidity, are maintained by the *natural ventilation* of the canyon unit. That is, the air is regularly replaced by (actually mixed with) "fresh air" — from outside the canyon, usually from above it — whose humidity is lower and whose temperature is higher. The ventilation is caused mostly by the regional windflow, but also in part by the fact that the air within the canyon moves in response to temperature differences — and thus density differences — across the canyon. In either case, the air isn't stagnant.

Conservatively, the ventilation rate of this wide-open space would be between 50 and 100 volume changes per hour. Dividing this rate into the "stagnant rate", Δ, gives

(30 deg per hour) ÷ (50 - 100 per hour) = (0.3 - 0.6) deg — a much smaller, and more realistic, cooling effect.

Thus it is that the cooling of urban canyon volumes by the transpirational effects of street trees is not very large, in terms of human comfort. However, street trees reduce the solar radiant heat load at street level in the canyon, so that *trees have a cooling effect by shading rather than by evaporation*.

Climate and air conditioning

As suggested in **Chapter 8**, we may discern — on a broad, regional scale — three major options for air conditioning: (i) passive cooling, (ii) active evaporative cooling, and (iii) active refrigerant cooling. In addition — as is also suggested in **Chapter 8** — the choice from among these three is usually obvious over large, geographical, climatic regions, at least on the basis of economics. Along the boundaries of these large regions, however, the choice may not be clear.

I offer a rudimentary analysis indicating how several concepts from atmospheric ecology might affect the decision about choice from among the three options. As a general proposition, the cost of air conditioning, *during one time period, such as an hour,* has several components:

(a) the amount — the number of units — of conditioning required, call it the *atmospheric function* (**A**) with dimensions of (conditioning units per hour);

(b) the energy efficiency *of a particular piece of equipment* in producing one unit of the required conditioning, call it the *efficiency function* (**E**) with dimensions of (energy units per conditioning unit);

(c) the monetary cost of delivering one energy unit, call it the *cost function* (**C**) with dimensions of (cost units per energy unit); and

(d) the probability that (**A**) conditioning units will be required at that hour on that date at that location, call it the *probability function* (**P**) with no dimensions, since probability is always just a number between zero and one.

Multiplying these four functions yields a product, call it the *total cost function* (X) with dimensions of (cost units per hour). Thus

$$(\text{X}) = (\textbf{P}) \ (\textbf{C}) \ (\textbf{E}) \ (\textbf{A}).$$

We need to note one additional factor before we consider the potential contribution of the atmospheric consultant to solving the regional planning problem. An air conditioner has an upper limit to the number of conditioning units it can deliver; call that the *mechanical limit* (**M**). The point is that, if the number of conditioning units *required* exceeds the number of conditioning units *possible*, then the result will be *incomplete* conditioning during that hour. We indicate the presence of this upper limit by saying that (**A**) cannot exceed (**M**): (**A** not > **M**).

What are conditioning units? Referring to **Fig. 5-5** showing the TRe chart of temperature, relative humidity, and vapor pressure (also see McGuinness *et al*, 1980, page 95) note that the standard human "comfort zone" is centered on the point at (T = 75°F) — which is (23.8°C) — and (RH = 50%). The effective temperature (T_e) is 72°F at that point.

If you move, on the chart, to any point of higher effective temperature, you are moving toward the hyperthermic side of the comfort zone, and air conditioning is required. As discussed by McGuinness *et al* (1980, page 226), you may attain comfort by the appropriate combination of changes in sensible heat (usually cooling) and in latent heat (usually drying). The thermodynamics of these changes is beyond the scope of this book, but we will note that, conceptually, there is an appropriate combination of changes in sensible heat and in latent heat required to attain comfort. In a hot, humid climate both cooling and drying are required; while in a hot, arid climate only cooling may be required.

We will indicate that two kinds of modification of air are required — change in *sensible* heat and change in *latent* heat — by using the symbol Δ for each kind of change, and noting that the sum of the two kinds of change is the number of conditioning units required: $A = (\Delta_S + \Delta_L)$. Each hour has a value of (T_e) associated with it, and with each value of (T_e - 72) goes one value of (**A**).

What is the probability function? In this case it is slightly different from that described for the case of the large rehab commercial block. Not every hour of every day, at the location being studied, will require exactly (**A**) conditioning units per hour — proportional to (T_e- 72). Some would require less than (**A**) units and some more. A great many would require no conditioning units — (**A** = 0). You could construct, from hourly weather observations taken near your location, a *frequency distribution* (see **Chapter 8**) of the number of conditioning units required for each hour of the record at that location.

In the sample of hours included in the records for that location, the value of (**A**) starts with zero and covers the whole range up to its maximum value. Exactly (**A**) units would be required at different hours on different days and in different years. At a particular hour on a particular day of the year — say 10 a.m. on July 18 — different years will probably have different numbers of conditioning units required, and so on.

You can estimate the probability that exactly (**A**) conditioning units are required at any one hour in the year by dividing the number of hours requiring exactly (**A**) conditioning units by the total number of hours in the record from which you calculated the frequency distribution. Call that probability **P(A)**.

Now we put together the discussions about the mechanical limit, the definition of (**A**), and the probability function to get the *working equation*:

$$X(A) = P(A) \times (C) \times (E) \times [\ (\Delta_S + \Delta_L)\ \text{not} > M\ ]$$

which tells us the cost of conditioning, over a year at that location, for each value of (**A**), for one particular piece of conditioning equipment. To get the total cost over a year, for one particular piece of conditioning equipment, add up the values of X(**A**) for all the values of (**A**) — each one proportional to $(T_e - 72)$. When you do the same for each kind of equipment, you have the basis for comparing equipment for that location.

Clearly, in any location each type of conditioning method — from passive to the various models for active refrigerant conditioning — can do some or all of the required job. *Passive cooling* can, at no operating cost, do all of the job some of the time and some of the job all of the time, but can it do enough? *Evaporative coolers* — called "desert coolers" — are popular and effective in desert regions. They reduce sensible heat and increase latent heat contents of environmental air. *Refrigerant conditioners* — by far the most common kind — reduce both the sensible and the latent heat contents of environmental air. With different rate schedules for power in different climatic regions, different kinds or different combinations of equipment will prove optimal.

Values of (**C**) come from the local power utility, while values of (**E**) and (**M**) come from the equipment supplier. A competent atmospheric consultant can supply the required information on $(\Delta_S + \Delta_L)$ and **P(A)**.

Appendix D.
Flow Rates in Figure 9-7

OUTSIDE THE ECOSPHERE

FROM ...		TO ...		
	Absorbed by Green Plants	Ecosphere	Respiration	Heat/ Space
Solar at Surface	280,000	-	-	4,820,000
Absorbed by Green Plants	X	12,843	-	267,157
The Ecosphere				
Living Plants	-	X	7188	-
Plant Eaters	-	X	842.5	-
Flesh Eaters	-	X	75.9	-
Decomposers	-	X	4736.6	-
Respiration	-	-	X	12,843
Totals	280,000	12,843	12,843	5,100,000

INSIDE THE ECOSPHERE

FROM ...			TO ...			
	Respiration	Plant Eaters	Flesh Eaters	Decomposers	Dead Materials	Heat/ Space
Living Plants	7188	981	-	-	4674	-
Plant Eaters	842.5	X	89.1	-	49.4	
Flesh Eaters	75.9	-	X	-	13.2	
Dead Organic Materials	-	-	-	5262.9	X	
Decomposers	4736.6	-	-	X	526.3	
Respiration	-	-	-	-	-	12,843
Totals	12,843	981	89.1	5262.9	5262.9	12,843

References

Anthes, R.A., J. Cahir, A. Fraser, and H. Panofsky (1975), *The Atmosphere*, (Third Edition), Charles E. Merrill, New York.

Architectural Graphic Standards, Robert Packard, Editor, (Seventh Edition, 1981), John Wiley & Sons, New York.

Ahrens, C.D. (1982), *Meteorology Today*, West Publishing, St. Paul, Minnesota.

Baldwin, J. (1973), *Climates of the United States*, NOAA, Department of Commerce, Washington, D.C.

Bates, M. (1969), Chapter 1 in *Resources and Man, A Study and Recommendations by the National Academy of Sciences and the National Research Council*, W. H. Freeman, San Francisco.

Bligh, J., and R. Moore (Editors), (1972), *Essays on Temperature Regulation*, Elsevier, New York.

Byers, H.R. (Fourth Edition, 1974), *General Meteorology*, McGraw-Hill, New York.

Changnon, S.A. (1962), "A climatological evaluation of precipitation patterns over an urban area", in *Air Over Cities, U.S. Public Health Service Publication SEC A62-5*, Cincinnati, Ohio.

_____ (1968), "The LaPorte weather anomaly: fact or fiction?", *Bulletin of the American Meteorological Society, v.49*, pages 4-11.

_____ (1980), "More on the LaPorte Anomaly", *Bulletin of the American Meteorological Society, v.61*, pages 702-717.

_____ *et al* (1977), "Summary of METROMEX, Volume I: Weather Anomalies and Impacts", *Bulletin 62*, The Illinois State Water Survey, Urbana, Illinois.

Critchfield, H. (1974), *General Climatology (Third Edition)*, Prentice-Hall, Englewood Cliffs, New Jersey.

Defant, Fr. (1951), "Local Winds", in *Compendium of Meteorology*, American Meteorological Society, Boston.

Fuller, R. Buckminster (1978), *Operating Manual for Spaceship Earth*, E.P. Dutton, New York.

Gates, D.M. (1965) in *Agricultural Meteorology*, Meterological Monograph 28, American Meteorological Society, Boston.

_____ (1980), *Biophysical Ecology*, Springer-Verlag, New York.

Geiger, R. (1965), *The Climate Near the Ground*, Harvard Press, Cambridge.

Givoni, B. (1969), *Man, Climate, and Architecture*, Elsevier, New York.

Goody, R. and J. Walker (1972), *Atmospheres*, Prentice-Hall, Englewood Cliffs, New Jersey.

Hubbert, M. K. (1969), Chapter 8 in *Resources and Man, A Study and Recommendations by the National Academy of Sciences and the National Research Council*, W. H. Freeman, San ·Francisco.

Huff, F. and S.A. Changnon (1972), "Climatological Assessment of Urban Effects on Precipitation", *Final Report for Contract GA-18781* to the U.S. National Science Foundation.

_____ and D. Jones (1975), "Precipitation Increases in the Low Hills of Southern Illinois", *Monthly Weather Review, v.103*, pages 823-836.

Jauregui, E. (1986), "Tropical urban climates: review and assessment", page 26 in *Urban Climatology (in Tropical Areas)*, Publication No. 652 of the World Meteorological Organization, Geneva, Switzerland (for address, see Appendix B).

Kalkstein, L. and K. Valimont (1986), "An Evaluation of Summer Discomfort in the United States Using a Relative Climatological Index", *Bulletin of the American Meteorological Society, v.67*, pages 842-848.

_____ (1987), "An evaluation of winter weather severity in the United States using the Weather Stress Index", *Bulletin of the American Meteorological Society, v.68*, pages 1535-1540.

Kemeny, J. G., J. Snell, and G. Thompson (1966), *Introduction to Finite Mathematics*, Prentice-Hall, Englewood Cliffs, N. J.

Kleiber, M. (1961), *The Fire of Life: an Introduction to Animal Energetics*, John Wiley and Sons, New York.

Kondratjev, K. (1969), *Radiation in the Atmosphere*, Academic Press, New York.

Landsberg, H.E. (1979), *Urban Ecology, v.4*, pages 53-81.

_____ (1981), *The Urban Climate*, Academic Press, New York.

Loftness, V. (1982), *Climate Data Applications in Architecture*, Publication WCP-30, World Meteorological Organization, Geneva, Switzerland (for address, see Appendix B).

Lovelock, J. E. (1979), *GAIA: a new look at Life on Earth*, Oxford University Press, New York.

_____ (1986), "Geophysiology: a new look at Earth Science", *Bulletin of the American Meteorological Society, v.67*, pages 392-397.

Lowry, W.P. (1967), "The Climate of Cities", *Scientific American*, August, page 15.

_____ (1969), *Weather and Life: An Introduction to Biometeorology*, Academic Press, New York.

_____ (1977), *Journal of Applied Meteorology, v.16*, pages 129-135.

_____ (1978), *A Biophysical Energy Balance for the Earth-Atmosphere System*, Research Report No. 4, Institute for Environmental Studies, University of Illinois, Urbana.

_____ (1979), "Interactions between Cities and Their Local and Regional Weather and Climate", in *Western European Cities in Crisis*, (M. Romanos, Ed.), D.C. Heath and Company, Lexington, Massachusetts.

_____ (1980), *Direct and Diffuse Solar Radiation: Variations with Atmospheric Turbidity and Altitude*, Research Report No. 6, Institute for Environmental Studies, University of Illinois, Urbana.

Lutgens, F. and E. Tarbuck (First Edition, 1979), *The Atmosphere: an Introduction to Meteorology*, Prentice-Hall, Englewood Cliffs, New Jersey.

_____ (Second Edition, 1982), *The Atmosphere: an Introduction to Meteorology*, Prentice-Hall, Englewood Cliffs, New Jersey.

McDonald, J. E. (1952), "The Coriolis Effect," *Scientific American*, May, page 2.

McGuinness, W., B. Stein, and J. Reynolds. (Sixth Edition, 1980), *Mechanical and Electrical Equipment for Buildings*, John Wiley & Sons, New York.

Mendelsohn, E. (1964), *Heat and Life: the Development of the Theory of Animal Heat*, Harvard University Press, Cambridge.

Miller, D. H. (1977), *Water at the Surface of the Earth: an Introduction to Ecosystem Hydrodynamics*, Academic Press, New York.

_____ (1981), *Energy at the Surface of the Earth: an Introduction to the Energetics of Ecosystems*, Academic Press, New York.

Monteith, J. L. (1976), *Vegetation and the Atmosphere, Volume 2*, Academic Press, New York.

Oke, T.R. (1978), *Boundary Layer Climates*, John Wiley & Sons, New York.

_____ (1981), *Journal of Climatology of the Royal Meteorological Society, v.1*, pages 237-254.

_____ (1982a), Review of *The Urban Climate*, by H.E. Landsberg, *Bulletin of the American Meteorological Society, v.63*, pages 535-536.

_____ (1982b), "The Energetic Basis of the Urban Heat Island", *Quarterly Journal of the Royal Meteorological Society, v.108*, pages 1-24.

_____ (1987a), "Street design and urban canopy layer climate", *Energy and Buildings, v.11*, pages 103-113.

_____ (Second Edition, 1987b), *Boundary Layer Climates*, Methuen, New York.

Olgyay, V. (1963), *Design with Climate*, Princeton University Press, Princeton, N.J.

Platt, R.B. and J. Griffiths (1964), *Environmental Measurement and Interpretation*, Reinhold, New York.

Reiquam, H. (1970), *Science, v.170*, pages 318-320.

Rosenberg, N. R. *et al* (Second Edition, 1983), *Microclimate: The Biological Environment*, Wiley and Sons, New York.

Schaefer, V.J. and J.A. Day (1981), *A Field Guide to the Atmosphere*, Houghton Mifflin, Boston.

Scientific American Editors (1970), *The Biosphere*, W.H. Freeman, San Francisco.

Sellers, W.D. (1965), *Physical Climatology*, University of Chicago Press, Chicago.

Smithsonian Meteorological Tables (Edition of 1966), The Smithsonian Institution, Washington, D.C.

Steadman, R. G. (1984), *Journal of Climate and Applied Meteorology, v.23*, pages 1674-1687.

Stern, A. C., *et al* (1973), *Fundamentals of Air Pollution*, Academic Press, New York.

_____ (Second Edition, 1984), *Fundamentals of Air Pollution*, Academic Press, New York.

Tarbuck, E. and F. Lutgens (1976), *Earth Science*, Merrill, Columbus, Ohio.

U. S. Committee for an International Geosphere-Biosphere Program (1986), *Global Change in the Geosphere-Biosphere: Initial priorities for an IGBP*, National Academy Press, Washington, D.C.

Watson, A. J. and J. E. Lovelock (1983), "Biological homeostasis of the global environment: the parable of daisy world", *Tellus, v.35B*, pages 284-289.

Wendland, W. M. (1987), *Bulletin of the American Meteorological Society, v.68*, pages 616-619.

Author Index

Subject Index

Atrium, as a design component 265, 402 et seq
Automobile, as a source of air pollutants 362-363
Average (see Mean)
Average daily temperature, calculated for degree days 122

B

Baguio (see Tropical cyclones)
Balance, energy (see Energy balance)
Balance, mass 6-8
Balance of nature 272, ecological overview 223, 383
Balanced motion, theory governing 319-320, 352
Barometric pressure (see Pressure)
Beaufort scale of wind speed 25
Behavioral thermoregulation 112, 172
Bell-shaped curve, related to turbulence 26
Bergeron-Findeisen process of cloud drop growth 86
Biogeography 201-204
Biological energy, compared with geophysical energy 287, distribution within the
    ecosphere 287-289
Biological response to temperature, in animals, various measures 175
Birds 156
Black body, defined 54, physical laws describing 55, radiative characteristics 54-55,
    the earth as a 57, the sun as a 57
Blizzards 338
Boiling 15
Boiling temperature of water, related to pressure/altitude 67
Boundary layer 27, microscale 29, 164-166, of Earth 29, 38, sizes 29, urban 134
Box model, for air pollutant concentration 373-376, 387
Budget, heat and energy, related to balance 91
Budget, radiation 91
Building-climate interactions 151
Buildings, compared with clothing as thermal resistance 121
Buoyancy 28, 30, related to air pollution 361

C

Canopy, plant 87, cityscape as a 134
Cap cloud 87
Capillaries, blood, related to thermoregulation 172
Carbon dioxide, canopy profiles related to photosynthesis 88, profiles in plant
    canopies 88, related to plant physiology 192, and the Greenhouse effect 293,
    related to global climatic change 384
Cardiovascular resistance to heat flow 164, thermoregulation 112
Carnivores 161, related to energy distributions in the Ecosphere 288
Carrier molecules, in Photosynthesis 186
CDD (see Cooling degree days)
Celsius temperature 15
Centigrade temperature 15
Central tendency, as a statistical concept 251-253
Chill requirement, as thermal preconditioning in animals 176
Circle of illumination 278
Circulation cells in fluid motion 79, local 79, size scales 79, of Earth's atmosphere
    296-297
Cities, related to cloud and rain formation 146 (also see Climate, urban, and Urban)
Clay 72, 75
Climate space 172
Climate, urban, data available 224, 247-249, 393-400, defined 133, proper use of
    data 224

Degree days, annual averages, problems for planning 126, calculations 122, cooling 122, growing 179, heating 122, 404 , maps of annual averages 127

Designer/planner, the professional 209-213, atmospheric expertise required 227, 272, direct use of air pollution concepts 373-380, formulating the atmospheric component 226-228, 269, interactions with atmospheric ecologist 221, 268, 272, presentations of atmospheric components 268-271, required view of the Balance of nature 234, sample problems with atmospheric components 264-268, 402 et seq, relationships to air pollution control 353eq, utility of knowledge about global scale climates 295, utility of knowledge about urban climatic effects 150

Development vs. growth, in organisms 177

Development, in plants 195, rate of, related to degree days 180

Developmental subperiod (see phenology)

Dewfall, in dry soil 77

Dewpoint depression, defined 71

Dewpoint temperature 271, annual patterns 23, defined 70, diurnal patterns 23, versus relative humidity 22

Differential heating, cause of motion 79, related to global scale motion 297

Differential transmissivity, and the Greenhouse effect 293

Diffusion factor, related to air pollutant plume 366

Diffusion vs. dispersion 365

Dispersion, as a statistical concept 253-254, vs. diffusion 365

Distance, scales 15

Diurnal thermoperiodism 196

Drain size, as a concept, related to air pollution control 357-362

Dry adiabatic lapse rate, related to vertical motion 361

E

Earth, estimated average temperature 56, average surface temperature, factors affecting 283-285, distance from the sun 277, general circulation of the atmosphere 296, geometry of insolation seen from space 277-279, orbital geometry 277

Earth's atmosphere, as a compressible fluid 282, composition of dry air 281, gaseous composition 281-282, logarithmic relationship of pressure/altitude 283, 294, pressure versus altitude 282-283, trace gases 282, variability of water vapor 281

Economics of externalities, related to air pollution 382

Ecosphere, energy distributions 287-289, energy flows 414

Ecosystem, global 295, related to the Balance of nature 234, 257

Eddy, turbulent 28, large scale, 325-327

Effective heat load, defined for animals 164

Effective temperature (see Environmental temperature)

Efficiency, related to temperature-energy relationships 125

El Cordonazo (see Tropical cyclones)

Electromagnetic spectrum 54-57, shortwave vs. longwave 55, 57

Elevation angle, solar 46

Emission spectra of atmospheric gases 58

Emissivity 54

Energy in biological processes on Earth 287-289

Energy 6-9, 59-60, 90, changes in urban demand with changes in architecture 132, comparative contents of atmospheric phenomena 317, conservation and conversion 59-60, envelope 20-21, metabolic, allocation by animals 162, partitioning by surfaces, rules of thumb 97, per capita use related to urban form 130, per unit urban area, use related to typical insolation rates 131, per unit urban area, use related to urban form 130, stored in plant canopies 98

Energy balance 6-9, microclimatic 89, 91-94, 107, 128, 139, 154, 194, 206 391-392, 402-406, of Earth and its atmosphere 285-287, 293

G

Gaia hypothesis 182, 289-293, related to global climatic change 383
Game of environmental design, best play 211-213, 218, game board 210, players 208, spatial scales 211, who wins? 208, 219, the Environmentalist as a player 233, the goal of 235
Gases, atmospheric, radiative characteristics 58
Gaussian plume, model 386, for air pollutants 373, utility for planning 376-377
GDD (see Growing degree days)
Generalist vs. specialist 388-390
Geophysical energy, compared with biological energy 287
Geostrophic flow 319-320
Global climatic change, interactions with air pollution 383-385, related to carbon dioxide 384, related to fossil fuels 383-385, related to solid particulates 385, related to the Greenhouse effect 383-385l, related to water vapor 385
Gradient, defined 26, 62, of temperature 62, vertical, of wind speed 26
Gradient flow 320 (also see Balanced motion)
Gravitational attraction, effects on planetary atmospheres 284
Gravity waves 323
Greenhouse effect, defined 285, related to energy balance of Earth's atmosphere 285, related to global climatic change 383-385
Growing Degree Days 179, 181, 202, 206, related to Heating and Cooling degree days 181, related to phenology, 202, 206
Growth, in plants 195, related to net photosynthesis 189, vs. development, in organisms 177
Guard cells 184, 192
Gust fronts 86
Gyres, in ocean currents 312

H

Habitat 275
Hadley cell, of Earth's atmosphere 301
Hailstorms 85
Haircoat 164, 174, color, related to climate space 174
HDD (see Heating degree days)
Heat vs. temperature 14
Heat, direction and rate of flow 61, related to temperature 60, storage 60
Heat budget, related to heat balance 91
Heat capacity 64-65, 73, 105, soil 73, of soil, as a weighted mean 105, related to soil temperature behavior 64-65
Heat energy 6, kinds of 90, related to biological energy 287
Heat flow in soil, air, and water 61-64
Heat load, effective, defined for animals 164, metabolic for animals 164, radiant for animals 164, leaf 194
Heat storage in soil, air, and water 64
Heat trapping, as a method for energy management 237
Heat Units Concept, defined 179, 198, related to plant growth and development 198, utility 179, 199, 206
Heating cables, soil, as a method for energy management 238
Heating degree days, defined 122, 404
Hedgerows 32
Height-to-length ratio 30
Herbivores 161, related to energy distributions in the Ecosphere 288
High (pressure area) 320-327, related to motion in a mountain stream 325-327
Hills, effects on wind patterns 33-34
Histogram, as a statistical tool 256
Homeotherms 110, 155, core temperature related to body function 110
Homo sapiens, as primate specialist 120

Hot capping, as a method for energy management 238
HU Concept, HU Model (see Heat Units Concept)
Human being, as a physical system 6, as homeotherm 110, as primate specialist 120
Human comfort 1-3, 21, 107, 113, 116-118, 140, 162, in midday urban canyon 140, on global scale 116-118, energy management in microenvironments 235, related to urban form 268
Human energy balance, as management problem 107
Human mortality and morbidity, related to air pollution 372
Human population, increase 387, increase in use of fuels 387, 389
Humidity, 21-23, 66, 266
Humidity, relative 19-23
Humidity, urban patterns 143, urban, compared with precipitation 143
Humiture index, or sultriness index of human comfort 115
Hurricane (see Tropical cyclones)
Hyperthermia in human beings 114
Hypothermia in human beings 114
Hypothesis, related to the scientific method 292-293

I

Ice storms 338
Impaction, an air pollution removal process 357
Index cycle, of atmospheric flow 326-327
Insects 155, related to a design/planning problem 265
Insolation, variables involved 11-12, definition 11, on local terrain 14, geometry of, seen from space 277-279, pathways through the atmosphere 51-52, rate defined 51, relationships to urban form 135, related to latitude and date 279-281, related to path of solar beam through the atmosphere 279, 292, sources of data 42
Instability, as a concept in atmospheric motion 360-362
Insulation, in animals 164
Interactions of buildings and climate 151
Interquartile range, as a statistical tool 254
Intertropical convergence zone (ITCZ) 302
Inverse square law 292
Inversion, temperature 59, 246-247, related to air pollution 361, Trade Wind 362
Involuntary thermoregulation 112
Irrigation, as a method for energy management 238, 244-245
Isobar 318, related to Highs and Lows 320-321
Isohyet, defined 145
Isotherm, defined 17, global means 310-313
ITCZ, of Earth's atmosphere 302

J

Jet stream, related to the Index cycle 326-327
Joint probability, as a statistical concept 257

K

Kelvin temperature 43
Kirchhof's law 55

L

LAI (see Leaf Area Index) 38-39
Lakes, temperature patterns, annual 19-20
Lake-shore breeze 33
Land masses, related to general circulation of the atmosphere 302
Land-sea breeze 33
Latent heat 90

Latitude, as a climatic control 304
Law of Conservation 7-9, 89
Leading edge, effects on windstream 29
Leaf area density 38
Leaf Area Index, of a perennial plant 40, of an annual plant 40, related to canopy structure 38-40, related to wind speed 40, various forms 38-39
Leaf, energy balance 194, radiative characteristics 58
Length, scales 15
Life, evidence for, related to the Gaia hypothesis 289-292
Life cycles 177
Life zones of Earth 275
Light reaction of photosynthesis 186
Light saturation in leaves 187
Limits, thermal, in animals 175
Loam 72, 75
Local wind systems 33
Logarithmic nature of profiles 27, 38, 43
Longitude, as a climatic control 304
Longwave radiation 95
Low (pressure area) 320-327, related to motion in a mountain stream 325-327

M

Macroscale dimensions of environmental design, the region 212
Management, of microenvironmental energy balance 100, of the human energy balance 107, of soil characteristics 17, decision making 231, 272, of energy, the central design and planning problem 222, feedback in 228, formulation of the problem 229, of a system 228, 272, of energy, in human microenvironments 235, of energy, in plant microenvironments 236-247, role of a system model in 229, strategies for 235, techniques in air pollution control 377
Manometer 104
Markov chain, as a statistical tool 258-261, 273-274
Mass 6, balance 6-9, 90, 102, 104
Materials, architectural, effects on microclimate 2, 135
Matric potential, of soil moisture 74
Maximum drain size (see Drain size)
Mean (average), as a statistical tool 251, weighted 105, 253
Measure, related to frequency, as a statistical tool 250
Mechanical turbulence 28
Median, as a statistical tool 252
Meridional flow, related to global scale motion 297
Mesophyll 184, 192
Mesoscale dimensions of environmental design, the area 212
Metabolic energy, allocation by animals 162, rate in human beings 107, related to energy balance analyses 99, urban sources, related to climate 135
Metabolic heat production, rate related to animal body size/mass 159, 206
Metabolic rate 155, homeotherms, comparison by size 155, 157, 204, plants and animals compared 155, related to climate space 174
Meteorological range, related to visual range, defined 387
Microbursts 86
Microclimate, four variables for description 24, related to urban form 128 (also see Energy balance)
Microscale boundary layer 29, resistance 164-166
Microscale dimensions of environmental design, the site 212
Midlatitude wave cyclone, life history 332-334, model 332-337, related to motion in a mountain stream 332
Migration, human, related to climate and energy demand 126
Milky Way 276

Millibar, defined 69
Mixing ratio, defined 71
Mode, as a statistical tool 252
Model, of energy balance, utility of 154, animal energy balance 163-167, essential
    elements of 229, regression, as a statistical tool 261, role in management 229,
    utility of 230
Model, mathematical, for air pollutant concentration 363-366, box 373-376, 387,
    limitations for planning 367, puff 377-379
Moisture, soil 24, urban relationships 143
Momentum, conservation of angular 300-301
Monsoons 79, 302, related to thunderstorms 339, and comparative energetics of
    atmospheric phenomena 317
Morbidity and mortality, Human, related to air pollution 372
Mortality, human (see Morbidity)
Motion, accelerated 352, first law of 352, in the general circulation of the
    atmosphere 296, related to the Coriolis effect 299, subsidence, related to air
    pollution 358-359
Mountains, effects on wind patterns 33-35
Mountain-valley winds 33
Mulching, as a method for energy management 238, 240-243

N

Naming air masses 328-329
Naming clouds 83, 104
Negative feedback 9-10
Net photosynthesis 189
Net radiation 95
Neutrality, as a concept in atmospheric motion 360-362
Newton's first law of motion 352
Niche, ecological, defined 157
Normalization of profiles 39
Nurse cropping, as a method for energy management 238

O

Observation vs. theory 13
Obstacles to windflow 30
Occluded front 332
Oceans, global currents, related to climate 308-309, 312-313, related to general cir-
    culation of the atmosphere 302, and shorelines, as a climatic control 304
Omnivores, related to energy distributions in the Ecosphere 288
Optimum temperature, in animals 175, in plants 198
Oxygen, related to plant physiology 192

P

Panting 172
Parks, urban, related to temperature contrasts 142
Partitioning of energy by surfaces, rules of thumb 97
Passive solar heating 265
Patchiness of surfaces, effects on windstream 30
Persistence, as a statistical concept 259
Phenology, 201-206, related to Growing Degree Days 202, 206
Photosynthesis 89, 185, diurnal behavior 190, related to atmospheric variables
    185-190, related to energy balance analyses 99, related to gas exchange in leaves
    185, related to limiting factors 188, related to respiration 185, related to transpi-
    ration 192
Physical systems 6

Physical units, four basic 42
Physics, as related to environmental design 220, laws related to weather and climate 295
Physiology, responses of animals to temperature 169, plant, roles of water 193
Planck's law 55
Planets of our solar system 276
Planner, the professional 209-213, atmospheric expertise required 227, 272, formulating the atmospheric component 226-228, 269, interactions with atmospheric ecologist 221, 268, 388-390, presentations of atmospheric components 268-271, required view of the Balance of nature 234, sample problems with atmospheric components 264-268, utility of knowledge about global scale climates 295, direct use of air pollution concepts 373-380, relationships to air pollution control 353
Planning, problems using annual average degree days 126, related to urban heat island 141, related to tropical cyclones 345, limitations of air pollution plume models 367-368, regional, related to air pollution 375-376
Plants, varied roles in atmospheric ecology 182, anatomy 183, 193, canopies, energy stored in materials 98, canopies, measures of structure 38-40, canopies, profiles of atmospheric variables 41, physiology, roles of water 193
Players in the game of environmental design 208
Plowing, as a method for energy management 238
Plume, from surface patchiness 29-30
Plume, air pollutant, centerline 363, mathematical model 363-366, mathematical model, Gaussian 373, rise 365
Poikilotherms 155
Polar cell, of Earth's atmosphere 302
Polar circles 277-278
Polar frontal zone 325
Pollutants, concentration, related to visual range 371
Pollution, air 259, 353-387, control, access to literature 386, potential 379, receptor effects 369-372, removal processes 357, sources, access to literature 386, sources, classification of configurations 363, strategies for control 357, 359, transport, access to literature 386, transport, diffusion vs. dispersion 365
Pollution, environmental 355
Population dynamics 156
Porosity, soil 73
Positive feedback 9-10
Potential, soil moisture, 74
Power, defined 318, of various atmospheric phenomena 318
Precipitation 37, 81, 144-149, 152, belts in the general circulation of the atmosphere 296, "depth" as a measure of rate 145, compared with humidity in mapping 143, descriptions vs. explanations of observed urban patterns 146, detection of urban effects on totals 149, explanations of observed urban patterns 146, expressed by isohyets 145, first urban mappings 144, global means 307, global scale types 313-314, related to global scale vegetation 314-317, LaPorte anomaly 145, mapping by radar 144, mapping of totals 144, requirements for 81, seasonal, global means 308, totals, related to terrain 152, totals, time trends as evidence of urban effects 152
Preconditioning, thermal, in animals 176, in plants 197
Predator-prey relations 157
Presentations of atmospheric variables, insolation 12-14, moisture 22-24, 41, temperature 15, 41, wind 27, 41
Pressure, defined 104, barometric, related to altitude 78, belts in the general circulation of the atmosphere 296, constant, maps of large scale atmospheric surfaces 321, effect on boiling temperature of water 67, global means 303-306, in Earth's atmosphere 282-283, related to altitude in Earth's atmosphere 282-283, sea level mapping 323, surfaces related to circulation cells 297-299, turgor 193
Pressure gradient 318, force 318, force vector 318

Prevailing westerlies 267, 269, 312
Primary consumers of the Ecosphere 288
Primary producers, plants as 183, 288
Probability, as a statistical concept 257-258, 405-407, 411-413
Process lapse rate 361, 385
Producers, in food chains 183
Profile, of wind speed 27, 41, of temperature 16-17, 41, in plant canpies 41, 88
Programs of air pollution control 371-372
Proof, related to the scientific method 292-293
PSN (see Photosynthesis)
Psychrometric diagram 68, 104
Public, as a player in the Game of Environmental design 208, health planning 259
Puff model, for air pollutant concentration 377-379

Q
Quasi-range, as a statistical tool 254

R
Radiant flux density on Earth, compared with other planets 284, factors affecting
    283-285
Radiant heat load, animal 164, leaf 194, human 107
Radiation, as an energy transfer process 90, infrared 55, net, defined 95, shortwave
    vs. longwave 55, 57, ultraviolet 55, visible, 55
Radiation budget 91
Radiation physics, behavior of "infrared film" 58, behavior of atmospheric gases
    58, behavior of leaf materials 58, Kirchhof's law 55, laws 55, Planck's law 55,
    Stefan-Boltzmann law 55, related to temperature inversions 59, the atmospheric
    "window" 58, Wien's law 55
Rainfall, urban effects on 146
Randomness in turbulent flow 26
Range, as a statistical tool 253
Rankine temperature 43
Rate of development, in organisms, related to degree days 180
Rate of flow, graphical indication of magnitude 7
Reducers (see Decomposers)
Reflectivity 54
Regional planning, related to air pollution 375-376
Regression, as a statistical concept 261
Relative cumulative frequency, as a statistical tool 250-251
Relative frequency, as a statistical tool 250-251
Relative humidity 21-23, 68, 271, diurnal patterns 22-23, related to human comfort
    21, versus dewpoint temperature 22-23
Reservoir, in a system model 8, 257
Residence time, as a statistical tool 255-257
Resistance, cardiovascular 112, 164, stomatal 191, thermal, related to human com-
    fort 111, related to architecture 122, to heat flow in animals 164
RESP (see Respiration in plants)
Respiration in plants 185, related to atmospheric variables 185-190, related to
    energy balance analyses 99
Resultant vector 352
Return period, as a statistical tool 255-257
Roughness, effects on windstream 38, effects on general circulation of the
    atmosphere 302
Ridge, on map of constant pressure 322
Running (moving) mean, as a statistical tool 252-253

S

Salinity, effects on soil moisture tension 77
Sand 72, 75
Saturation deficit, defined 71
Scales of windspeed 25
Scattering, of light waves, related to color 53, solar radiation by the atmosphere 51, related to visual range 369
Scattergram, as a statistical tool 261
Sea breeze 33, related to air pollution 368
Secondary consumers of the Ecosphere 288
Secondary producers, herbivores as 183
Sensible heat 90
Separation, surface of 31, 35
Sequence, as a statistical concept 254-255, 258-261
Severe storms, hailstorms 339, related to a midlatitude wave cyclone 335, thunderstorms 338-340, times of occurrence 344-345, tornadoes 340-345, tropical cyclones 345-350, winter 338
Shading, as a method for energy management 237
Shadow area of a human being 107
Shadows related to solar geometry 49-50
Shear, wind 86
Shelterbelting, as a method for energy management 32, 238, 247
Shore breeze 33
Shorelines, and wind patterns 33
Shortwave radiation (also see Solar radiation and insolation) 95
Silt 72
Sky view factor 141
Sky, why it is blue 53
Snow storms 338
Snowflakes, related to precipitation 85
Soil, packing, as a method for energy management 238, physical characteristics 17, 24, porosity 73, temperature patterns, annual 16, decay of amplitude 17, diurnal 16-17, management 17, rules of behavior 17, time delay 17, temperature profiles, curvature and temperature change 62-63, texture 72, water ratio 73, water, forces acting on 73, 77
Soil heat capacity, as a weighted mean 105
Soil moisture 72-77, as experienced by humans 24, as experienced by plants 24, characteristic curve 75, effects of salinity on tension 77, local patterns 24, movement of liquid 76, movement of vapor 77, movement under field conditions 76, self-mulching of dry soil 77, tension 74-75, 77
Solar constant, defined 51
Solar elevation angle 46
Solar energy, and comparative energetics of atmospheric phenomena 317 (also see Insolation)
Solar geometry 12, 42, 44-51, 277-279, angles defined 47, calculations 49,
Solar heating, passive 265, 402, 404
Solar path length 51
Solar radiation, absorption and scattering by the atmosphere 51, direct and indirect compared 52, pathways through the atmosphere 51-52 (also see Insolation)
Solar system 276, planets 276
Solar zenith angle 46
Solstice 279
Source strength 363, 374
Sources and sinks, defined 88, related to profile curvature 88
Spatial dimensions of environmental design 212
Specialist, use of consultants 4, vs. generalist 388-390
Specific humidity, defined 71

Specific surface, related to soil moisture tension 75
Spectrum, of absorption and emission of atmospheric gases 58, of Black Body
    54-55, of leaf materials 58
Speed vs. velocity 25
Spraying, water, as a method for energy management 238
Squall line, and comparative energetics of atmospheric phenomena 317, related to a
    midlatitude wave cyclone 335
Stability, as a concept in atmospheric motion 360-362, related to formation of
    tropical cyclones 348
Stack height, air pollutant 364
Stationary front 334
Statistics, arithmetic mean (average) 251, central tendency 251-253, conditional
    probability 258, contingency 258, correlation 261, dispersion 253, 263, double
    mass analysis 132, frequency 250, 263, frequency distribution 250-251, 263,
    histogram 256, interquartile range 254, joint probability 257, kinds of variables
    249-250, Markov chain 258-261, mean (average) 251, "measure" related to fre-
    quency 250, median 251, mode 251, odds related to probability 257, persistence
    259, probability 257, 405-407, 411-413, quasi-range 254, range 253, regression
    261, relative cumulative frequency 251, relative frequency 250, residence time
    255-257, return period 255-257, running (moving) mean 252, scattergram 261,
    sequence 252, 263, threshold values 255, time series 255, turnover time
    255-257, weighted mean 105 253, wind rose 262
Steering, or storm systems, related to the jet stream 326-327
Stefan-Boltzmann law 55
Stomata (singular "stoma") 184, 191, resistance 191
Storms, as generalized atmospheric phenomena 317, maps in urban environs 148,
    urban patterns of variables related to 148, and weather changes, midlatitudes
    317, related to motion in a mountain stream 325-327
Strategies for energy management 235, vs. tactics 235
Streamlines 28, 81, related to air pollutant plume 367
Streamlining 30
Streams, of energy transfer 91
Subsidence, belts in the general circulation of the atmosphere 296, related to air
    pollution 358-359
Suction tension (see Soil moisture tension)
Sultriness index, or humiture index of human comfort 115
Sun, estimated average temperature 57
Sun path diagram 12, 42, 44-46, coordinate systems 44-45, input and output data
    44 -45, Southern Hemisphere 44, three dimensional 46, sources, 42
Sunrise and sunset 14, times of 49-50, why they are red 53, related to net radiation
    96
Supersaturation of water vapor 82
Surface of separation 31, 35
Surface-to-volume ratio 160
Sweating 172
Symbiosis 157
Systems, physical 6

T

Tactics vs. strategies 235
Technical consultant 4, 210-213 (also see Atmospheric consultant)
Technology, related to air pollution 382
Temperature, annual patterns in soil 16, average daily, calculated for degree days
    122, average of Earth 56, average of surface value on Earth, factors affecting
    283-285, conversion formulas 43, core 110, 164, critical 170, daily mean, calcu-
    lation of 225, diurnal patterns in air 15-16, 22-23, diurnal patterns in soil 15-16,
    diurnal patterns in water 18, diurnal range, calculation of 273, envelope 20-21,

Vector 352, addition of 352, of pressure gradient force 318, related to atmospheric motion 352

Vegetation, global scale types 314-317

Velocity, vs. speed 25, average in turbulent flow 26, turbulent components 26, related to air pollutant plume 367

Virga 86

Visibility, related to visual range, defined 386

Visual range, defined 386, related to air pollution 369-371

Voluntary vs. involuntary theroregulation 112

Vortex, large scale, related to motion in a mountain stream 325-327, related to wave cyclone 332

Vorticity 350

W

Wall view factor 141

Warm front 332

Washout, an air pollution removal process 357

Water, change of phase 66, density related to temperature 19, latent heats 66, roles in plant physiology 193, temperature patterns 18-20, three physical states 66, urban vapor sources compared with rural 143

Water ratio, soil 73

Water vapor, urban sources 143, related to global climatic change 385

Watershed 355

Waterspouts 336

Waves, as generalized atmospheric phenomena 317, gravity, on large scale atmospheric surfaces 323, on large scale atmospheric surfaces 327, thermal, on large scale atmospheric surfaces 322-326

Wave cloud 87

Weather, data available 224, 247-249, 338, 393-400, and midlatitudes storms, 317, tropical 351, modification 86, proper use of data 224, 344, types 222, vs. climate 295

Weather maps 222, of large scale atmospheric surfaces 322, related to a midlatitude wave cyclone 336-337

Weather stress index 17, 271

Weighted average (see Weighted mean)

Weighted mean, as a statistical tool 105, 253

Wet bulb, depression 71, temperature 70

Wien's law 55

Willy Willy (see Tropical cyclones)

Wind, effects on animals 167, effects on transpiration 194, global means 305-306, related to Highs and Lows 320-321, machines, as a method for energy management 238, 246-247, profiles, logarithmic 27, 38, 41, 43, profiles in plant canopies, 41, related to climate space 174, scales 25, Beaufort scale of speed 25, daily cycle in mountains 33-34, directional naming 25, effects on leaf energy balance 194, generalized flow around obstacles 30-31, gust fronts 86, microbursts 86, responses to different surfaces 29-30, 41, speed vs. velocity 25, the "prevailing westerlies" 267, 269

Wind chill index of human comfort 115

Wind rose, as a statistical tool 262

Wind shear 86

Windflow, effects of building clusters 31-32, 42, effects of hedgerows 32, effects of obstacles 30-31, effects of shelterbelts 32, effects of terrain 32-36

Window, radiative, in the atmosphere 58

WSI (see Weather Stress Index)

Z

Zenith angle, solar 46